Natural Resource Management in Transition 1

For further volumes:
http://www.springer.com/series/13274

Carsten Schmitz-Hoffmann • Michael Schmidt •
Berthold Hansmann • Dmitry Palekhov
Editors

Voluntary Standard Systems

A Contribution to Sustainable Development

With contribution by David J. Smyth

Springer

Editors
Carsten Schmitz-Hoffmann
Berthold Hansmann
Deutsche Gesellschaft für Internationale
Zusammenarbeit (GIZ) GmbH
Eschborn, Germany

Michael Schmidt
Dmitry Palekhov
Department of Environmental Planning
Brandenburg University of Technology
Cottbus, Germany

ISBN 978-3-662-51254-8 ISBN 978-3-642-35716-9 (eBook)
DOI 10.1007/978-3-642-35716-9
Springer Heidelberg New York Dordrecht London

Printed on acid-free paper

Springer is part of Springer Science+Business Media (www.springer.com)

Foreword

Just a decade or so ago, you would never have found a book delving into sustainability standards and their diverse use around the globe. There were only a handful of examples of successful certification initiatives and the so-called first generation of standards such as Fairtrade International and the Forest Stewardship Council, among others, were just starting to consolidate and scale up their programmes. Certified products on store shelves were niche items, and mainstream companies such as Unilever, McDonald's and IKEA were not yet fully engaged in transforming their supply chains for sustainability. Today, studies, conferences, training programmes and even social media platforms overflow with information on certification and standards. Global companies have announced major commitments to source certified products and governments are increasingly choosing to reference voluntary standards through co-regulation, public procurement, or support to producers for implementing sustainable practices.

Despite the many successes in the last two decades, awareness of sustainability standards is still surprisingly low. In particular, people working in sustainability need to understand what makes a standard truly credible, what the business and producer cases are for certification, what positive impacts have been seen to date and what still needs to be demonstrated, and how stakeholders can use sustainability standards to achieve their social and environmental objectives. This book will help to build that awareness and thereby aid our future sustainability leaders in strengthening effective standards and related initiatives for the benefit of all.

London, UK
April 2013

Karin Kreider

We all know it: unsustainable patterns of production and consumption are having serious and long-term detrimental economic, environmental and social impacts worldwide. To start changing the current situation, we need game changers. Game changers that will address both consumption and production by mobilising consumers, producers and intermediaries on an unprecedented scale—making sustainable consumption and production the norm, the easy choice and the mainstream. Game changers that address—and interact with—green growth and provide benefits from sustainable development for those most in need.

But will VSS be one of the game changers? Through my work at Consumers International, I have become more and more involved in the VSS community and space. It is an area where I see potential for a game changer, in particular where there is strong focus on the potential value of standards as tools for developing countries to achieve their sustainable development goals.

As a consumer advocate credible VSS has certainly made my life easier as they have started to populate the vacant space between 'official' standards/eco-labels and often dubious self-declared claims with very little standard behind them. But for me as an 'ethical consumer', I see the front end of the systems, more often than not in the form of labels, logos, colours and statements, attempting to help me navigate (and persuade me) of my consumption choice of products and services. And there seem to be more and more of them. At the launch of the new United Nations Forum for Sustainability Standards (UNFSS), the Committee on Sustainability Assessment (COSA) counted 435 seals and standards. For me that illustrates one of the risks of VSS not becoming a game changer as the VSS landscape develops—proliferation of (often overlapping) standards and the associated communication to end users with a focus on niche production, niche issues and niche consumers.

The reality is that we need VSS to work and to be a game changer in particular in terms of mainstreaming. And as this book illustrates there are a lot of positive case studies that can be scaled up and learned from, and there are VSS that have already made a real difference and are able to demonstrate real impact. In some specific sectors such as coffee, bananas, tea and forests, some would rightly argue that we are getting closer to the 'tipping point' for mainstreaming with market shares of certified products climbing steadily. Other areas are ripe and ready for innovation and initiative.

This book sets a milestone in the VSS debate with a state-of-the-art overview of the VSS landscape. It charts the development of VSS, their potentials and challenges and interaction with legal instruments. The various practitioner contributors provide invaluable insights as well as their outlook on the future of VSS, creating a must read for VSS practitioners, the standards community and policy makers alike.

Kuala Lumpur, Malaysia
June 2013

Bjarne Pedersen

Preface

Voluntary Standard Systems (VSS) are a promising and rapidly evolving concept with considerable potential to promote 'green economy'. They encompass the three pillars of sustainability—social, environmental and economic aspects, and consequently they can be considered as a tool, which makes sustainable development visible. Currently, VSS are becoming a significant element in international trade and in the promotion of sustainable development strategies, especially in the context of globalised markets and supply chains.

This book is divided into five parts, which provide a comprehensive overview of the current VSS concepts and contains numerous examples of their implementation in different sectors of the economy. Part I introduces the concept and nature of VSS and discusses various issues related to their functioning. Part II highlights the difference between formal and private standards, their complementary characteristics and their co-functioning. Part III places VSS in the broader context of global development issues and challenges, including development policy and international sustainability commitments, progress towards achieving, green economy, and meeting climate protection targets. Part IV presents a representative selection of case studies to aid in demonstrating their wide range of applications and effectiveness in contributing to development objectives. Part V closes our publication with discussion of the current challenges related to the development of VSS and the future outlook.

The completion of this volume leaves us indebted to many people. First of all we wish to thank all authors from across the globe for their valuable article contributions, which made possible the preparation of this comprehensive publication. We are very grateful to David J. Smyth for his help in preparing this volume and for the excellent work on proofreading of all chapters and authoring introductions. We also wish to express our gratitude to Tania Wang and Bradlie Martz-Sigala for their assistance in the initial stages of preparing the manuscript.

We hope that practitioners within the field of standards application, the wider business community and policy makers, in addition to academic researchers, teachers and students will find the enclosed material valuable in their respective work and research endeavours.

Eschborn, Germany Carsten Schmitz-Hoffmann
Cottbus, Germany Michael Schmidt
Eschborn, Germany Berthold Hansmann
Cottbus, Germany Dmitry Palekhov
September 2013

Contents

Part I Concepts and Operation of Voluntary Standard Systems (VSS)

1 **Introduction to Voluntary Sustainability Standard Systems** 3
Kristin Komives and Amy Jackson

2 **Standards for Standards: The Foundations for Voluntary Standards for Sustainability** 21
Robert B. Gibson

3 **Conceptual and Practical Aspects of Indicators for Measuring Sustainability of Certified Products and Producers** 39
Oscar David Matallana-Tost and Michael Schmidt

4 **Evolution of VSS: From Niche to Mainstream** 49
David Ovando Jeria and Miguel Araque Vera

Part II Formal and Private Standards: The Added Value

5 **VSS and Legal Standards: Competition or an Added Value?** 61
Eike Albrecht

6 **VSS Where Formal Regulations Are Missing: Potential Study on Example of Nanotechnologies** 77
Joel Goebelbecker and Eike Albrecht

7 **Recognition of Private Sustainability Certification Systems for Public Regulation (Co-Regulation): Lessons Learned from the EU Renewable Energy Directive** 99
Martina Gaebler

Part III Voluntary Standard Systems: The Development Dimension

8 The Role of Voluntary Standards in German Development Policy 115
Evita Schmieg

9 Voluntary Sustainability Standards: Measuring Their Impact 133
Carsten Schmitz-Hoffmann, Berthold Hansmann, and Sophie Klose

10 Environmental Standards and Embedded Carbon in the Built Environment 145
Callum Hill and Andrew Norton

11 Exploring Market Strategies Based on Voluntary Environmental Certification in a Post-Soviet Transition Economy 165
Ludmila Palekhova and Gennadiy Pivnyak

Part IV Implementation and Impact of VSS

12 Evaluation of the Interrelation Between Voluntary Standard Initiatives and Regulatory Approaches Relevant to Forest Management 179
Berthold Hansmann, Stefan Essel, and Sophie Klose

13 Voluntary Standard Systems and Regulatory Processes for Timber Products: Analysis of Green Procurement in Germany . . . 195
Eike Albrecht, Franziska Rückert, and Michael Schmidt

14 The Role of Sustainability Standards in the Energetic Use of Palm Oil Plantation Residues: Case Study of Cameroon 211
Michael Schmidt, Berthold Hansmann, and Pia Dewitz

15 The Adoption and Impact of Forest Stewardship Council Standards in the Congo Basin Forestry Sector 229
Mercy Nambu Diangha and Gerhard Wiegleb

16 Issues and Opportunities for Implementation of VSS in China 243
Ni An and Eberhard Schaller

17 A Feasibility Study of Utilising Voluntary Sustainability Standard (VSS) Systems in Paper-Making Enterprises in Liaoning Province, China 253
Xiaoying Gu and Gerhard Wiegleb

18 Voluntary Environmental Certification in Ukraine: Experience and Issues 261
Dmitry Palekhov and Michael Schmidt

19 Comparative Analysis of Environmental and Social Impacts of Cocoa Production: Case Study Cameroon 275
Eric Ambe Asoh, Ambe Emmanuel Cheo, Michael Schmidt, and Hans-Jürgen Voigt

20 Transnational Initiatives to Promote Sustainable Cocoa Production and Trade: The Case of the German Initiative on Sustainable Cocoa (GISCO) 287
Elena Rueda, Ulrich Helberg, Vera Morisse, and Eberhard Krain

21 VSS and Climate Change in the Coffee Sector: The 4C Climate Module 305
Kerstin Linne

22 The Role of VSS in Enhancing the Contribution of Fisheries and Aquaculture to Sustainable Development 315
Mark Prein and Uwe Scholz

23 Measuring Sustainability in the Construction and Real Estate Sector: A Case Study of the DGNB Certification System 345
Christine Lemaitre

Part V Challenges and Future Trends

24 Corporate Social Responsibility and the Role of Voluntary Sustainability Standards 359
Daniele Giovannucci, Oliver von Hagen, and Joseph Wozniak

25 Voluntary Standards and Approaches for Sustainable Communities . . . 385
John Blewitt

26 Small-Scale Farmers' Involvement in Ecolabelling: Limitations and Conflicts 403
Carolin Möller, David Smyth, and Michael Schmidt

27 Tariff Preferences for Sustainable Products: A Summary 419
Philipp Schukat, Jenny Rust, and Julia Baumhauer

28 The Political Challenge of Voluntary Standard Systems 431
Günther Bachmann

About the Authors 441

Index 453

Contributors

Eike Albrecht Department for Civil and Public Law with References to the Law of Europe and the Environment, Brandenburg University of Technology (BTU) Cottbus, Germany

Ni An Department of Environmental Meteorology, Brandenburg University of Technology (BTU) Cottbus, Germany

Miguel Araque Vera ThyssenKrupp Industrial Solutions, Bad Soden, Germany

Eric Ambe Asoh Department of Environmental Planning, Brandenburg University of Technology (BTU) Cottbus, Germany

Günther Bachmann German National Council for Sustainable Development, Geschäftsstelle c/o GIZ GmbH, Berlin, Germany

Julia Baumhauer Julia Baumhauer Give Something Back To Berlin, Berlin, Germany

John Blewitt Aston Business School, Aston University, Birmingham, UK

Ambe Emmanuel Cheo Department of Environmental Geology, Brandenburg University of Technology (BTU) Cottbus, Germany

Pia Dewitz Department of Environmental Planning, Brandenburg University of Technology (BTU) Cottbus, Germany

Programme for Social and Environmental Standards, Deutsche Gesellschaft für Internationale Zusammenarbeit (GIZ) GmbH, Eschborn, Germany

Mercy Nambu Diangha Department of General Ecology, Brandenburg University of Technology (BTU) Cottbus, Germany

Stefan Essel Programme for Social and Environmental Standards, Deutsche Gesellschaft für Internationale Zusammenarbeit (GIZ) GmbH, Eschborn, Germany

Martina Gaebler Programme for Social and Environmental Standards, Deutsche Gesellschaft für Internationale Zusammenarbeit (GIZ) GmbH, Eschborn, Germany

Robert B. Gibson Department of Environment and Resource Studies, University of Waterloo, Waterloo, ON, Canada

Daniele Giovannucci The Committee on Sustainability Assessment (COSA), Philadelphia, PA, USA

Joel Goebelbecker Department for Civil and Public Law with References to the Law of Europe and the Environment, Brandenburg University of Technology (BTU) Cottbus, Germany

Xiaoying Gu Department of General Ecology, Brandenburg University of Technology (BTU) Cottbus, Germany

Berthold Hansmann Programme for Social and Environmental Standards, Deutsche Gesellschaft für Internationale Zusammenarbeit (GIZ) GmbH, Eschborn, Germany

Ulrich Helberg Programme for Social and Environmental Standards, Deutsche Gesellschaft für Internationale Zusammenarbeit (GIZ) GmbH, Eschborn, Germany

Callum Hill Norsk Institutt for Skog og Landskap, Ås, Norway

JCH Industrial Ecology Limited, Gwynedd, Wales, UK

Renuables, Gwynedd, Wales, UK

Amy Jackson ISEAL Alliance, London, UK

David Ovando Jeria Department of Environmental Planning, Brandenburg University of Technology (BTU) Cottbus, Germany

Sophie Klose Programme for Social and Environmental Standards, Deutsche Gesellschaft für Internationale Zusammenarbeit (GIZ) GmbH, Eschborn, Germany

Department of Environmental Planning, Brandenburg University of Technology (BTU) Cottbus, Germany

Kristin Komives ISEAL Alliance, London, UK

Eberhard Krain Programme for Social and Environmental Standards, Deutsche Gesellschaft für Internationale Zusammenarbeit (GIZ) GmbH, Eschborn, Germany

Christine Lemaitre German Sustainable Building Council (DGNB), Stuttgart, Germany

Kerstin Linne Green Line Consulting, Bremen, Germany

Oscar David Matallana-Tost Department of Chemical Engineering and Environmental Management, University of Engineering and Technology (UTEC), Lima, Peru

Carolin Möller Department of Environmental Planning, Brandenburg University of Technology (BTU) Cottbus, Germany

Vera Morisse Programme for Social and Environmental Standards, Deutsche Gesellschaft für Internationale Zusammenarbeit (GIZ) GmbH, Eschborn, Germany

Andrew Norton Renuables, Gwynedd, Wales, UK

Dmitry Palekhov Department of Environmental Planning, Brandenburg University of Technology (BTU) Cottbus, Germany

Ludmila Palekhova Department of Marketing, Institute of Economics, National Mining University (NMU), Dnepropetrovsk, Ukraine

Gennadiy Pivnyak Rector of the National Mining University (NMU), Dnepropetrovsk, Ukraine

Mark Prein Rural Development and Agriculture Division, Programme on Sustainable Fisheries and Aquaculture, Deutsche Gesellschaft für Internationale Zusammenarbeit (GIZ) GmbH, Eschborn, Germany

Franziska Rückert Department of Environmental Planning, Brandenburg University of Technology (BTU) Cottbus, Germany

Elena Rueda Programme for Social and Environmental Standards, Deutsche Gesellschaft für Internationale Zusammenarbeit (GIZ) GmbH, Eschborn, Germany

Jenny Rust Social and Environmental Standards Programme, Deutsche Gesellschaft für Internationale Zusammenarbeit GmbH (GIZ), Eschborn, Germany

Como Consult GmbH, Hamburg, Germany

Eberhard Schaller Department of Environmental Meteorology, Brandenburg University of Technology (BTU) Cottbus, Germany

Michael Schmidt Department of Environmental Planning, Brandenburg University of Technology (BTU) Cottbus, Germany

Evita Schmieg Stiftung Wissenschaft und Politik (SWP), Berlin, Germany

Carsten Schmitz-Hoffmann Programme for Social and Environmental Standards, Deutsche Gesellschaft für Internationale Zusammenarbeit (GIZ) GmbH, Eschborn, Germany

Uwe Scholz Rural Development and Agriculture Division, Deutsche Gesellschaft für Internationale Zusammenarbeit (GIZ) GmbH, Eschborn, Germany

Philipp Schukat Social and Environmental Standards Programme, Deutsche Gesellschaft für Internationale Zusammenarbeit GmbH (GIZ), Eschborn, Germany

David Smyth Department of Environmental Planning, Brandenburg University of Technology (BTU) Cottbus, Germany

Hans-Jürgen Voigt Department of Environmental Geology, Brandenburg University of Technology (BTU) Cottbus, Germany

Oliver von Hagen The International Trade Centre (ITC), Geneva, Switzerland

Gerhard Wiegleb Department of General Ecology, Brandenburg University of Technology (BTU) Cottbus, Germany

Joseph Wozniak The International Trade Centre (ITC), Geneva, Switzerland

Part I
Concepts and Operation of Voluntary Standard Systems (VSS)

At the outset of our investigation into voluntary standard systems, as with any endeavour, we require a secure basis from which to begin. Those new to the subject require foundation for thought, whilst those revisiting may need some reinforcement before tackling headlong into the later material. Part I sets out to frame VSSs within their contextual environment, outlining basic principles, introducing the commonly used technical language and tools, and revising the forces which conceived them and how VSS development progressed with time. Acronyms run rampant within discussions of sustainability, in an attempt to acknowledge and include all the contributing organisations and schemes involved within this relatively 'new' management arena.

Chapter 1 acts as a guide through essential terminology, linking to the structural components of a standards system and elaborating on these building blocks. A brief history follows, with reference to the various organisations spawned during VSS development and uptake from the early twentieth century to present day. Trends are highlighted relating to system models and assurance, covering topics ranging from collaboration of standard setting bodies to the alternative uses for auditors in verification as a step towards encouraging eventual certification. Concerns linked to VSS credibility and effectiveness are tackled, where obstacles are represented in the form of 'green-washing', exclusion issues and accessibility to standards for producers, and the realities of claims versus actual deliverables.

Chapter 2 features discussion upon humanity's short-comings when dealing with our environment, with the core challenges to sustainability being listed; including topics such as global carrying capacity, meeting basic needs of populations and distribution of wealth. The implications for a transition from 'unsustainability to sustainability' are also realised within this section. There is the summarised 'eight basic requirements for progress towards sustainability', tempered by the recommendation that such generic guidelines should be adapted and elaborated upon according to the context in which they are to be applied. Further insight is achieved with the acknowledgement of the practical difficulties to sustainability, exemplified in cases of resolving disagreements or how best to represent future generations in decision making, etc. The implications for sustainability based VSS feature, including a comparison of ISEAL and IISD project standards (International Institute for Sustainable Development) with their respective merits and short-comings.

Indicators dominate Chap. 3, along with issues relating to their development and implementation. Indicators can serve as incredibly powerful tools through which a particular change can be viewed; however their complexity and sensitivity make their use an expert activity. Readers are shown how developers and users can be aided in their application of indicators and the need to acknowledge the inherent weaknesses of such tools so that they can be used effectively. Definitions of data and indicators are coupled with data gathering methods, feeding into the usefulness of indicators for application in measuring sustainability. The presentation of indicators using analytical aids and tools is discussed, where suggestions are made regarding the appropriate use of comparators, thresholds, targets and baselines, depending on the situation at hand.

Chapter 4 rolls out an expanded timeline for VSS development and use, detailing the pedigree of VSS throughout three notable periods of the twentieth century, and revealing the driving forces acting upon events at these times. The emergence of sustainability movements (1960–1990) such as IFOAM (International Federation of Organic Agriculture Movements) and Fairtrade are acknowledged, through to the 1992 Rio Earth Summit and the current age of mainstream voluntary sustainability standards being established. This retrospective view of VSS is important to ensure that those carrying development forward have experience of past events and can use that to better guide future decision making, as characterised by the obstacles being recalled e.g. proliferation of different labelling schemes, trade barriers obstructing producers, etc.

Chapter 1
Introduction to Voluntary Sustainability Standard Systems

Kristin Komives and Amy Jackson

1.1 Introduction

Private voluntary sustainability standard systems are an innovative market-based approach to promoting sustainable production and business practices. Adoption of these sustainability standards is intended to be voluntary: the standards are not created, run, or required by governments or government regulation. Instead, voluntary sustainability standard systems are non-government initiatives that seek to drive sustainable production and consumption by creating market demand for sustainable products, and a supply to meet that demand. They help buyers (both consumers and businesses) identify sustainably-produced products, and they guide producers, forest managers, mine and tourism operators, and factory owners and others in the choice of sustainable practices.

Voluntary sustainability standard systems have become important tools for moving production in some sectors toward sustainability. Some of the best known sustainability standards—e.g. Fairtrade International, the Forest Stewardship Council (FSC), and the Marine Stewardship Council (MSC)—are now well-known brands in many countries, and consumers rely on associated 'eco-labels' to inform buying decisions. Business attention to sustainable procurement has grown, increasing both supply and demand for products produced in accordance with sustainability standards. A 2010 study of the market presence of voluntary sustainability standards found that, as of 2009, 18 % of globally managed forests were certified to the FSC or PEFC (Programme for the Endorsement of Forest Certification Schemes) standards, 17 % of global coffee supply was produced in compliance with a sustainability standard, and sustainable bananas made up about 20 % of global banana exports (Potts et al. 2010). Parallel to the growth in these sectors with long experience with certification, new voluntary standard systems are emerging in

K. Komives (✉) • A. Jackson
ISEAL Alliance, The Wenlock Centre, 50-52 Wharf Road, London N1 7EU, UK
e-mail: kristin@isealalliance.org; amy@isealalliance.org

C. Schmitz-Hoffmann et al. (eds.), *Voluntary Standard Systems*, Natural Resource Management in Transition 1, DOI 10.1007/978-3-642-35716-9_1,

sectors such as bio-trade, mining, energy, water and sports (e.g. Union for Ethical Biotrade, Responsible Jewellery Council, Initiative for Responsible Mining Assurance, Alliance for Responsible Mining, Roundtable for Sustainable Biomaterials, Golf Environment Organisation, Alliance for Water Stewardship and more). New uses for voluntary sustainability standards are also emerging. Actors such as governments and financial institutions are now employing standards to help implement policy objectives and assess portfolio risk.

Voluntary sustainability standard systems are an increasingly important market mechanism for driving sustainability, and the world of standards is rapidly innovating and evolving. In the first half of this chapter we explain what private voluntary sustainability standard systems are and how they are structured. We briefly review the history of these standards and examine recent trends in the evolution of standard systems. In the second half of the chapter, we examine concerns expressed about the credibility and effectiveness of sustainability standards and how the voluntary standards movement is addressing these concerns.

1.2 Voluntary Sustainability Standard Systems: Structure, History, and Evolution

1.2.1 Terminology

Voluntary sustainability standard systems are known by many different names. Even within this book, authors use various different terms to refer to these market-based instruments, including 'voluntary sustainability standards', 'private standards', 'standard systems', 'certification', and 'eco-labels'. Often authors employ these terms interchangeably, without attaching a particular meaning to one term. However, there are subtle differences in the terminology that are important to understand.

'Private' highlights the non-governmental nature of these systems. It does not mean that they are business-driven initiatives; often private voluntary standards are developed and managed by multi-stakeholder groups or even groups dominated by non-governmental organisations. The use of 'eco' or 'sustainability' in the name differentiates voluntary sustainability standards from other similar non-governmental, market-based initiatives that are not focused on addressing sustainability concerns.

The term 'system' highlights that these instruments rely on more than just the standard itself (the list of required practices or performance levels) to drive change. Below we describe each of these pieces of the system. 'Certification' is a reference to one piece of a standard system—the assurance process—and to one particular approach to assuring that products are actually produced in accordance with the standard.

In general, our discussion in this chapter focuses on 'voluntary sustainability standard systems'—multi-faceted, market-based systems with sustainability goals.

1.2.2 Structure

At the heart of any sustainability standard system is a standard that defines good social and environmental practices or performance in an industry or product (see Chap. 2, "Standards for Standards" for an in-depth discussion of the content of these standards). But a standard alone would not be sufficient to create a market for sustainable products. The market mechanism behind standards relies on four other important components of a standard system: assurance, labels and claims, traceability, and capacity building.

- Producers and other businesses seeking to meet a standard (e.g. farms, fisheries, forests, factories, or operations) are assessed to determine whether they meet the standard. This is done through the *assurance system* set up by the standard-setting organisation. Assurance of compliance has traditionally been based on an independent, third-party audit leading to *certification*, though new approaches are emerging. The assurance systems gives buyers the confidence that they are buying sustainably produced products.
- Many standard setting organisations offer corporate buyers of certified products the right to use a consumer-facing *label or claim* on product packaging (e.g. Fairtrade-certified coffee or the Rainforest Alliance green frog label on certified products). Others permit only business-to-business claims. Labels and claims are appealing to buyers and consumers and thus help increase demand for products produced in accordance with the standard.
- *Traceability* systems trace the 'chain of custody' of products, from where they were produced, through the full supply chain, and through to the final product, to provide proof of the origin of products carrying a label or a claim.
- Some standard setting organisations provide *capacity building* services to help producers, operators, or enterprises come into compliance with their standard. Others work with partner organisations that provide this training service.

By combining these five elements (the standard, assurance, labels and claims, traceability, and capacity building), voluntary sustainability standard systems provide incentives for many different actors to support and implement more sustainable practices. Consumers rely on standard systems to identify products that were produced using practices they value and want to support. For businesses seeking to source sustainably, the standard systems provide assurance that they are in fact buying products produced using responsible practices. Together, consumers and purchasing businesses build a demand for sustainable production. For suppliers of this product, standard systems offer guidance on how to improve production and meet sustainability goals and connect them to a market for sustainable products (which often provides higher prices than conventional markets).

1.2.3 History and Evolution

The first private voluntary sustainability standards date from the first half of the twentieth century. These early sustainability standards were private organic standards for agriculture, for example the Soil Association in the United Kingdom. Organic standards were developed locally, each with somewhat different criteria and different required practices. IFOAM (International Federation of Organic Agriculture Movements) was established in 1972 to lead, unite and assist the organic movement. Their work continues, for example with the publication of the consolidated Principles of Organic Agriculture, an international guideline for certification criteria, in 2005.

The fair trade certification movement followed a similar path, starting with one national standard, Max Havelaar in the Netherlands in 1988, and being replicated in several other markets across Europe and North America. These national standards then came together under one umbrella organisation, Fairtrade Labelling Organisations International (today Fairtrade International, or FLO), in 1997. In identifying the need for international coordination very quickly, Fairtrade provides an interesting transition from the organic movement to the next generation of sustainability standards which took a global approach from the beginning of the standard-setting process.

These standards, which emerged in the 1990s, aimed to develop global consensus on sustainable practices for particular industries and sectors. Early examples of this are the FSC, the MSC, the Rainforest Alliance's Sustainable Agriculture Network (RA-SAN), and Social Accountability International (SAI), which set standards for forestry, fisheries, agriculture, and labour respectively. All four set their standards through multi-stakeholder processes that brought NGOs, businesses, and other stakeholders to the table, and they still use this approach today.

These systems were created at a time when market mechanisms for environmental protection were gaining in popularity due to disillusionment with the effectiveness of government regulation and legislation to address sustainability challenges. The principle reason for creating the systems was to offer a non-governmental tool for achieving social and environmental change. NGOs and other actors who promoted the creation of these systems saw them as important vehicles for changing consumer buying patterns, business purchasing decisions, and production practices. Using the market was a powerful approach for using consumer demand for sustainability in one country to drive change in production practices in others. For example, the first product sold with a Fairtrade label was coffee from Mexico sold into Dutch markets.

A third generation of standard systems emerged after the turn of the century—commodity-based 'roundtables' bringing together stakeholders from industry, NGOs, and government to develop standards for commodities with known negative impacts on the environment. The roundtable standards were an initiative of WWF, seeking to use market forces to make sweeping changes in these sectors.

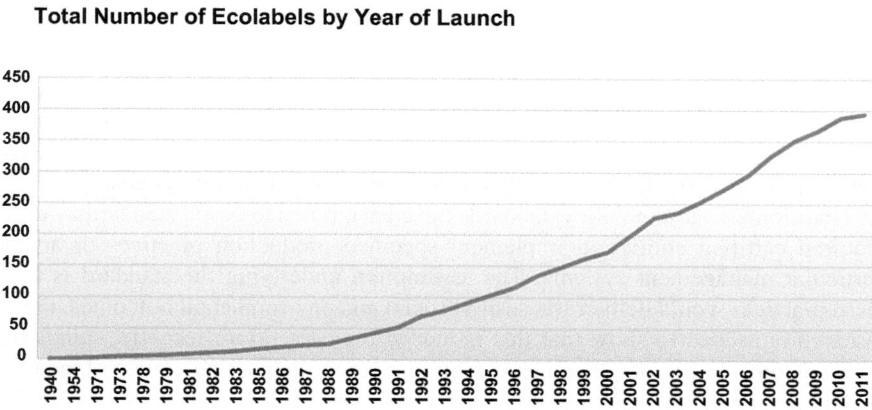

Fig. 1.1 Growth in labelling initiatives (*Source*: Ecolabel Index 2013)

Roundtable for Sustainable Palm Oil (RSPO), Roundtable for Responsible Soy (RTRS) and Bonsucro are three examples.

Since the early 1990s there has been a dramatic increase in the number of standards and eco-labels. As of July 2013, the Ecolabel Index is tracking 435 ecolabels in 195 countries (Ecolabel Index 2013; see also Fig. 1.1); more standards and labels emerge every year. The number of industry sectors employing voluntary standards as an approach for achieving sustainability is also growing, with standards in sectors like mining (e.g. Responsible Jewellery Council) and oil and gas (e.g. Equitable Origin) being established. More detailed information on the history of voluntary standards systems is available in Chap. 4, “The Evolution of Voluntary Standards Systems – From Niche to Mainstream”.

The first signs of organisation within the voluntary standard system industry itself appeared at the end of the 1990s when the Forest Stewardship Council (FSC), the International Federation of Organic Agriculture Movements (IFOAM), Fairtrade International and Marine Stewardship Council (MSC) came together to discuss the feasibility and benefits of working in closer collaboration. By 2002 four more organisations—International Organic Accreditation Service (IOAS), Marine Aquarium Council, Rainforest Alliance (RA-SAN) and Social Accountability International (SAI)—joined to form the International Social and Environmental Accreditation and Labelling Alliance (now just referred to as the ISEAL Alliance). The aim of the newly formed ISEAL Alliance was to enable collaboration between its members and coordinate and represent their common interests to government and other key stakeholders. In June 2013, the ISEAL Alliance had 14 full members and 7 associate members, all international bodies involved in standard setting or accreditation.

1.2.4 Trends in Standard System Models

Over the course of the relatively short history of voluntary sustainability standards, standard setting bodies have sought to meet the changing expectations and needs of stakeholders by innovating and adopting new standard and assurance models. This has led to increasing diversity in the types of voluntary standard systems.

Traditionally sustainability standards have been practice-based standards—they required certified entities to implement specified production practices or adopt particular management systems. The assumption underlying the standard is that these practices would deliver the desired social and environmental outcomes. Field research is needed to show that this is indeed the case. More recently, standards with outcome or performance-based requirements are emerging. These standards require certified entities to meet specified performance targets (e.g. water quality or carbon measures). This approach to standard setting has the advantage that certified entities can themselves decide which practices or systems to adopt in order to meet the goal. Performance based standards also provide more immediate evidence that the standards are achieving desired outcomes. However, some stakeholders feel that these types of standards may not be appropriate in all scenarios, for example in social auditing, and performance standards are too new to fully evaluate the feasibility of implementing a performance based standard on a global scale or the relative effectiveness of process versus performance based standards. In Chap. 2, author Robert B. Gibson argues that no one solution is applicable to all sustainability challenges. Diversity in the structure of standards is likely to increase over time as standard-setters seek to improve the effectiveness of their standards.

Just as the diversity of standards models is increasing, so is diversity in the objectives and geographical base of standard systems. Early voluntary standards were developed in Europe and the United States and generally aimed to set a high bar for sustainability and focused on a niche market where price premiums were a likely benefit. More recently we have seen the emergence of standards based in Latin America and Asia (e.g. RTRS and RSPO; see also Chaps. 16 and 17 for a discussion of standards in China) and the creation of standards that aim to improve mainstream industry practice or eliminate the most egregious practices. The 4C Association, for example, developed a baseline standard for coffee, which is intended to offer a first step in moving towards sustainable production in the coffee sector (see Chap. 8, Sect. 8.3 for more information about the 4C Association). Some standard systems permit certification at different levels of performance, such as the various Green Building Council's globally which offer Platinum, Gold, Silver, and Bronze certifications for green buildings (see Chap. 23 for a detailed review of one example, the German Sustainable Building Council's (DGNB) certification system), and many standard systems build requirements for performance improvement over time into their standards (e.g. Fairtrade and UTZ Certified standards in agriculture).

Another important trend in the standards' world is the emergence of collaboration across standard setting bodies as they recognise that they do not operate in isolation. For example, 4C Association now works in collaboration with Rainforest

Alliance, Fairtrade International, and UTZ Certified to develop "stepping up" programs that would move producers from the baseline 4C standard to one of the other more rigorous standards. New standards such as the Roundtable for Sustainable Biomaterials (RSB) have benchmarked their standard to existing agricultural standards so as to accept compliance with one of these standards as proof of compliance with part of the RSB standard. Cooperation across standards is making it possible for standard systems to specialise (for example, by focussing on one sector—Alliance for Water Stewardship) and still provide standards users with the option of addressing a broad range of sustainability issues in their supply chains. One example of this is joint Fairtrade/FSC certification for non-forest timber products. These forms of cooperation could ultimately help address the complex and multi-faceted nature of sustainability challenges (see Chap. 2).

1.2.5 Trends in Assurance

Parallel to the evolution in the standards themselves have been innovations in two other components of standard systems—assurance and traceability. The traditional approach to assurance has been, and remains, the use of independent, third-party auditors checking that an enterprise complies with the standard, which results in a certificate being issued. These independent auditors are normally from a conformity assessment body or *certification body*. The ability and quality of work done by the certification body can be checked in various ways, a process generally referred to as *accreditation* or *oversight*. In the past this oversight has normally been done by the standard-setting organisations themselves, or with reliance on National Accreditation Bodies. A more recent trend is for the standard-setting organisation to appoint an independent, international body—IOAS, Social Accountability Accreditation Services (SAAS), or Accreditation Services International (ASI)—to do the accreditation for their scheme and help ensure global consistency in the performance of certification bodies. For example, this approach has been taken up by SAI, MSC, FSC, RA-SAN, a large proportion of the organic movement, and newer entrants like Aquaculture Stewardship Council (ASC) and RSPO, among others.

At the same time, alternatives to audits as a means for checking compliance are also emerging. For example, while it is generally recognised that auditors should not give advice during an audit, some standards have identified ways to take advantage of the presence of auditors in the field or on the factory floor to help improve practices, without compromising the impartiality of the audit. Some entry-level standards with a focus on performance improvement, such as the 4C Association, use what they call 'verification' rather than certification. In this approach, the 'verifiers' may provide advice about how to improve practices during the audits. The companies that are successfully 'verified' do not receive a certificate, and companies buying these products cannot use an eco-label on their packaging, however the general level of practice is improved in the sector. Another set of emerging assurance models rely heavily on transparency and peer review to provide assurance of compliance. One example is the Participatory Guarantee Systems (PGS), in which producers check the performance of their peers. These changes are motivated by a desire to reduce the cost and complexity of assurance and to facilitate scaling up the use of sustainability standards.

In order to substantiate sustainability claims, most standards systems employ a form of traceability, sometimes also termed 'chain of custody'. 'Identity preservation' is the most strictly controlled form of traceability, but this can require considerable work as it requires products to be 100 % traceable, from certified origins. 'Segregation' keeps certified products separate from non-certified but does not allow one to track a specific batch of product to origin. 'Mass balance' keeps track of certified volumes, but these may be mixed with non-certified products. An interesting approach that is used to connect remote producers with the market is the certificate trading system, where credits are sold that equal the amount of certified product produced, but the actual product is not shipped to the buyer of credits. This is mainly used in large commodities such as palm oil, or where traceability of the product itself is impractical.

The choice of traceability system can have important implications for the overall functioning and impact of the standard system, as illustrated with an example from the Renewable Energies Directive (RED) of the European Commission. RED establishes that mass balance approach to traceability is acceptable for their requirements to demonstrate responsible sourcing of biofuels, however, some stakeholders are expressing concerns that this could allow false claims to be made. Conversely, if traceability requirements are too high, it could decrease the uptake of the scheme and therefore threaten its broader impact. There is additional detail on the chain of custody considerations and their implications in one sector, forestry, in the second half of Chap. 10 "Environmental Standards and Embedded Carbon in the Built Environment".

Standard systems are seeking new solutions to these dilemmas, including how to use technology to increase accessibility whilst maintaining rigour of traceability systems. Online traceability is seen as a possible replacement or supplement to traceability systems, to decrease on-site audit time and cost. Technology can also be used to add valuable controls to combat fraud, which is a risk given the price margin sometimes available for certified goods. FSC is currently setting up an online system to verify and trace the use of FSC claims. Many systems provide access to databases to provide real time certification information (i.e. that a company is still certified to handle certified products) to help reduce fraud.

Traceability systems are one area where active collaboration between systems is already being seen. UTZ Certified, for example, hosts the 'Good Inside Portal' which tracks UTZ Certified products, and has begun outsourcing this to other standards, beginning with RSPO. The ASC has accessed the MSC chain of custody standard and certification methodology and hosts their data within the same database as MSC. This improves accessibility for participants in their shared seafood supply chains and leads to more efficiencies within the newer ASC to allow for faster growth.

1.2.6 Trends in the Uses of Voluntary Sustainability Standard Systems

Sustainability standard systems offer a global approach to reward and encourage actors to stop harmful practices and adopt practices and systems that will improve sustainability outcomes. The standard systems themselves and many of their key stakeholders see this as the primary goal and long-term benefit of standard systems. And yet, as the standard system industry has matured, other actors have begun to use sustainability standard systems for other purposes as well.

Although initially conceived of as an alternative to government regulation, many governments now rely on voluntary standards systems to help enforce or implement their own policies. For example, LEED standards have been incorporated in green building policies of numerous state and local governments in the United States. Tunisia based its national organic agriculture policy on IFOAM standards, which it saw as global best practice (Carey and Guttenstein 2008). These are examples of what a recent report (*Toward sustainability: The roles and limitations of certification*) characterises as 'superseding' private standard systems by incorporating them as part of legislation (Steering Committee 2012). In 'hybrid' models of interaction between governments and standard systems, there is a division of labour and function between governments and sustainability standard systems. For example, the European Union has recognised voluntary sustainability standards as a mechanism for verifying compliance with EU biofuels regulation and its requirements on forest product legality verification.

Sustainable public procurement is another example of government use of sustainability standards. The crucial advantage of sustainability standards systems for public procurers is that they 'outsource' the identification of sustainability hotspots in the supply chain and the verification of compliance with sustainability criteria. This is particularly significant when public purchasers are procuring in multiple categories, with limited time and expertise available to assess each and every product category in detail. In some regions, legal frameworks place limits on the extent to which sustainability standards can be included or referred to in public procurement processes.

Like governments, retailers, manufacturers, and brand managers use procurement of certified products to help meet their own sustainability commitments. The appeal of standard systems to businesses, however, goes beyond sustainable procurement. Partnering with standard systems helps reduce the risk of exposure of unsustainable practices in their supply changes, where one example of bad practice highlighted in the media can significantly damage brand value. Standard systems with consumer facing labels also offer the potential to market certain values to consumers, and to potentially recover higher costs of procurement. Global brands use different sustainability standards in different products in order to appeal to different consumer groups.

One key driver for the growth in standards has been some businesses' recognition of the value of sustainability standards as an approach to securing long-term supply of products and ingredients. For example, Unilever joined forces with WWF to form the MSC shortly after the collapse of the Grand Banks cod fishery, as this lack of supply was a clear threat to the success of their Birds Eye and Iglo frozen fish business. This trend is also now visible in agriculture where climate change, land use pressures, and aging farming populations threaten to reduce supplies. Companies operating in these sectors see certification helping them secure future supply in three ways. The first is that by addressing sustainability issues (e.g. environmental problems that limit production, vulnerability to climate change, and farmer incomes) they will make the production more viable into the future and help ensure continued supply. The second is that the standard system itself creates a link between producers and buyers, which can help ensure a particular company access to the supply they need. Thirdly, having made this link it can help to drive change at the production level, where the longer term relationships can act as reassurance that investments in improved practices have an interested customer, and so will be repaid at a later date.

1.3 Addressing Concerns, Defining Credibility, Demonstrating Effectiveness

Growth in the supply and demand for certified products, increasing diversity in standards and standard systems, and the emergence of new actors and uses for standards are all signs that the world of voluntary sustainability standards is maturing. With this maturity comes also more attention and scrutiny of standard systems in general and of individual systems. Both standards proponents and sceptics have raised important concerns, and standard systems are working to address them.

What are the major concerns that proponents and critics of standard systems have raised about voluntary sustainable standards? And how are standard systems seeking to address those concerns? Three broad areas of concern discussed in the second part of this chapter are growth and market, accessibility and exclusion, and impact and claims.

A first set of concerns relates to the market for sustainable products created by voluntary standard systems. There is concern that these systems will not grow quickly enough—that they will not develop enough supply for the market, or conversely, not generate rapid uptake in the market when certified supplies are available. Finding a balance between demand and supply is tricky. The oversupply of certified products that do not end up being sold with the relevant claim is referred to as leakage. Leakage reduces any margins or benefits from securing customers as a result of certification and can reduce the incentive of producers, factories, or other operations to seek to demonstrate compliance with a standard. Where there is not sufficient supply of certified product available, it can prevent larger customers from making commitments to sourcing from a particular standard because they do not think these commitments can be met. This in turn reduces the market demand for the products produced according to sustainability standards. For this reason, anticipating market demands, developing a market, providing connections between buyers and suppliers, and ensuring that capacity building and assurance services are available to help boost supply are important functions of standard owners. Some standard-setting organisations take on this challenge themselves with strong market development departments and capacity development areas, while others seek partnerships to address the issue. Regular discussions with potential buyers about their sourcing needs helps standard systems and their partners determine where to invest in building new supply. Partnerships with capacity building organisations and donor institutions provide farmers, factories, and enterprises in target commodities and markets support to prepare for certification.

A related concern is that voluntary standard systems are not well-suited to mainstream production, that they are appropriate only for select producers, operators, factories, and fisheries, and not for reaching the vast majority of entities with unsustainable practices. Not only does this threaten to limit the growth of standard systems, it also raises concerns about equal access to standards and the markets they create.

The accessibility of standards is of particular concern regarding smallholder farmers and fishers, manufacturers, and other operators in the developing world. The concerns about accessibility stem from both the costs of achieving compliance (needed investment) and demonstrating compliance (auditing processes). Improving practices in order to meet the sustainability standard can be seen as an investment in the future of the business. However, there are concerns that those who cannot afford to make the initial investment in improvement, or lack the know-how to do so, will be excluded. The cost of audits or verification to demonstrate compliance also creates a possible source of exclusion. When the assurance process includes a site visit, this is typically done at the cost of the entity being certified. Although participation in sustainability standard systems is voluntary, standards compliance is a condition for some buyers. This means that an inability to make investments to come into compliance with voluntary sustainability standards or to cover the cost of compliance checks could exclude producers from certain markets. On the other hand, voluntary sustainability standard systems are actually, at their core, a mechanism to connect producers with new markets, which might not have been accessible to them without the standard system mechanism. Moreover, standards systems and buyers of certified products can provide or facilitate provision of capacity building activities and credit to producers that do not otherwise have access to this assistance.

Sustainability standard systems are addressing concerns about exclusion and accessibility in many different ways. Expanding capacity building and facilitating finance for improvements is one approach. Another area of intervention is in the standard setting process itself: including stakeholder groups in different countries and from different production models in standard development and revision processes and seeking strategies to make global standards nationally relevant and applicable. The on-going revision of Sustainable Agriculture Network

(RA-SAN), for example, will include stakeholder workshops in 20 countries in Europe, the Americas, Africa, and Asia. As described earlier, new models of standards are also emerging to offer a less demanding entry into a standard system, and new models of assurance seek to reduce costs of the compliance check process. At the same time standard systems are investing in building monitoring and evaluation systems, which will give them better information about the entities they are reaching with their standards, those who might be left out, and why. This type of business intelligence will help standard owners improve the reach and inclusiveness of their systems.

The third set of concerns about voluntary sustainability standard systems relates to what they deliver and which claims are based on the use of standard systems. With more and more actors relying on standards to meet their own sustainability objectives, stakeholders and standard system users increasingly want to know whether standards really result in sustainability improvements and/or stop harmful practices (see, for example, Chap. 9 on measuring the impact of standard systems). Expectations that standards should be able to reliably demonstrate that they are making a difference are growing. This demand to prove results is not unique to the standards world: demands for data and evidence are growing in public policy and development work in general. It takes on a special importance for voluntary sustainability standard systems however, because these systems are meant not only to deliver sustainability results but also assure buyers that their purchases are supporting sustainable production. In response to this concern, many standard systems have ramped up efforts to evaluate their systems and collaborate with independent researchers to conduct independent research about their systems effectiveness and impacts. As described earlier, they are also building monitoring systems to track key performance indictors over time. Public access to study and monitoring results is improving, as standard systems increasingly make findings available on their websites.

A related concern is greenwashing. Some critics worry that standard systems and/or their users are making false or exaggerated claims. These false claims could simply not be true (for example, a claim that a process meets a standard when it does not) or claims could be vague or difficult to verify (for example, a claim that the product is 'natural'). False or exaggerated claims are seen with regard to all environmental and social issues and are not isolated to voluntary sustainability standards. In fact, the benefit of products, processes or services making a claim about performance against a standard is that the claim can be independently verified, and the standard owner itself can take steps to prevent false claims about use of its system.

Lack of transparency on who is behind a particular standard can also raise concerns of greenwashing. Sustainability standard systems seek to address this concern with balanced multi-stakeholder standard setting and governance. The FSC, for example, is governed by its members, who represent environmental NGOs, the timber trade, community forest groups and forest certification organisations. Members are organised into three chambers—social, environmental, and economic—and each chamber is divided into north and south sub-chambers. Voting rules ensure balanced input from north and south thereof the respective chambers. Similarly, Fairtrade International has a board with representatives of producer networks, labelling initiatives, and certified traders.

Given these general concerns about sustainability standard systems and the large number of new systems coming into the market, several actors in the standards landscape are developing tools to help standards users identify the individual systems that best meet their needs and address their primary sustainability and credibility concerns. The International Trade Centre (ITC) has created a database (www.standardsmap.org) with detailed information about many sustainability standard systems, and this database feeds tools for public procurement officers and for producers looking for standards that meet their needs. Ecolabelindex.com covers more ecolabels, but with fewer data fields. GoodGuide is a product-specific database and Ekobai provides a central database of companies certified to one of over 400 different sustainability standards. Industry initiatives, such as the Global Sustainable Compliance Program, and other organisations interested in using standards to achieve sustainability objectives are developing benchmarking tools to enable users to compare standards content and processes against each other and/or against a benchmark.

These benchmarking approaches have met with some criticism from many of the sustainability standards systems themselves, and from other stakeholders who are concerned with how benchmarking or equivalency is conducted. For example, there is concern that benchmarking or equivalency tools show two standards are 'equal' when in reality they handle issues very differently in their standards, have very different assurance or traceability practices, or are more or less able to prove their effectiveness. One way to address this challenge is to ensure that the criteria used to compare standards are based on a broad multi-stakeholder agreement about the critical factors that define the credibility of a sustainability standards system.

The ISEAL Alliance works within the community of standard systems, standards users, and other stakeholders to reach this global view of what constitutes credibility and best practice for voluntary sustainability standard systems and then builds this global understanding into its Codes of Good Practice for sustainability standard systems: the Standard Setting Code (2004), the Impacts Code (2010), and the Assurance Code (2012). To become a member of the ISEAL Alliance, sustainability standard systems must commit to achieve compliance with these codes. Code compliance is one of several strategies that the ISEAL Alliance uses to improve the effectiveness and impact of voluntary sustainability standard systems.

The ISEAL Alliance has just conducted an intensive, multi-stakeholder global consultation process, including in-person workshops on four continents, to update its understanding of credible practices for sustainability standards. In June 2013, the Alliance published the Principles for Credible and Effective Sustainability Standards Systems: the ISEAL Credibility Principles[1]—ten principles widely understood to make a standard system credible and to increase the likelihood that the

[1] www.iseal.org/credibilityprinciples.

system will deliver positive sustainability impact (Annex 1). These principles include *impartiality*, *transparency* and *truthfulness* in claims. *Efficiency* and *accessibility* are also critical, but cannot come at the expense of the *rigour* of the assurance process. Also important is *engagement* of affected and interested stakeholders in setting *sustainability* objectives and creating a *relevant* standard that will address the most critical sustainability concerns in a variety of different local settings. Once sustainability objectives are clear and the standard is set, then the systems need to be put in place to track what impacts a standard is having on the ground and to enable sustainable standards to make *improvements* to their systems over time. These Credibility Principles will guide ISEAL's work with sustainability standard systems and are also intended to provide a common framework for wider discussion, assessment, and comparison of the characteristics, strengths and weaknesses of individual sustainability standard systems.

1.4 Conclusion

Though the first examples were seen in organic farming as many as 70 years ago, voluntary sustainability standard systems have really only been around as a significant presence for the last two decades. The ways of using standards and the nature of the standards themselves are evolving rapidly. Different models are emerging to address new needs and particular concerns. Pressure is on the standard systems to innovate to reduce costs and make their standards more accessible, without reducing rigour. Standard systems are also increasingly asked to prove that they do make a difference for sustainability and meet their standard-specific objectives.

Given recent trends, we can expect voluntary sustainability standard systems to continue to grow and evolve. As they are only one tool in the sustainability toolbox, standard systems will increasingly be used to complement other initiatives in combined solutions to the complex sustainability challenges our world is facing. To remain credible and effective in this evolving space, voluntary sustainability standards will need to embrace the ideas embodied in the ISEAL Credibility Principles and strive to continuously improve their performance and effectiveness over time.

Annex 1. Principles for Credible and Effective Sustainability Standards Systems: The ISEAL Credibility Principles

The ultimate aim of sustainability standards systems is to bring about positive social, environmental and economic impacts while decreasing negative impacts. Impacts can be difficult to demonstrate, particularly in the short-term. Integrating these principles increases the likelihood that a standards system will achieve its intended positive impacts.

Sustainability

Standards scheme owners clearly define and communicate their sustainability objectives and approach to achieving them. They make decisions that best advance these objectives.

Improvement

Standards scheme owners seek to understand their impacts and measure and demonstrate progress towards their intended outcomes. They regularly integrate learning and encourage innovation to increase benefits to people and the environment.

Relevance

Standards are fit for purpose. They address the most significant sustainability impacts of a product, process, business or service; only include requirements that contribute to their objectives; reflect best scientific understanding and relevant international norms; and are adapted where necessary to local conditions.

Rigour

All components of a standards system are structured to deliver quality outcomes. In particular, standards are set at a performance level that results in measurable progress towards the scheme's sustainability objectives, while assessments of compliance provide an accurate picture of whether an entity meets the standard's requirements.

Engagement

Standard-setters engage a balanced and representative group of stakeholders in standards development. Standards systems provide meaningful and accessible opportunities to participate in governance, assurance and monitoring and evaluation. They empower stakeholders with fair mechanisms to resolve complaints.

Impartiality

Standards systems identify and mitigate conflicts of interest throughout their operations, particularly in the assurance process and in governance. Transparency, accessibility and balanced representation contribute to impartiality.

Transparency

Standards systems make relevant information freely available about the development and content of the standard, how the system is governed, who is evaluated and under what process, impact information and the various ways in which stakeholders can engage.

Accessibility

To reduce barriers to implementation, standards systems minimise costs and overly burdensome requirements. They facilitate access to information about meeting the standard, training, and financial resources to build capacity throughout supply chains and for actors within the standards system.

Truthfulness

Claims and communications made by actors within standards systems and by certified entities about the benefits or impacts that derive from the system or from the purchase or use of a certified product or service are verifiable, not misleading, and enable an informed choice.

Efficiency

Standards systems refer to or collaborate with other credible schemes to improve consistency and efficiency in standards content and operating practices. They improve their viability through the application of sound revenue models and organisational management strategies.

References

Carey C, Guttenstein E (2008) Governmental use of voluntary standards: innovations in sustainability governance. ISEAL Alliance, London

Ecolabel Index (2013) Ecolabel Index – global directory of ecolabels. http://www.ecolabelindex.com/. Last accessed 05 July 2013

ISEAL Alliance (2004) ISEAL code of good practice for setting social and environmental standards. ISEAL Alliance, London

ISEAL Alliance (2010) Code of good practice for assessing the impacts of social and environmental standards. ISEAL Alliance, London

ISEAL Alliance (2012) Code of good practice for assuring compliance with social and environmental standards. ISEAL Alliance, London

ISEAL Alliance (2013) Principles for credible and effective sustainability standards systems: ISEAL credibility principles. ISEAL Alliance, London

Potts J, van der Meer J, Daitchman J (2010) The state of sustainability initiatives review 2010: sustainability and transparency. International Institute for Sustainable Development (IISD) and the International Institute for Environment and Development (IIED), Winnipeg and London

Steering Committee of the State-of-Knowledge Assessment of Standards and Certification (2012) Toward sustainability: the roles and limitations of certification. RESOLVE, Inc., Washington

Chapter 2
Standards for Standards: The Foundations for Voluntary Standards for Sustainability

Robert B. Gibson

2.1 Introduction

Twenty years ago Paul Hawken introduced *The Ecology of Commerce* with a story about a moment of personal enlightenment. It had occurred as he was accepting a major award for his company's efforts to be more environmentally responsible. As he reached the podium to receive the award, he "suddenly realised two things: first, that my company did not deserve the award, and second, that no one else did, either". Despite all the positive steps that had been taken, the effects were at best marginal. Moreover, he concluded, "If every company on the planet were to adopt the best environmental practices of the 'leading' companies . . . the world would still be moving toward sure degradation and collapse" (Hawken 1993, pp. xi–xiii).

Since then the standards of best practice in environmental behaviour, social responsibility and overall contributions to sustainability have risen substantially. There have been innumerable laudable initiatives in the past 20 years. They have been undertaken in most sectors and jurisdictions, and have included not only innovations driven directly by the usual legislative and economic imperatives but also a wide range of other more or less voluntary efforts engaging a rich diversity of motives, drivers and participants. A now vast literature documents these initiatives and while there have certainly been disappointments, the achievements merit celebration.

Nevertheless, Hawken's statement from 1993 remains largely valid. The sustainability initiatives undertaken so far are still marginal phenomena in a world where the entrenched practices of the dominant forces of political economy remain highly problematic—devoted to growth based on expansion of material and energy demand and demonstrating no effective inclination or ability to deliver the bulk of the benefits to those who most need them. Whether or not current best practices, if

R.B. Gibson (✉)
Department of Environment and Resource Studies, University of Waterloo, 200 University Ave. W., Waterloo, ON, Canada N2L 3G1
e-mail: rbgibison@uwaterloo.ca

C. Schmitz-Hoffmann et al. (eds.), *Voluntary Standard Systems*, Natural Resource Management in Transition 1, DOI 10.1007/978-3-642-35716-9_2,

universally adopted, could be adequate to reverse continuing trends towards deeper unsustainability is perhaps an open question. Best practice standards are now both higher and more clearly demonstrated in particular concrete examples. But the challenges are also greater. The past two decades have been characterised by worsening reports on many if not most of the key indicators of sustainability—higher atmospheric concentrations of greenhouse gases, more losses of biodiversity, further depletion of groundwater supplies, declines in wild fisheries and other ecological resources, increased reliance on more fundamentally risky technologies, and a rising conjunction of financial and ecological debt—all of which compromise the prospects of future generations.

In this situation, more numerous and more advanced sustainability initiatives of all sorts, including voluntary standard systems, are surely needed. Moreover, it is clearly important to design, apply and assess these initiatives with ambition as well as creativity, mobilising our best understanding of the challenges to be faced, while recognising the lessons from past experience, the imperatives for progress towards sustainability, and the practicalities of acting on these imperatives in the great diversity of particular cases and contexts.

The purpose of this chapter is to provide an initial basis not only for thinking about sustainability-oriented voluntary standard systems but also for designing and implementing them. Perhaps a little too grandly, the chapter title promises a set of standards for the design and implementation of voluntary standard systems for sustainability. Given how little we know about how the world works or about how to foster positive transitions in human institutions and behaviour, we should not anticipate much more than a provisional working understanding of how best to move forward. But we have now had over a quarter century of thought and experimentation with sustainable development and can claim to have learned a little.

With that in mind, this chapter now turns to an examination of the essential rationale and requirements for a transition to sustainability and the implications for voluntary standard systems. Section 2.2 considers the challenges posed by the deepening unsustainability of current practices globally and in most jurisdictions. Section 2.3 reviews the general implications—the core generic requirements for reversing direction and moving towards sustainability, and the key considerations in deciding how best of act on the requirements in particular contexts. Section 2.4 then sets out the evident implications for sustainability-based voluntary standard systems, including implications for their design and implementation, and also their place as interactive components in a larger set of initiatives for change towards sustainability. The final Sect. 2.5, reviews the argument and points to the promise of higher standards for voluntary standards.

2.2 Deepening Unsustainability

There is much that is good in this world. The human record includes long chapters featuring the stupid and brutal, but also a great and diverse legacy of worthy traditions and collective accomplishment. The recent centuries of modern economics and science have brought especially significant improvements as well as larger perils. Compared to previous generations, most people today live longer, with more material security and broader opportunities. While the distribution of benefits has been highly uneven, and whole categories of people have lost more than they have gained or are otherwise still lacking the basics for wellbeing, the starting point for thinking about sustainability is that there is much to be sustained in the socio-economic realm as well as in the ecological one.

The second key point is that what has been achieved for human wellbeing so far, and all prospects for maintaining and extending those achievements, are ultimately and inevitably dependent on the viability of the biosphere—as our home as well as our source of material wherewithal for existence and advance. Especially since we have seen the astronauts' haunting photos of Earth as a lonely blue ball, it is easy enough to grasp the basic logic of a limited planet that cannot support indefinitely expanding demands on its capacities to supply, assimilate and withstand. Adherents to cornucopian views argue quite rightly that the planet's carrying capacity for humans is not fixed; it can be and has been expanded through technological and managerial innovations that deliver greater efficiencies, mitigate damages, permit exploitation of previously inaccessible resources or supply substitutes for depleting commodities. But it does not follow that the planet's capacities are infinite, or that the needed innovations can keep up as rising demands press harder against what remains. Indeed the most fundamental evidence before us is that the innovations have not kept up, that sustainable carrying capacities have been exceeded in several crucial areas, and that few of the benefits of growth are going to sustain what has been achieved or to address the most serious continuing needs.

In that context, we can turn to the three fundamental factors that define the core challenges for sustainability:

1. Global demands on the biosphere's carrying capacity for humans at our present level of technology and managerial competence are almost certainly well over what might be sustainable in the long run. The WWF's biennial calculations of the global human footprint suggest that we crossed the line into unsustainability in the 1970s, and that the effects of rising demands for energy and materials, assimilative capacity and other ecosystem services are now 50 % above the sustainable level and still climbing (WWF 2012, pp. 39–51). While the specifics are debatable, the key effects in biodiversity decline, greenhouse gas loadings, desertification, soil loss, and depletion of other key ecological resources are well documented (e.g. MEA 2005; IPCC 2007; UNEP 2012). These effects have already reduced our foundations for wellbeing, and the damages and risks are still rising faster than our improvements in efficiency and mitigation.

2. At the same time that we are demanding too much of the planet's capacities, a very large portion of the human population suffers from basic material deficiencies. UNICEF estimates that a billion children "are deprived of one or more services essential to survival and development" (UNICEF 2009, p. 18). Roughly 870 million people—one person in eight—now suffer from chronic malnutrition (FAO 2012). Over 780 million have to rely on unsafe drinking water sources (UNICEF 2013). And so on. Some of these numbers represent considerable improvements over past decades, but there are still many people whose basic needs have yet to be met, even though we are already over-exploiting the resource base. We are using too much and at least a billion people do not have enough.
3. Very few of the benefits derived from the long post-war record of economic growth and biophysical stress have been going to those most in need. On the contrary, the richest 10 % of the world's population gets about 67 % of total global income while the poorest 10 % of people get less than 0.2 % of income (Milanovic 2011, pp. 152–153).[1] The hypothesis that the fruits of increasing wealth will trickle down to the poor is not disproven by these data, but clearly the trickle is very small and its effects have been insufficient. Moreover, reliance on continued conventional economic growth to deliver a trickle to the poor is not viable when that growth is tied to increasing material and energy demand and rising stress on already impaired biophysical systems.

Many other factors add detail to this picture. It is, for example, important to consider the highly uneven distribution of these global phenomena, the crucial role of wellbeing and security (especially for women) in reducing human population growth, the added perils and promises of technological advances and movements for change, and the evident wobbliness of financial systems wresting with debt problems that have arisen from another form of ill-advised borrowing from the future. But the thrust of these considerations is to clarify and strengthen the quite obvious implications for sustainability represented by the three points above.

2.3 Implications for a Transition from Unsustainability to Sustainability

2.3.1 The Basics

The core implication is that we must reduce overall demands on the biosphere substantially if we are to maintain the foundations for human and other life while

[1] The gap between the incomes of the richest and poorest is a little less extreme if recalculated to take into account the differences of purchasing power from place to place. The richest 10 % then get 56 % of global income and the poorest 10 % still get close to 0 % (Milanovic 2011, p. 152).

also ensuring material sufficiency, security and opportunity for those currently deprived. That is a tall order. If the WWF is right that our demands are already 50 % over what might be sustainable, and if a billion or more people are in more or less serious material need, then a 60–70 % cut is probably a conservative estimate of what is needed. And for some parameters, for instance greenhouse gas emissions with delayed adverse effects that build over time, the needed reductions are higher.[2] At the same time we must reduce dramatically the grotesque and in many places still widening inequities of wealth, influence and opportunity that drive excess, maintain deprivation, waste resources, encourage grievance and deepen insecurity.

The easy part is probably finding technically feasible ways to achieve these reductions and redistributions. The harder parts are conceptual and institutional. As the Brundtland Commission emphasised a quarter century ago, the problems and the requirements for change are fully intertwined (WCED 1987). Reducing our demands on the biosphere, establishing a more equitable world and ensuring essential wellbeing for all, need to be pursued together—not merely at the same time, but as a package of mutually reinforcing initiatives—so that gains in one aspect foster, complement and support gains in the others. Approaching issues that way is not often attempted in a decision-making world more often typified by defined expertise, specified mandates and fragmented authority. The required transitions will require better integration of action as well as understanding. The task, after all, is not just to invent better ways of living. It is to reconfigure and redirect a vast and firmly entrenched structure—the great suite of assumptions, institutions and habits that underlie an economic system based on ever expanding demands for materials and energy, stress accommodation and waste assimilation—and to do so in ways that protect the most vulnerable and serve the most neglected.

While the transition required is both sweeping and fundamental, it must also be gradual. The changes involved are not merely in methods and behaviour, but also in structures and eventually cultures. To be lasting, such changes must be widely embraced and fully absorbed into institutional form and substance. It is true that we have seen quite rapid change in the recent past. The great post-World War II rise of production and consumption, sometimes called the 'great acceleration' (UNEP 2012, p. 22) brought transformative changes in ideas and cultural practices as well as in economies and ecologies globally over a few decades. But as the concept of acceleration suggests, these changes merely sped up and expanded long-standing trends driven by already well-established interests. In contrast, a transition to sustainability requires a reversal in direction. It will take time.

---

[2] According to Weaver et al. (2007), even if greenhouse gas emissions reductions of 90 % relative to 1990 levels were achieved before 2050, overall global warming would eventually exceed the 2 ºC threshold likely to bring significant adverse effects.

2.3.2 Transition, Reassertion and Redirection

Not everything has to change. The characteristics of a more sustainable society reflect basic values—stewardship, collaboration, precaution, equity, creativity, foresight, etc.—that have been celebrated at least since the dawn of literature, and are preserved and practiced still in various ways in most communities. The path to sustainability does not require the invention of new values so much as the reassertion of old ones.

The need to reverse direction does not imply an obligatory end to economic expansion. The requirement is only that such expansion be pursued through means that reduce stress on the biosphere and that a much larger portion of the benefits from remaining extractive activities and other biospheric stresses go to those most in need. These are the main elements of the long-standing distinction between development and growth and the main grounds for defending the concept of sustainable development. Economic expansion can contribute to lasting wellbeing, but only if it combines reduction of biospheric stress while ensuring enough for everyone. Practically, this entails transition to economies (and societies, cultures, ways of thinking and living) that are much more equitable in the delivery of material sufficiency, much more efficient in the use of materials and energy (in the order of ten times more benefit per unit of resource), and much more focused on delivery of life-enriching opportunities and satisfactions that rely minimally if at all on material or energy consumption. Sustainability respects biophysical limits, moral obligations and the practical imperatives for just and viable socio-ecological arrangements. But it also requires creativity and innovation, the expansion of opportunities, the enrichment of understanding and the open-ended pursuit of enhanced (and lasting) wellbeing.

Similarly, the needs for transition do not entail an end to significant reliance of government and market institutions, though both will have to depart from business as usual. Accountable institutions of public government will remain indispensible as venues for deliberation and authoritative action on matters of collective interest. And the market mechanism, suitably informed by prices that reflect the full costs of goods, will continue to be crucial as our most ingenious tool for matching supply with demand. By themselves, however, government authorities and market players cannot reasonably be expected to deliver a transition to sustainability. Their established inclinations and practices have been driving us in the opposite direction, and if by some magic that were suddenly to change, the basic agenda for a transition to sustainability would still be ambitious and complex beyond the capacities of any particular authority or mechanism. The skills and capacities of all potential participants—civil society organizations and citizens as well as governments and private sector bodies—will be needed. Moreover, the practical feasibility of a transition to sustainability will depend heavily on the understanding, acceptance and positive involvement of those who will be affected. Not surprisingly therefore, the literature on decision making in the pursuit of sustainability centres on governance, how to engage multiple bodies (public, private and civil society organizations and less formal groups and individuals) at multiple interacting scales (local to global) in collective decision-making arrangements that are transparent, accountable and effective (Adger and Jordan 2009; Kemp et al. 2005; Lafferty 2004).

Finally, a commitment to sustainability does not impose obligations to follow a single path or a strict agenda. This is in part because we are working in and with complex socio-ecological systems with dynamic interactions at many scales and no fixed ends. Not surprisingly we do not understand these systems well. Even where we have invested massive funding and effort (e.g. in climate change science), our understanding even of the biophysical aspects of how key systems function, interact and deal with stress and opportunity, is highly imperfect. When we add the further complexities of predicting, or even explaining, the behaviour of human institutions, the difficulties multiply. With such limited grounds for describing sustainability-related situations and the possibilities, we are in no position to claim much confidence about how best to nudge system change in desirable directions. Instead, the key lessons from studies of sustainability and complexity favour diversity and experimentation, multiplicity of participants and approaches, pilot projects with careful monitoring and adjustment, a precautionary bias preferring low risk options, and iterative review of desired ends and attempted means (Gunderson et al. 1995; O'Brien 2000; Walker and Salt 2006). These are characteristics at the opposite end of the spectrum from a single fixed path.

Taken together, these characteristics of transition towards sustainability make the challenge both daunting and potentially within our reach. The changes required are extensive and fundamental. They go well beyond mitigation of damage to reversal of direction. At the same time, strategies for transition can rely on longstanding values and entail extension rather than elimination of key existing mechanisms. While quick and effective action is needed on many particular fronts, the overall cultural transition must be gradual. And while there is an evident basic set of unavoidable general obligations if we are to move in the desired direction, the means of achieving progress are multiple, diverse and mostly still experimental. The door for creativity and innovation is wide open.

2.3.3 Generic Criteria and Particular Applications

The pursuit of sustainability needs both broadly applicable rules and sensitivity to the demands of particular circumstances. Unsustainability is a global problem and making a transition to sustainability is a global responsibility. Effective responses would seem to require at least some global guidance that is influential enough to foster basic consistency of direction. But, as suggested above, no single top-down solution is available, in theory or in practice. Moreover, aside from climate change and stratospheric ozone depletion, most threats to sustainability are the cumulative effects of multitudes of local and regional abuses. While these abuses and appropriate responses to them may have common features, they arise in different contexts with different ecologies and cultures, stresses and opportunities, capacities, aspirations, possibilities and priorities. Sustainability initiatives that do not respect these differences are likely to suffer the same unfortunate fate as military interventions and development aid projects that failed to grasp the significance of context.

There is endless room for debate about how best to foster overall consistency and context sensitivity at the same time. Probably the most effective solutions will vary—what works for biodiversity recovery may not work for women's empowerment. The most common general solution, however, is to define a limited set of basic shared criteria for universal application and to specify them for particular applications. This approach has been demonstrated, explicitly and implicitly, in the development of many initiatives meant to contribute to a transition to sustainability. Some of the best examples centre on efforts to design and apply voluntary standard systems, many of which involve establishing broadly applicable basic standards and defining how the generic requirements will be elaborated for particular applications. The Forest Stewardship Council, for instance, has set basic principles and criteria plus policies, procedures and other mandatory standards that provide the common foundation for specific applications—establishing national standards, setting expectations for particular forest types, and certifying forest owners/managers and others in the forest product chain for the labelling of sustainably harvested forest products (FSC 2012; 2013).

The starting point, then, is establishment of a broadly applicable generic set of criteria for designing and assessing undertakings to assist in the transition to sustainability. Logically, these should be based on the essentials of what we need to accomplish—the basic requirements to be met if we are to move towards sustainability. Identifying these is a task less difficult than might be expected. While the countless sustainability deliberations and experiments over that past few decades are remarkable for their variety, they easily provide a sufficient base for identifying the basic generic requirements for moving towards sustainability. The short list in Box 2.1 below represents a synthesis of the literature. It also reflects the considerations noted above.

Box 2.1: The Eight Basic Requirements for Progress Towards Sustainability

Socio-ecological system integrity

Build human-ecological relations that establish and maintain the long term integrity of socio-biophysical systems and protect the irreplaceable life support functions upon which human as well as ecological well-being depends.

Livelihood sufficiency and opportunity

Ensure that everyone and every community has enough for a decent life and opportunities to seek improvements in ways that do not compromise future generations' possibilities for sufficiency and opportunity.

(continued)

Box 2.1 (continued)

Intragenerational equity

Ensure that sufficiency and effective choices for all are pursued in ways that reduce dangerous gaps in sufficiency and opportunity (and health, security, social recognition, political influence, etc.) between the rich and the poor.

Intergenerational equity

Favour present options and actions that are most likely to preserve or enhance the opportunities and capabilities of future generations to live sustainably.

Resource maintenance and efficiency

Provide a larger base for ensuring sustainable livelihoods for all while reducing threats to the long term integrity of socio-ecological systems by reducing extractive damage, avoiding waste and cutting overall material and energy use per unit of benefit.

Socio-ecological civility and democratic governance

Build the capacity, motivation and habitual inclination of individuals, communities and other collective decision-making bodies to apply sustainability principles through more open and better informed deliberations, greater attention to fostering reciprocal awareness and collective responsibility, and more integrated use of administrative, market, customary, collective and personal decision-making practices.

Precaution and adaptation

Respect uncertainty, avoid even poorly understood risks of serious or irreversible damage to the foundations for sustainability, plan to learn, design for surprise and manage for adaptation.

Immediate and long term integration

Attempt to meet all requirements for sustainability together as a set of interdependent parts, seeking mutually supportive benefits.

–from Gibson et al. (2005, *ch. 5)*

The eight requirements for progress towards sustainability can serve as common criteria for evaluations and decisions on undertakings at any scale, in any sector or place, anywhere on the planet. They should be the core considerations in all potentially significant evaluations and decisions, including in the design and implementation of voluntary standard systems. Note that none of the listed requirements is expendable and all are interdependent.[3] Progress in all areas is necessary and the

[3] The basic substance of the requirements/criteria in Box 2.1 could be presented in many different ways. The approach taken here makes no attempt to express the criteria so that they fit conveniently into the commonly adopted three pillars of sustainability (economic, social, ecological). This is in part because the core requirements do not naturally fit into those categories. But the non-pillar categorisation also encourages attention to interconnections and discourages the habitual separation of sustainability initiatives (especially research and reporting) into separate solitudes of expertise and mandate. Avoiding the pillars opens more space for recognising the interactions and interdependencies that are at the heart of sustainability (see Gibson 2006).

desired results of application are multiple, mutually reinforcing, fairly distributed and lasting gains.

Generic criteria are, however, only the beginning. For all particular applications, they need to be specified and elaborated for the case and context. That too is not in principle enormously difficult. It entails identifying the key sustainability-related issues that are crucial to the application—the problems, stresses and concerns, valued qualities, assets and deficiencies, aspirations and opportunities, etc.—and combining them with the generic requirements in a consolidated package, framed as a set of evaluation and decision criteria. The character as well as the specifics of the case and context issues will vary from one application to another. Issues for standard setting in an industry sector will differ from those for planning decisions in an urban region. But for illustration, the following questions could help guide thinking about the core list of big issues in a rural or remote area needing to evaluate economic development options:

- What qualities are most highly valued in the relevant communities, and associated natural and built environments?
- What resources, ecosystems, populations, traditions and other assets are already stressed or otherwise vulnerable to damage or loss?
- Are new or expanded or more diverse livelihood opportunities needed and if so where and for whom?
- Where are the greatest needs and openings for greater efficiencies and less waste in the use of environmental and other assets?
- Where are the greatest needs and openings for more equitable distribution of livelihood opportunities and for fairer distribution of influence in decision making, risks of adverse effects, etc.?
- Where are the greatest needs and openings for building greater community and regional self-reliance and adaptive capacities (greater ability to take advantage of new opportunities, and reduced vulnerability to outside pressures, unexpected problems, etc.)?
- Where are the greatest needs and openings for shared learning, collaborative action and mutual assistance?

Answers to these and similar questions will be better if informed by the perspectives of diverse stakeholders, as well as those with professional or experiential expertise in the relevant matters. The prospects for understanding and adoption of the resulting criteria will also depend heavily on the extent to which the criteria development process, including the issues identification work, is transparent, broadly participative and credibly rigorous. Some important insights will involve recognition of what is not known, what may be at risk or what may be worth trying. The information base will always be imperfect and identifying the main uncertainties will help in specifying criteria related to the generic requirement for precaution and adaptability.

Together, the generic requirements and the key issues for the particular application provide the makings of a set of criteria (sometimes in the form of a set of sustainability-based standards) for evaluations and decisions. How best to structure these criteria depends chiefly upon the needs of the audience—those who are meant to adopt and apply the criteria and to live with the results. The criteria must address all of the generic requirements, including recognition of their interdependencies, but it does not follow that the generic requirements should be adopted as the framework and the issues used only to specify the most important concerns. Usually it is better to adopt framework categories, or at least category names, based on concerns and language familiar to the audience and to use the generic requirements list only to ensure that no big considerations are neglected. The priority, after all, is to establish well-founded sustainability-based criteria that will be understood, embraced and applied.

All this is, as noted above, easy enough in principle. The essential criteria are now reasonably evident and the basic approach to specification is not far beyond common sense. Unfortunately, there are major challenges in practice. Because the unsustainable trends of the present are rooted in the underlying ideas as well as the entrenched structures and behaviours of dominant institutions and dependent individuals, resistance is inevitable. Moreover, implementation involves a host of practical difficulties:

- Who should be at the table in identifying the key relevant issues and specifying the criteria? And how should the interests of those who cannot be at the table (e. g. future generations) be represented?
- How should we deal with tensions between the actual complexity of overlaps and interactions, and the understandable desire for simple, distinct and measurable criteria?
- To what extent should the criteria development be tied to identification of indicators? To what extent is it important that indicators be quantitative? What is to be done when appropriate data are not available?
- Given the great gap between current practice and what might be sustainable, how high should the bar be set? Is it better for the criteria to be a genuine test of sustainability or for the standard to be within the reach of most current players who make an effort?
- What means will best encourage broad adoption, effective implementation and further learning?
- How can consistency of criteria for longer term planning be combined with continuous improvement of the criteria as new understandings, challenges and possibilities emerge?
- How should disagreements be resolved?
- How can all of these difficulties be addressed with good information, careful deliberation, fair processes and defensible results without being frustratingly slow and expensive?

All of these matters and many more have been confronted and examined in the literature and in practical initiatives. There are good answers to many of the questions, though predictably the answers are multiple and varied, largely because different cases demand different responses. In all cases, however, the means of implementation need to respect the substance of the core objectives. The generic criteria for sustainability along with the specifications for case and context need to be applied in designing implementation strategies as well as in defining the criteria and standards themselves.

In the end, we are left with a reasonably clear overall imperative and set of basic requirements for moving towards sustainability, evident needs to specify these generic requirements for particular cases and contexts, and a rich set of more or less well-tested approaches to implementation. These three interrelated components constitute a package with a solid centre and necessary flexibility. They serve well as a base for the design and evaluation of voluntary standard systems intended to assist the transition to sustainability.

2.4 Implications for Sustainability-Based Voluntary Standard Systems

Voluntary standard systems have emerged as one of the most promising, and already influential, tools for fostering and guiding more sustainable behaviour in a wide range of applications, sectors and places around the world. Current standards follow no rulebook and take may forms. Perhaps ironically there is no standard name or firm boundaries for the concept. The 'voluntary' aspect generally excludes requirements imposed by law. But many 'standards', such as those for formal certification purposes (e.g. fair trade coffee or sustainably harvested seafood), are binding and linked to provisions for proof of compliance. Other standards are merely guidelines designed to be applied in different ways to suit the circumstances or to be adopted to the extent the relevant player or players find feasible (e.g. GRI and other efforts to encourage and guide corporate sustainability reporting). Some standards are meant to define what is actually needed to justify claims to sustainability. Many others aim lower—to educate by defining current best practice, or to entice their audience onto the first tiny steps in the right direction. And some reflect little more than the manipulative self-interest of powerful participants. Almost all of these, perhaps even a few in the mostly greenwash category, can be helpful so long as they do not detract from more useful action. Considering the greatness of the need and the range of organisations and activities to be moved, diversity of form and strategy is mostly an advantage.

Moreover, voluntary sustainability standards are not merely tools; they are (and are participants in) larger systems. Most obviously they depend on motivations that are rooted in and arise from many aspirations and pressures affecting institutions and individuals in corporate, public sector and civil society life. In the corporate world, for example, the outside pressures influencing adoption of voluntary sustainability standards and associated initiatives include fear of regulatory action or liability claims, anticipation of new tax burdens or new market opportunities, resource price and availability worries that encourage attention to energy and materials efficiency, pressures from other companies in a supply chain, requirements of bankers and insurers, demands of consumers, complaints of neighbours, expectations of current and potential employees, and personal commitments of individuals including but not limited to corporate leadership.

Inevitably, the factors surrounding voluntary standards for sustainability also include opposition, resistance and other limiting pressures. The positive incentives noted above face entrenched behaviours and powerful counter influences. Many of the desirable benefits are not effectively valued (not priced at all or not priced adequately) in the current marketplace. Demands for immediate results dampen interest in gains that may not be visible soon enough to maintain support. The supply of low hanging fruit—initiatives that clearly promise short-term profitability as well as longer-term future benefits—can be quickly exhausted. Means of ensuring accountability are not automatically present. Voluntary initiatives may be presented as substitutes for enforceable regulation or used as rationales for dismantling regulatory capacity even though fear of strict and inflexible regulation is a significant driver of participation in voluntary measures. As a result, sustainability-based standards many be embraced more to enhance reputation than to guide serious implementation.

The positive and negative factors are a complex and dynamic mix of interacting pressures, responses and further expectations. They are also clearly important determinants of the potential effectiveness of voluntary standards for sustainability. Perhaps most significantly, the influences overlap and interact. The positive factors build upon one another as a set of mutually supportive encouragements that expand adoption and implementation of sustainability standards. In addition, as adherence to voluntary sustainability standards becomes more common and more evidently feasible and effective, they raise the bar for what is acceptable and what merits recognition for excellence. The results include more positive pressures on regulatory, fiscal and market behaviour, as well as higher expectations among suppliers, investors, customers, neighbours, employees and peers. This is the upward spiral of desirable systems interactions that is needed to overcome the barriers to effective sustainability standard implementation and ultimately to reverse the downward spiral of unsustainability.

To date the interacting positive aspects, and accompanying strategies for overcoming the barriers and limitations, have come together more or less accidentally in particular applications. No doubt the diversity, flexibility and dynamism of voluntary standard systems and their associated drivers have been crucial contributors to their growing influence. Nevertheless, greater shared understanding of the bigger picture would help. That understanding would include lessons from experience about the main drivers of effective adoption and application and how to link them, about the main barriers and limitations and how to overcome them, and about the resulting desirable characteristics of voluntary standards. Many of these have been ably documented by academics, practitioners and associated organizations. But the most fundamental component of the needed big picture for designing and implementing voluntary sustainability standards is centred on the basic objectives—whether they address the essential requirements for progress towards sustainability and how well these requirements have been specified for application in and through particular voluntary standards.

A basic set of sustainability-based criteria for voluntary standard development and application can serve two functions. It can provide core direction for the design of particular standards—ensuring that all key transition-to-sustainability considerations are included in each standard and guiding how application of the standard will amount to application of a comprehensive sustainability-based framework for identifying problems, evaluating options, making decisions and assessing results. At the same time, the basic sustainability criteria can be used to test the core substance of existing and proposed voluntary standards.

There are now many hundreds of serious sustainability-based voluntary standards addressing a host of very different sectors, commodities, practices and places. Many are well documented in accessible media and the documentation on the best examples often reveals significant efforts to establish credible sustainability criteria, along with appropriate indicators and protocols for monitoring, review and reporting. Learning from those initiatives, including their deficiencies as well as their triumphs, has also led to a sizeable literature. While the bulk of these studies have focused on particular initiatives or sets of initiatives in certain fields, some have attempted overview assessments of the state of the art (e.g. IISD 2010) and others have offered general guidance on how to establish credible and effective voluntary sustainability-based standards (e.g. ISEAL Alliance 2010a, b). Both the broad assessments and the general guidance documents devote considerable attention to process concerns—matters of governance inclusivity and transparency, credible monitoring of compliance, provisions for dispute resolution, respect for regional and local differences, attention to the needs of small players, etc.—but also address the substantive requirements for contributions to sustainability. Implicitly in the case of the assessments and explicitly in the case of the general guidance, these projects propose standards for voluntary sustainability standards.

The resulting standards for standards differ. The ISEAL guidance on general sustainability concerns, for example, identifies 13 issue categories (from labour rights to value chains) divided into the usual social, environmental and economic pillars of sustainability. The 13 are "intended to represent the most important issues for societal well-being and environmental resilience at a global level" and to provide "a common basis for standards systems to assess their contributions," but the section is described as "informative" rather than mandatory (ISEAL Alliance 2010b, p. 8). Sustainability criteria selection and elaboration are mostly left to the defined processes for particular standards. In contrast, the IISD project examines whether or not and to what extent the reviewed standards include attention to 55 more specific environmental, social and economic concerns (IISD 2010, pp. 147–149), plus a similarly lengthy list of other, largely process-related factors. The analysis considers only what the reviewed standards expect, not what is delivered on the ground by those organizations that adopt the standards. The implication, however, is that all voluntary standards ought to address all these concerns.

Both approaches have merit. Recognising the crucial importance of credibility among diverse interests, both have focused on issues identified by respected global governance bodies with multi-stakeholder processes, and have emphasised procedural best practice in the deliberations of particular standards organizations. They have also accepted the need for reliable data for monitoring impacts. Unfortunately, the price of conventional action on these decisions is reliance on a reductionary approach to criteria and indicators that focuses on individual issues, relies on isolated data categories following the three pillars, and seeks linear improvements. The criteria based on individual issues encourage positive steps in important areas. But they do not provide direct or comprehensive means of addressing the key requirements for progress towards sustainability.

In view of the distance to be travelled and the practical challenges of going beyond individual issues, conventional data and linear gains, the current approaches may often represent the best we can do in the circumstances. And in the end, efforts focused on individual issues may often combine well enough to deliver gains that support each other and promote further advances. But the complex, intertwined and daunting requirements for progress towards sustainability identified in Box 2.1 remain. Those requirements do not fit in tidy boxes and are not likely to be satisfied by adding up a list of individual accomplishments where we have managed to improve wellbeing prospects for some people here and mitigate ecological damage in some areas there. As in most initiatives to pursue a transition towards sustainability, the development of voluntary sustainability standards would gain from a strengthening of our abilities to respect and make use of complex interactions—to find better means of designing our activities to deliver an expanding foundation of mutually reinforcing gains.

For voluntary standards, the advantages of a more complex, integrative and ambitious approach to sustainability criteria parallel the long-recognised advantages of an integrated approach to mobilising the motivations, drivers and associated tools for change towards sustainability. The many potentially available motives for 'voluntary' change (e.g., for private sector participants, anticipation of cost savings, reputational enhancement, lower risk of liabilities, market expansion, and avoidance of inflexible regulatory imperatives) and related drivers (customers and suppliers, regulators, bankers and insurers, employees, neighbours and peers, etc.) serve more powerfully and more reliably if combined. Similarly, combined use of multiple tools (voluntary measures supported by regulatory action, plus market pricing changes, product chain links, transparency expectations and educational initiatives, etc.) works far better than use of any one of them alone. Applying the same principles to the development and application of sustainability criteria would be equally fruitful. Certainly the integrative approaches are more demanding. They require broader collaboration, deeper innovation, and more creativity in implementation, none of which can be mobilised easily. However, it is unlikely that a transition to sustainability can ever be achieved without embracing the complexities and building positive interactions.

In sum, there is room for a higher overall standard to be set for and by voluntary sustainability standards. That higher standard would begin, as this chapter has done, with the basic realities of deepening global unsustainability and the fundamental requirements for reversing direction and fostering a transition towards sustainability. It would combine these general requirements with recognition of the specific big issues for voluntary standards and the specific considerations of particular contexts and applications. And it would focus not only on promoting measurable gains in individual categories, but also on building a positive spiral of interactive effects. Such practice may emerge and flourish more easily in local scale standard-setting initiatives than in global efforts. But the core imperatives for change—the list of requirements in Box 2.1—are most visible from a global perspective. The advantage of a general standard for standards is that it can take the big picture and draw out the framework implications for developing the more specified criteria needed for particular standards. We have not done enough of that yet.

2.5 Conclusions

The central argument in this introductory chapter has been quite simple:

- What we are doing on this planet is increasingly unsustainable and dangerously unfair in linked ways that threaten a downward spiral of impaired biospheric capacity and constrained well-being.
- Redirecting our structures and practices towards more desirable and lasting futures will entail a substantial set of changes affecting most realms of activity.
- The basic requirements for transition towards sustainability are easily identifiable but they interact in complex ways, apply in a world of uncertainty, and need to be specified for each of the great diversity of particular applications.
- Specification of sustainability requirements and translation of them into criteria for evaluating options and designing solutions is easy in principle, though richly complex in practice.
- Our best hopes lie in initiatives that can foster multiple, mutually reinforcing, fairly distributed and lasting gains that combine into an upward spiral of movement towards sustainability.
- In a world where governments and markets cannot reasonably be expected to deliver all of the needed changes, voluntary sustainability standards that mobilise diverse players, incentives and tools offer a promising way to link broad engagement with a serious commitment to sustainability (though there are plenty of barriers to overcome).
- Making good on that promise requires not only the careful attention to credibility demonstrated in the current best guidance on voluntary standards, and the combined engagement of many motives and drivers, but also adoption of a more critical, complex and integrated understanding of overall sustainability requirements and how to specify them for particular applications.

The simple argument unfortunately entails more complex challenges for those already doing heroic and insufficiently celebrated work at the leading edges of establishing, implementing and improving voluntary sustainability standards and their applications today. Current sustainability standards vary greatly in quality and ambition as well as in character. Some are mere marketing devices adopted to cloak unsustainable business-as-usual practices. The best of them, however, are exemplars of new models of participatory, transparent and progressive governance that offer a necessary supplement to the old vehicles of government authority and market choice. As such they represent valuable openings for the application of new understandings—especially about sustainability and complexity—at a time of pressing global needs to change direction. Arguably the leading voluntary sustainability standards initiatives are among the brightest prospects for defining and promoting more hopeful paths to lasting well-being. With this great potential come heavy responsibilities and daunting complexities but also the satisfactions of work that is of the highest public significance and at the frontiers of concept and practice.

References

Adger NW, Jordan A (2009) Governing sustainability. Cambridge University Press, Cambridge

FAO – Food and Agriculture Organization of the United Nations (2012) The state of food insecurity 2012. FAO, Rome. http://www.fao.org/docrep/016/i3027e/i3027e00.htm. Last accessed 09 Mar 2013

FSC – Forest Stewardship Council International Center (2012) FSC International Standard: FSC principles and criteria for Forest Stewardship, FSC-STD-01-001 (V5-0) EN. https://ic.fsc.org/principles-and-criteria.34.htm. Last accessed 12 Mar 2013

FSC – Forest Stewardship Council International Center (2013) FSC certification. https://ic.fsc.org/certification.4.htm. Last accessed 12 Mar 2013

Gibson RB (2006) Beyond the pillars: sustainability assessment as a framework for effective integration of social, economic and ecological considerations in significant decision making. J Environ Assess Policy Manage 8(3):259–280

Gibson RB, Hassan S, Holtz S, Tansey J, Whitelaw G (2005) Sustainability assessment: criteria and processes. Earthscan, London

Gunderson LH, Holling CS, Light SS (1995) Barriers and bridges to the renewal of ecosystems and institutions. Columbia University Press, New York

Hawken P (1993) The ecology of commerce. HarperCollins, New York

IISD – International Institute for Sustainable Development (2010) The state of sustainability initiative review 2010: sustainability and transparency. IISD, Winnipeg. www.iisd.org/pdf/2010/ssi_sustainability_review_2010.pdf. Last accessed 14 Mar 2013

IPCC – Intergovernmental Panel on Climate Change (2007) Climate change 2007 synthesis report. IPCC, Geneva. http://www.ipcc.ch/publications_and_data/ar4/syr/en/contents.html. Last accessed 09 Mar 2013

ISEAL Alliance (2010a) ISEAL Code of Good Practice for Setting Social and Environmental Standards v5.0 (Standard-Setting Code). ISEAL Alliance, London. http://www.isealalliance.org/our-work/codes-of-good-practice/standard-setting-code. Last accessed 14 Mar 2013

ISEAL Alliance (2010b) ISEAL Code of Good Practice for Assessing the Impacts of Social and Environmental Standards (Impacts Code). ISEAL Alliance, London. http://www.isealalliance.org/our-work/codes-of-good-practice/impacts-code. Last accessed 15 Mar 2013
Kemp R, Parto S, Gibson RB (2005) Governance for sustainable development: moving from theory to practice. Int J Sustain Dev 8(1/2):12–30
Lafferty WM (2004) Governance for sustainable development: the challenge of adapting form to function. Edward Elgar, Cheltenham
MEA – Millennium Ecosystem Assessment Board (2005) Current state and trends assessment. http://www.millenniumassessment.org/en/Condition.aspx. Last accessed 07 Mar 2013
Milanovic B (2011) The haves and the have nots. Basic Books, New York
O'Brien M (2000) Making better environmental decisions: an alternative to risk assessment. MIT Press, Cambridge
UNEP – United Nations Environment Programme (2012) Global Environment Outlook 5: environment for the future we want. UNEP, Nairobi. http://www.unep.org/geo/geo5.asp. Last accessed 07 Mar 2013
UNICEF – United Nations Children's Fund (2009) The state of the world's children, special edition celebrating 20 years of the convention on the rights of the child. UNICEF, New York. http://www.unicef.org/publications/index_51775.html. Last accessed 07 Mar 2013
UNICEF – United Nations Children's Fund (2013) Water, sanitation and hygiene. UNICEF, New York. http://www.unicef.org/wash/. Last accessed 09 Mar 2013
Walker B, David Salt D (2006) Resilience thinking: sustaining ecosystems and people in the changing world. Island Press, Washington
WCED – World Commission on Environment and Development (1987) Our common future. Oxford University Press, Oxford
Weaver AJ, Zickfield K, Montenegro A, Eby M (2007) Long term climate implications of 2050 emission reduction targets. Geophys Res Lett 34:L19703. http://wikyonos.seos.uvic.ca/people/alvaro/Emi_2050.pdf. Last accessed 09 Mar 2013
WWF International (2012) Living planet report 2012: biodiversity, biocapacity and better choices. WWF, Gland. http://wwf.panda.org/about_our_earth/all_publications/living_planet_report/2012_lpr/. Last accessed 07 Mar 2013

Chapter 3
Conceptual and Practical Aspects of Indicators for Measuring Sustainability of Certified Products and Producers

Oscar David Matallana-Tost and Michael Schmidt

3.1 Introduction

This chapter aims to discuss the issues related with the development and implementation of indicators. It displays a structured overview on the conceptual and practical aspects of indicators for measuring sustainability of certified products and producers. A description of some of the analytical aids and tools for the presentation of indicators is also part of this chapter.

There are no lists of indicators which could serve for all purposes. Products and supply chains should be assessed to understand their negative and positive impacts relative to established goals in order to develop suitable indicators. Besides, such assessment should include economic, environmental and social aspects. In other words, the theoretical background and applicability of the indicator should actually meet the ultimate objective of sustainability, reliably indicating the degree of compliance of products to such principles. In addition, each production system has its own distinguishing features, namely a set of properties which can be monitored to gather data on the interaction between the activities related to the fabrication of its particular products and the environment to be preserved or improved. In turn, the definition of that environment is not only determined by its inherent characteristics but by the special perspective of the observer (who wants to follow changes according to an already defined aim).

The main focus of this chapter is to reflect on this manifold complexity and serve as a guide for developers and users of indicators. This chapter argues why, in order

O.D. Matallana-Tost (✉)
Department of Chemical Engineering and Environmental Management, University of Engineering and Technology (UTEC), Av. Cascanueces 2221, Santa Anita, Lima, Peru
e-mail: omatallana@utec.edu.pe

M. Schmidt
Department of Environmental Planning, Brandenburg University of Technology (BTU), 101344, 03013, Germany
e-mail: michael.schmidt@tu-cottbus.de

C. Schmitz-Hoffmann et al. (eds.), *Voluntary Standard Systems*, Natural Resource Management in Transition 1, DOI 10.1007/978-3-642-35716-9_3,

to develop instruments for measuring sustainability of certified products, it is also necessary to understand indicators as ad hoc models of reality. This means that the indicators include some of the quantifiable aspects of reality and are made to serve a singular purpose. Indicators are a careful simplification of the complex system to be evaluated, which give enough information to make sound decisions in a cost-effective way. Consequently, the worth of an indicator rests on the quality of its primary data, the scale and context where it is being applied, its degree of adaptability, the soundness of its interpretation, and its relative simplicity (Segnestam 2002).

This chapter starts with Sect. 3.2 by defining what data means within the context of indicators development, explaining what kind of data is needed, and how they should be analysed. Section 3.3 is dedicated to describing the role of data collection and how it supports the indicator developer when deciding how the necessary data is going to be gathered. Section 3.4 uses the lessons learned in the previous sections to offer a solid theoretical foundation along with empirical information to support the indicator developer and user. A key element of this section is the explanation of all the factors which determine the usefulness of indicators. Section 3.5 serves as an introduction to the possible ways of presenting indicators once they have been developed or selected, as well as an explanation of the significance of the selected analytical aids and tools for presentation.

The chapter ends with conclusions and recommendations in Sect. 3.6 where the most significant outcomes of the preceding sections are summed up and analysed, highlighting the importance of aspects such as credibility, plausibility, applicability, explicability, communicability and admissibility for the development and implementation of indicators for measuring sustainability of certified products.

3.2 Data and Indicators

In theory, data is factual information that could serve to characterise an observed phenomenon. However, a set of data is not always mutually consistent or logical, rather it can sometimes be confusing, dissonant or completely apart from the phenomenon to be assessed (Dubois et al. 2000). In order to receive valuable information from the collected data, the results from a monitoring process should display a clear relationship with the observed reality to allow further processes of organization, analysis, ranking and/or evaluation. Furthermore, the availability, quantity and quality of data constitute key components to be optimised because they are related to the degree of confidence, extent and detail by which a particular system is being assessed (Giudici 2003).

Ideally, the development of indicators for measuring sustainability of certified products and producers would start with an extensive data collection phase where primary data on all the activities related to the production of the good and their effects on a precisely defined surrounding are gathered (Hendrikson et al. 2006).

Depending on the scale and level of analysis, this primary data should be appropriate in terms of extension and detail to give reliable and useful information.

Once the proper amount of quality data is collected, the second phase would be the full analysis of the primary data. Such analysis should be performed in a way that reduces complexity while enhancing the quality of information. The third phase includes the final implementation of indicators based on the reliable outcomes of the second phase. At this stage it is important to select appropriate names and units that take into account both the theory behind the resulting indicator and its potential to be understood by the stakeholders. A further step would be the development of indices based on the analysis of a set of indicators.

The information pyramid shown in Fig. 3.1 summarises the formerly described theoretical steps in its first case 'Theory', where the development of indicators rests on the primary data as explained before. On the other hand, the second case 'Common Practice' illustrates how indicators are developed in practice due to the lack of primary data. The pyramid is inverted in this case because many indicators are developed using pre-existing data which is rather limited in extension and quality (Segnestam 2002). Such a lack of reliable data could be explained by technical, institutional and political limitations that arise when trying to gather the necessary data to develop indicators, given that the data gathering could be not only resource-demanding and technically difficult, but also strongly dependent on people's preferences (Bateman 2002).

3.3 Data Gathering

The collection of the data can follow many different arrangements, depending on the scale or level of analysis. The term scale represents the observational level of the assessment for both spatial and temporal components, so that it defines the quantity and quality of data that will be the basis for the decision making process (Joao 2002; Gontier 2007).

At a project level, for example, modifications in the chosen scale could have important repercussions on the results of an environmental impact assessment, not only regarding the determination of impact significance and the measurement of environmental parameters, but also concerning the scope of the overall process and the relevance of the monitoring procedures (Patil et al. 2002; Joao 2002; Therivel and Ross 2007).

More to the point, whether the indicator initiative is at the level of one product or a set of products, whether the activities of one producer are being analysed at a national or international level, will also matter for the methods used to gather the primary data. To be precise, the level at which the measured factor lies, ultimately determines the methods of primary data capture. For instance, even if a certified producer would carry out activities worldwide, the data may be gathered for a single country when the indicator is intended to be used at a national level.

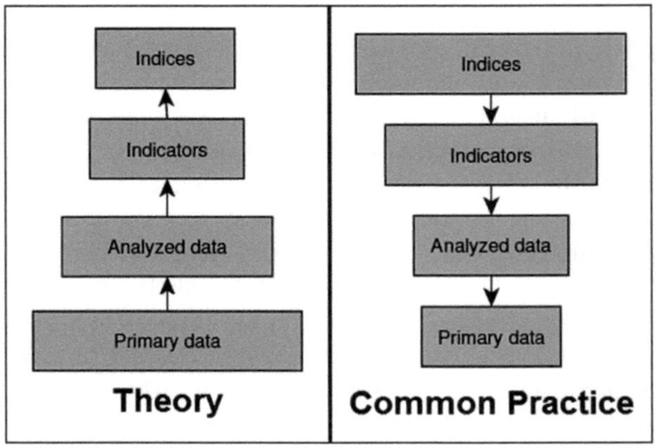

Fig. 3.1 Information pyramid related to indicators (adapted from Segnestam 2002)

Other issues such as credibility, cost efficiency, and incentives are also relevant for this data collection phase, since they establish the quality of the monitoring system and determine its potential to be sustainable and useful for making decisions based on the resulting indicators.

The word credibility makes reference here to both the trustworthiness and the capacity of the data collector, namely that the entity responsible for collecting the data should be reliable and have the capacity to manage the quality of data which is at stake. A data collector would be cost-efficient when it gathers only the primary data which is strictly necessary to develop the indicator, after using all the existing data that could be employed to the greatest possible advantage.

To address the problem of the costs and benefits associated with the selected data collection system, the aspect of incentives has to be taken into account. Monitoring to develop indicators for measuring sustainability of certified products and producers takes place only when both producers and consumers involved can realise the benefits attached to a particular data gathering arrangement. Producers have to be sure that they are going to be successful by investing money in monitoring while consumers want to be sure that the data gathered serves a purpose in resonance with their preferences (Bateman 2002).

3.4 Usefulness of Indicators for Measuring Sustainability

A crucial feature of an indicator is of course its usefulness. As illustrated in Fig. 3.2, the usefulness of indicators depends on factors such as scale, context and interpretation. These factors are related to the conceptual aspects previously explained and should be taken into account when applying indicators for decision making.

The spatial and temporal dimensions of scale determine how the indicator should be understood and implemented, since the quantities displayed by the same indicator could mean different things depending on the geographic perspective of the

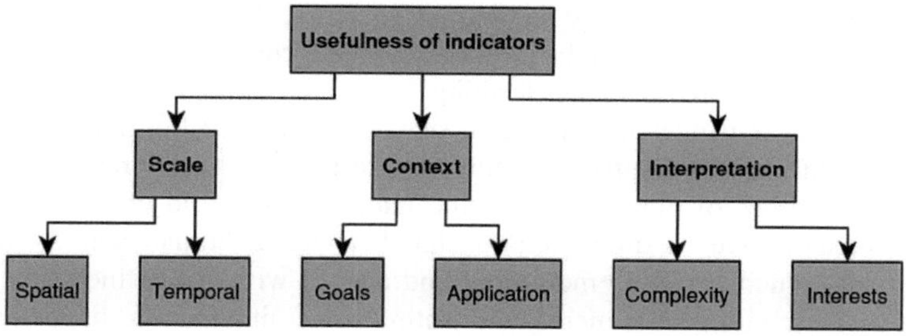

Fig. 3.2 Factors that determine the usefulness of indicators

decision maker and the considered time window. Hence, an indicator is not globally or locally important per se, but its significance is subordinated to the relative context in which is being applied. For example, the extinction of the Scarce Large Blue butterfly (*Maculinea teleius*) could be of grave concern for Germany but not that severe for the whole European Union if that species is proven out of risk in regions like France, Switzerland, Austria or Spain (Günter 1991).

Additionally, geographic considerations play an important role when deciding the spatial boundaries of the indicators which do not always coincide with the politically delimited areas. To ensure the value of indicators for decision making, it is crucial to select spatial limits that could be consistent with the explained conceptual background and the intended application. Likewise, the time scale of an indicator also influences how the indicator should be used and interpreted. Different indicators may imply different quantities of elapsed time needed to indicate a change in the status quo, thus the decision maker should select the time scale according to the goals that motivated the development of an indicator in the first place. A good example of this could be found in the forest products industry, where indicators for measuring sustainability of certified timber would imply different time scales depending on the process to be quantified, e.g., deforestation as well as land use changes should be addressed within shorter time lapses (day, month, year) than the increase of humus content or the carbon sequestration rate (year, decade, century) (Montagnini and Nair 2004).

Indicators for measuring sustainability could be also context-dependent in the sense that their development and application requires the understanding of the activities that pertain to a particular certified producer or product. To give an example, if a cheese producer wants to measure the sustainability of their certified cheese and include (amongst other considerations) an indicator to control pollution, one of the first issues to deal with would be what type of pollution is going to be considered. Depending on political, conceptual, practical and institutional factors, the cheese producer would decide to take into account all the practices of cheese production that may pollute in some way or just restrict such an indicator to the quantification of CO_2 emissions. Either to develop the indicator or to implement an existing one, the cheese producer would have to proceed according to the particular goals of the organization and its capacity, the current knowledge about pollution, the available measuring tools and techniques, the current developments within the cheese production industry, and the preferences of the customers. If the objective is to measure the sustainability of a farming system in Germany, for instance, an indicator such as 'Changes in population sizes of the Scarce Large Blue butterfly due to different mowing regimes' would be more appropriate to prevent biodiversity loss than 'Loss of Scarce Large Blue habitats' (Wätzold et al. 2008). The former indicator gives a direct relationship between a human activity and its environment, which could be monitored and altered within a defined time span, whereas the later indicator is merely recounting something that has happened.

The issue of interpretation is shown in Fig. 3.2 as a separated factor because the usefulness of an indicator is strongly related to the degree of understanding and the interests of the user, even if the indicator is already the appropriate one in terms of scale and context. In order to be effective, an indicator may include complex and extensive information that could not be understood by all sectors of society, possibly causing misunderstanding, non-acceptance (or overconfidence), misuse, miscalculations or premature abandonment. An example of this is the monitoring of changes in biodiversity due to the fabrication of a certified product. Such a task implies a deep comprehension of ecological processes and a huge amount of primary data. It also requires the development of new conceptual and technical tools to identify when the observed variations are a consequence of the certified production system and when just a normal trend explained by dispersal or random speciation (Hubbell 2001).

Finally, interests come into play when interpreting indicators because people could mix their preferences with the analysis of the outcomes. Thus a quantity expressed by an indicator may become value-loaded in the hands of politicians, private investors, activists or the general public, especially if the trends exhibited by the indicators have the potential to affect big markets. For instance, the development of indicators such as carbon footprint, virtual water or embedded energy, has had a strong impact on the food market because these indicators are being used to support claims against products like meat and its derivatives (Vegan Outreach 2011). Even experts could make wrong interpretations if they are biased to conservation or working under the pressure of a private sector (paying to see its interests reflected in a final report).

3.5 Analytical Aids and Tools for Presentation of Indicators

After the proper indicator for measuring sustainability of certified products and producers is developed or selected from an already available source, the next step towards successfully applying it is to decide how and by what means the indicator will be presented to the intended public (standard setter, producers, consumers, policy makers, NGOs, scientific community or general public).

Table 3.1 Basic analytical aids for sustainability indicators

Aid	Objective	Suggested use	Recommendations
Comparator	To give a clear reference and facilitate the understanding of the quantity expressed by the indicator Example: The water footprint of a meat product like beef (15,500 l/kg) compared to others like chicken (3,900 l/kg) (Raureif GmbH 2012)	When the indicator is rather unfamiliar To ensure the transmission of the proper message	The comparator should be as familiar and uncomplicated as possible It should be in resonance with the original goals of the indicator initiative
Threshold	Definition of upper and/or lower limits Example: Deforestation thresholds in tropical countries (Centre for International Forestry Research 2012)	Monitoring and control of environmental, economic or social impacts	The threshold should reflect the significant direct and indirect effects of the production activity
Target	Definition of aims to preserve the status quo or promote improvement Example: Indicators in the framework of the Millennium Development Goals (UN 2012)	Monitoring of changes in the status quo towards improvement	The target should represent the general and specific objectives of the indicator initiative
Baseline	To monitor the changes in the status quo due *only* to the activities related to the certified product or producer	Monitoring of environmental, social or economic impacts	Comprehensive approach It requires deep understanding and extensive data

An intelligent selection of both the analytical aid and the presentation tool could enhance the general value of the indicator, as the success of an indicator depends on how close it is to the real situation, how well it has been understood and how much it is being correctly used by the multiple stakeholders. The Table 3.1 summarises the typical cases for using basic analytical aids such as comparators, thresholds, targets and baselines. If a standard setter wants to develop sustainability standards, for example, the choice of the analytical aid would depend on the type of the standard, namely when they produce an outcome-based standard (Komives 2011) it is to expect that the related indicators would be managed in the form of thresholds and targets rather than by using baseline approaches.

With regard to the presentation tools it is important to remark that indicators could be shown in various ways depending on the expertise of the user, the current knowledge and the available technologies. Common presentational tools are data-oriented, like tables and text, and some others are visual-oriented, such as graphs, maps and charts. The final selection of a presentation tool should especially take

Fig. 3.3 Virtual water content of beef and one apple (adapted from Raureif GmbH 2012)

into account the type of indicator, its primary aim and the target audience (Segnestam 2002).

A good example of a simple but effective graph used to convey the message to a broad section of the public can be found in Fig. 3.3. Such a graph can give quick information about the virtual-water content of different products, making it easy for the user to perform visual recognition as well as qualitative and quantitative assessments such as comparison, ranking and evaluation (Raureif GmbH 2012).

3.6 Conclusions and Recommendations

As discussed in this chapter, there are several aspects that need to be taken into account when working with indicators for measuring sustainability of certified products and producers, belonging to both theoretical and empirical realms.

The quality of the data in terms of extent and detail, the reliability and capacity of the indicator developer, the nature of the necessary collection data arrangement, along with other factors such as scale, context and interpretation, all play a crucial role in developing and/or implementing indicators for measuring sustainability of certified products and producers. After the proper indicator is developed or selected, the success of the indicator initiative would depend on the fittingness of the chosen analytical aids and presentation tools, due to the fact that the worth of the indicator significantly depends on how close it is to the real situation, how well it has been understood, and how much it is being correctly used by the multiple stakeholders.

In a nutshell, aspects such as credibility, plausibility, applicability, explicability, communicability and admissibility have to be carefully evaluated when developing or implementing indicators for measuring sustainability of certified products and producers.

References

Bateman I (2002) Economic valuation with stated preference techniques. Edward Elgar, Cheltenham

Centre for International Forestry Research (CIFOR) (2012) Knowledge on forests and forestry. http://www.cifor.org/online-library/browse.html. Last accessed 21 Mar 2013

Dubois D, Hájek P, Prade H (2000) Knowledge-driven versus data-driven logics. J Logic Lang Info 9(1):65–89

Giudici P (2003) Applied data mining: statistical methods for business and industry. Wiley, Chichester

Gontier M (2007) Scale issues in the assessment of ecological impacts using a GIS-based habitat model – a case study for the Stockholm region. Environ Impact Assess Rev 27(5):440–459

Günter E (1991) Die Schmetterlinge Baden-Württembergs. Ulmer, Stuttgart

Hendrikson C, Lave L, Matthews H (2006) Environmental life cycle assessment of goods and services: an input–output approach. Resources for the Future, Washington

Hubbell S (2001) The unified neutral theory of biodiversity and biogeography. Princeton University Press, Princeton

Joao E (2002) How scale affects environmental impact assessment. Environ Impact Assess Rev 22(4):289–310

Komives K (2011) Voluntary sustainability standards: part 1: an introduction. Guest lecture, Brandenburg Technical University

Montagnini F, Nair P (2004) Carbon sequestration: an underexploited environmental benefit of agroforestry systems. Agroforestry Syst 61(1):281–295

Patil A, Annachhatre A, Tripathi N (2002) Comparison of conventional and geo-spatial EIA: a shrimp farming case study. Environ Impact Assess Rev 22(4):361–375

Raureif GmbH (2012) The Virtual Water Project. http://virtualwater.eu/#about. Last accessed 21 Mar 2013

Segnestam L (2002) Indicators of environment and sustainable development: theories and practical experience. Environ Econ Ser 89(1):17–36

Therivel R, Ross B (2007) Cumulative effects assessment: does scale matter? Environ Impact Assess Rev 27(5):365–385

UN – United Nations (2012) Millennium development goals. http://www.un.org/millenniumgoals/. Last accessed 21 Mar 2013

Vegan Outreach (2011) Resources and contamination. http://www.veganoutreach.org/whyvegan/resources.html. Last accessed 06 Feb 2012

Wätzold F, Lienhoop N, Drechsler M, Settele J (2008) Estimating optimal conservation in the context of agri-environmental schemes. Ecol Econ 68(1):295–305

Chapter 4
Evolution of VSS: From Niche to Mainstream

David Ovando Jeria and Miguel Araque Vera

4.1 Introduction

This following chapter aims to discuss the most important standards and historical developments that have led to the current situation facing voluntary sustainability standards (VSS).

During the 1970s, sustainability initiatives such as organic agriculture provided small groups of consumers with ecologically, socially and economically sound products (IFOAM 2013). Other groups developed interests in supporting the conditions in which farmers and craftsmen worked in. At that time, voluntary sustainability standards were regarded as being limited to 'niche' markets and often associated with novelty items (Sexsmith and Potts 2009). VSS are a group of non-obligatory schemes, codes and steps which focus on the social and environmental safety of consumer oriented practices (UNIDO 2010). They set the requirements of a number of today's products and production processes. VSSs are contributed to by consumers, non-governmental organisations (NGOs) and corporations.

Over the past few decades, there have been a number of movements and intergovernmental initiatives which have supported the development of VSS, which has resulted in a massive growth of initiatives extending to more than 400 standards today (Ecolabelindex 2012).

The development of VSS has positively influenced producers and consumers in several ways, for example by improving social and economic conditions of workers in developing countries and guaranteeing consumers healthy choices and environmentally sound practices. Alternatively, the increased pace with which private

D.O. Jeria (✉)
Department of Environmental Planning, Brandenburg University of Technology (BTU) Cottbus, P.O. Box 101344, 03013 Cottbus, Germany
e-mail: davidovandojeria@yahoo.com

M. Araque Vera
ThyssenKrupp Industrial Solutions, Bad Soden, Germany
e-mail: m.araquev@gmail.com

C. Schmitz-Hoffmann et al. (eds.), *Voluntary Standard Systems*, Natural Resource Management in Transition 1, DOI 10.1007/978-3-642-35716-9_4,

standards have developed and the high level of requirements that they impose, has set trade barriers upon small producers mainly in developing countries (FAO 2012). Organisations such as IFOAM (International Federation of Organic Agriculture Movements), FAO (Food and Agriculture Organization) and ISEAL Alliance (International Social and Environmental Accreditation and Labelling Alliance) have developed harmonisation strategies to address these issues.

According to the UNFSS (2012), in comparison to the growth of conventional food sales (2–4 %) the VSS markets can be considered to have reached mainstream status with growth in the following areas: organic products by 10–15 %, Fairtrade products with an increase of 35 % and forestry certification schemes with up to 105 %.

Taking this into consideration it is important to understand how this current phenomenon of sustainability schemes, and the demand for sustainable products, has historically evolved. In particular, with an emphasis on the movements that originated their foundation and the trends that have pushed their development into the mainstream.

In this context, Sect. 4.2.1 (1960–1990) begins by addressing the agricultural and Fairtrade movements. Section 4.2.2 (1992–2000) continues by examining the highlights of the 1990s where the environmental movement, in response to the failed negotiations of the GATT (General Agreement on Tariffs and Trade) but with the support of the 1992 Rio Earth summit, established the path for the creation of important initiatives; here we discuss what led to their foundation as well as other social factors. In Sect. 4.2.3 (2000–present time), the growth of coffee, social and fishery standards is analysed by discussing the reasons for their rapid development and the multi-stakeholder initiatives that have emerged aiming to harmonise these practices. In closing, a summary is presented of the most important historical developments of VSS, emphasising the need for harmonisation between the different systems and their potential role for the future.

4.2 Voluntary Sustainability Standards and Their Driving Forces

Voluntary sustainability standards belong to the family of non-mandatory or private standards that differentiate themselves within this family by seeking to ensure that materials, products, processes and services meet social, economic and environmental requirements (ITC 2010). Furthermore, they usually support the development of their guidance by basing themselves under technical standards such as ISO and CODEX Alimentarius (Salmon 2002).

The development of voluntary sustainability standards can be described by two main international and national factors: in the international arena, the proliferation of agreements aiming to protect the exploitation of natural resources and regulate international trade, at the national level, pressure from society driven by the lack of governmental regulation on key issues such as economic stability, social wellness and environmental protection. These two main international and national factors have triggered initiatives in the form of political movements, multi-stakeholder agreements, NGOs and corporative engagement to establish mechanisms of regulation.

The following subsections approach the historical development and evolution of VSS by discussing the driving forces that enabled these organisations to be formed and developed. Three time periods could be distinguished.

4.2.1 Emergence of Sustainability Movements (1960–1990)

The following section discusses the foundation and motivation of the first organised sustainability initiatives: Organic agriculture, forestry movements (health and environment) and the Fairtrade movement (socio-economic fairness).

Methods of environmentally friendly practice can be traced to the organic movements in northern Europe in the 1960s, where isolated agricultural groups provided local consumers with environmentally safe and healthy products through the development of small production/distribution/consumption networks constituted by cooperatives, box schemes and farmer markets (Raynolds 2004). These initiatives became strongly supported by the general population with the publication of "Silent Spring" in the early 1960s (Carson 1962), which highlighted the negative effects of intensive agricultural practices and the dangers of using hazardous pesticides. Over time, the concerns regarding the environment and human health within agricultural areas spread internationally, bringing together these movements into one organisation (RESOLVE Inc. 2012).

The unification of the agricultural movement began in Versailles, France, in 1972 with the foundation of IFOAM. Organised by Nature et Progrès of France, IFOAM united organisations from Great Britain, Sweden, South Africa and United States. The aim of the new organisation was to create a unified voice to promote organic food with the diffusion and exchange of information of the fundamental principles and practices of organic agriculture across the world (IFOAM 2013). Currently, IFOAM is an internationally recognised organisation and its establishment has supported the development of what we now know as 'bio standards and labels', with more than 750 affiliates around the world (Salmon 2002).

In the socio-economic area, handcrafted articles from developing countries are sold in USA and Europe in the so called 'world shops', aiming to support better trading conditions for the workers in these countries. This is known as the Fairtrade movement. Fairtrade can be tracked from several initiatives around the world, from shops of 'The Ten Thousand Villages' in the USA during the 1940s (Fairtrade Hub 2012[1]), to Europe with the emergence of Oxfam foundation, which provided food and essential supplies for starving women and children during World War II (Oxfam 2012). By 1988, the first successful attempt of implementing voluntary

[1] http://www.fair-trade-hub.com.

sustainability certification schemes was observed in the Netherlands. The first Fairtrade coffee was sold in Dutch supermarkets, branded under the fictional literary character of Max Havelaar, who opposed the exploitation of farmers in Dutch colonies. These initiatives were then replicated in several countries within Europe (Belgium, Switzerland, Denmark, Norway and France), North America (United States, Canada) and Asia (Japan) consolidating Fairtrade as an international movement (Fairtrade International 2012).

In the 1980s, NGOs such as Friends of the Earth, the Rainforest Action Network, and Greenpeace ran boycott campaigns against unsustainably harvested tropical timber and other environmentally harmful practices (RESOLVE Inc. 2012). In 1986, in response to the alarming rate of deforestation and species extinction, a group of environmentalists gathered for a small workshop with the purpose of discussing forest and biodiversity conservation measures, and established a non-profit organisation, the Rainforest Alliance (RA) (von Hagen et al. 2010). This gave rise to the first forest certification program, Smartwood (1989), with the aim of improving forest management by providing economic incentives to companies who chose to implement sustainable forestry practices (Rainforest Alliance 2012).

4.2.2 Environmental and Social Standards (1990–1999)

This subsection addresses the development of the environmental and social standards by emphasising the role of the 1992 Rio Earth Summit on the development of multi-stakeholder environmental standards. Furthermore, the social movements and campaigns against sweatshops and child labour that enabled the development of important social voluntary sustainability standards are discussed.

After the foundation of RA, the 1990s saw the strengthening of environmental and social lobbies with the launch in 1997 of the Fairtrade Labelling Organizations International (FLO) aimed at uniting all Fairtrade initiatives under one framework.

Forestry movements during this decade had a strong presence. The pursuit of mechanisms for regulating deforestation and species extinction could be observed in organised initiatives such as the Tropical Forest Action Plan (TFAP) and the establishment of the International Tropical Timber Organization (ITTO). However, the support at the time proved to be ineffective against the rising rates of deforestation (Lang 2006).

International recognition of environmental issues had become mainstream by 1992 during the Rio Earth Summit; here important discussions on climate change, species extinction and unsustainable growth were addressed. As a result the Kyoto Protocol, the Convention on Biological Diversity and the concept of environment management were developed. Many consider the 1992 Rio Earth Summit as a failure to produce binding agreements in respect to forest protection measures in the case of Agenda 21 (Lang 2006; Guéneau 2007; Hinrichs and Van Helden 2012).

However, it provided a forum for many organisations to come together and discuss independent and international forest certification schemes.

Discussions on biodiversity and forestry conservation evolved in 1993 to a positive outcome, with cooperation between the Rainforest Alliance and the World Wildlife Foundation (WWF) leading to the creation of the Forestry Stewardship Council (FSC), a multi-stakeholder organisation. The FSC promotes forestry management, focusing on the three pillars of sustainability, and serves as an international certification scheme that grants the trademark label to companies who comply with sustainable practices (Lang 2006).

Similar multi-stakeholder initiatives motivated by the 1992 Rio Earth Summit were established in 1994 with the Sustainable Forestry Initiative (SFI) in North America and the Programme for the Endorsement of Forest Certification schemes (PEFC) in 1999 as umbrella organisations with the purpose of mutually recognising national forest certification schemes.

In the mid-1990s sustainable forestry schemes increased in size and number. Considering this phenomenon, several organisations including the Rainforest Alliance, CyD, FIIT, Fundación Ambio and Imaflora united to form the Sustainable Agriculture Network (SAN) with the goal of coordination and collaboration between organisations (Sexsmith and Potts 2009).

On a more technical and official basis in 1996, the ISO 14000 norms emerged, inspired on a similar scheme as the ISO 9000 standards. The ISO 14000 standards family focuses upon Eco-Management and Audit Schemes (EMAS) designed to cover corporate requirements in auditing, labels and declarations, life cycle assessment, greenhouse gas (GHG) measurements and other factors (ISO 2010).

Besides the environment, corporate social responsibility discovered a new meaning with the pressure of social activism upon companies such as Nike and Home Depot for using sweatshops which employed child labour (Bartley 2007). In 1996 a meeting hosted by US President Bill Clinton with the support of corporations, NGOs and labour unions resulted with the foundation of Apparel Industry Partnership (AIP). This taskforce was later responsible in 1998 for the establishment of the Fair Labor Association (FLA), the first entity with the purpose of monitoring sweatshops, child labour and ecological degradation (Gereffi et al. 2001). Furthermore, following UN conventions and the International Labor Organization (ILO) guidelines, important VSS initiatives such as Social Accountability International (SAI) with the SA8000 standard and the Ethical Trade Initiative were created (SAI 2012).

Within the framework of corporate social responsibility in 1997, Unilever took one of the first steps in corporate VSS support by developing, in cooperation with the World Wildlife Fund for Nature (WWF), a sustainable management system for fisheries called the Marine Stewardship Council (MSC).

The 1990s were a decade of establishment, exponential growth of VSS and the introduction of multi-stakeholder initiatives, which focused on uniting and organising stakeholders with similar goals.

The evolution of forestry and social labelling schemes during the 1990s cleared the path for commodities based standards (cotton, coffee, cocoa and tea) such as:

UTZ (1997), Global Good Agricultural Practices (GlobalGAP) (1999) and FLP (Flower Label Program 1999).

4.2.3 Mainstream Consolidation (2000–Present Time)

In the previous subsections we observed how the influence of environmental conventions and social movements motivated the development of multi-stakeholder standards. The following section focuses on the mainstream establishment of voluntary sustainability standards, by discussing the collaboration of corporations and private organisations in the standardisation of greenhouse gas emissions accounting, the high growth in the number of standards and the private solutions available for harmonisation of criteria (meta-standards).

The increase in multi-stakeholder support for climate change initiatives composed of corporations and international institutions[2] could be clearly observed in 2001 with the launch of the Greenhouse Gas Protocol (GHG Protocol).[3] This standard aims to act as an accessible guide to business and organisations worldwide for accounting of greenhouse gas emissions (GHG Protocol 2012). Other initiatives for the assessment of GHG emissions are also available in technical institutions such as ISO (ISO 14064) and BSI (PAS2050).[4]

The growth of VSS with the demand for harmonisation and unification gave rise to meta-standards. After several meetings from voluntary standard setters such as FSC, MSC, FLO, IFOAM, RA-SAN and SAI concerning the content and scope of each respective standard, a common agreement was reached in 2002 with the foundation of the International Social and Environmental Accreditation and Labelling Alliance (ISEAL Alliance) (RESOLVE Inc. 2012). This established a framework and guiding principles for voluntary standard setters and coherence between standardisation mechanisms was achieved. Additionally, ISO contributed to the new model of meta-standards, developing the ISO 26000 standard for drawing up corporate guidance on social and environmental responsibility (Djama 2011).

Another multi-stakeholder approach was the establishment of commodity-based roundtables. Such initiatives started in 2004 with the WWF supporting the foundation of the Roundtable on Sustainable Palm Oil (RSPO), followed by the Round Table on Responsible Soy Association (RTRS) and sugar provided for by Bonsucro which also forms part of the ISEAL Alliance (de Man 2010). Roundtables differentiate themselves by qualifying certification systems and standard development in relation to the specific commodities that have the most impact on the environment, these being initially palm oil, soy and sugar respectively (RESOLVE Inc. 2012).

---

[2] World Resource Institute (WRI) and the World Business Council for Sustainable Development (WBCSD).

[3] GHG Protocol: Corporate Accounting and Reporting Standard.

[4] British Standards Institution.

The establishment of VSS on mainstream markets can be currently observed by the strong investment support which participating corporations are using to certify their supply chains. In the case of MSC, a considerable increase in the certification of sustainable fishing practices has been observed with new partners such as Wal-Mart (2006) and McDonald's (2011) submitting a significant proportion of their supply chain for certification (ISEAL Alliance 2012). The forest stewardship council, referred to some as "the mother of multi-stakeholder standards" (de Man 2010) had, by 2009, a market reach consisting of 113 million hectares of forest (Sexsmith and Potts 2009). Fairtrade being one of the biggest VSS institutions had, in 2010, approximately 27,000 certified products globally (ISEAL Alliance 2012). Finally, sustainable coffee is expected, for 2015, to have a worldwide market share of 20–35 % (ITC 2011), demonstrating the extensive reach that sustainable products will have on consumers and markets in the coming years.

A negative effect related to the high number of available standards has caused confusion amongst consumers and users regarding which standard can be adjusted to satisfy their needs. Web solutions such as the Ecolabelindex[5] and Ekobai[6] have provided assistance by supporting consumers with an online database, containing information on a great variety of VSS labels. More than 400 labelling schemes are currently available according to Ecolabelindex.

4.3 Conclusions

The purpose of this chapter aimed to address the historical development of VSS. Sustainability standards have supported society throughout the years by establishing bridges between consumers and producers, raising awareness of how health and natural resources are managed and providing the necessary platforms to address these issues. The agricultural and Fairtrade movements in the 1970s are a strong reflection of how unsustainable agricultural practices and deficient labouring conditions enabled consumers and organisations to work together to influence improvements in our productive systems.

As seen in the previous Sect. 4.2.2, the emergence of voluntary initiatives was strongly influenced by international support; this was the case for the internationalisation of global warming, deforestation and the increased rate of endangered species addressed in the 1992 Rio Earth Summit. This gave rise to further international collaboration between countries to develop agreements in the sustainability arena such as Agenda 21, the Kyoto Protocol and the Convention on Biological Diversity. Based on these conventions, environmental and social organisations developed schemes and systems for compliance with sustainability goals. Recent ventures such as the Rio + 20 in 2012 and the United Nations Forum on Voluntary

[5] http://www.ecolabelindex.com/.

[6] http://www.ekobai.com/.

Sustainability Standards (UNFSS) promise to maintain this tendency in providing support for sustainable development.

The current situation of VSS is seen to be favourable, where consumers are increasingly aware of the impacts linked to their daily choices, and with VSS providing a platform that encourages sustainable choices. In contrast, consumers can be confused and uninterested by the overflow of labelling schemes in the market. This has also affected small producers by generating trade barriers where sometimes up to eight standards must be complied with in order to have the possibility to access global markets (RESOLVE Inc. 2012). As a solution, harmonisation mechanisms have been developed under meta-standards as well as web based databases, which are now available to provide assessment on the available standards.

Trends in economic growth and population suggest that the increase in resource consumption and its resulting scarcity will probably influence VSS standards to shift from regulative mechanisms to the improvement of productivity by the application of sustainable mechanisms. This is predicted considering that the influences of climate change, in the case of agriculture, will have potential negative effects upon yields of today's main producers (RESOLVE Inc. 2012). Implementing sustainable practices that ensure the resilience and endurance of crops adds additional economic value and priority for VSS.

The lessons of the past have showed us that a lack of regulation by policy makers, and abuse and negligence by corporations, are important weak points that VSSs have addressed. The development of these events has also led to governments and corporations now working together with private initiatives to address these issues (Djama 2011). Public regulations now exists to control environmental, health and social problems, furthermore corporations now see sustainability as an innovative way of attracting consumers. VSSs serve as monitoring and supporting tools for these two entities and it is this function that is expected to be strengthened in the years to come.

References

Bartley T (2007) Institutional emergence in an era of globalization: the rise of transnational private regulation of labor and environmental conditions. Am J Sociol 113(2):297–351

Carson R (1962) Silent spring. Houghton Mifflin Co., Boston

de Man R (2010) Land issues in voluntary standards for investments in agriculture: a discussion paper. In: The World Bank annual bank conference on land policy and administration, Washington, 26 and 27 April 2010

Djama M (2011) Articulating private voluntary standards and public regulations. Perspective No. 11, CIRAD, Paris

Ecolabelindex (2012) Global directory of ecolabels, operated by Big Room Inc., Canada. http://www.ecolabelindex.com. Last accessed 07 Sept 2013

Fairtrade Hub (2012) History of Fair Trade. http://www.fair-trade-hub.com/history-of-fair-trade.html. Last accessed 07 Sept 2013

Fairtrade Labelling Organization International (2012) Fairtrade labelling international history. http://www.fairtrade.net. Last accessed 06 Sept 2013

FAO – Food and agriculture Organization (2012) Current FAO activities on sustainability standards. http://www.youtube.com/watch?v=6wUkrb6ZcDM. Last accessed 07 Sept 2013
Gereffi G, Garcia-Johnson R, Sasser E (2001) The NGO-industrial complex. Foreign Policy 125:56–65
GHG Protocol – The Greenhouse Gas Protocol (2012) About the GHG Protocol. http://www.ghgprotocol.org/about-ghgp. Last accessed 13 Sept 2013
Guéneau S (2007) Certification as a new private global forest governance system: the regulatory potential of the Forest Stewardship Council, IDDRI (Institut du développement durable et des relations internationales). In: Paper for the conference: non-state actors as standard setters: the erosion of the public-private divide, Basel Institute on Governance, Basel, 8–9 February 2007
Hinrichs A, van Helden F (2012) Can the FLEGT action plan and voluntary forest certification reinforce each other? ETFRN News 53, April 2012
IFOAM – International Federation of Organic Agriculture Movements (2013) About IFOAM. http://www.ifoam.org/en/about-us-1. Last accessed 13 Sept 2013
ISEAL Alliance (2012) Growth of the sustainability standards movement. http://www.uncsd2012.org/content/documents/Growth%20of%20the%20Sustainability%20Standards%20Movement%20timeline%20online.pdf. Last accessed 13 Sept 2013
ISO – International Organization for Standardization (2010) International standards and "private standards", Geneva, Switzerland. http://www.iso.org/iso/private_standards.pdf. Last accessed 13 Sept 2013
ITC – The International Trade Centre (2010) Voluntary standards in developing countries: the potential of voluntary standards and their role in international trade. International Trade Forum – Issue 3/2010. http://www.intracen.org/Voluntary-Standards-in-Developing-Countries-The-Potential-of-Voluntary-Standards-and-their-Role-in-International-Trade/. Last accessed 13 Sept 2013
ITC – The International Trade Centre (2011) Trends in the trade of certified coffees, Geneva, Switzerland. http://www.intracen.org/Trends-in-the-trade-of-certified-coffees/. Last accessed 13 Sept 2013
Lang B (2006) Experiences with voluntary standards initiatives and related multi-stakeholder dialogues. Deutsche Gesellschaft für Technische Zusammenarbeit (GTZ) GmbH on behalf of the German Federal Ministry for Economic Cooperation and Development (BMZ), June 2006
Oxfam International (2012) History of Oxfam International. http://www.oxfam.org. Last accessed 09 Sept 2013
Rainforest Alliance (2012) 25 years of conservation – highlights of our history. www.rainforest-alliance.org. Last accessed 07 Sept 2013
Raynolds LT (2004) The globalization of organic agro-food networks. World Dev 32(5):725–743
RESOLVE Inc., Steering Committee of the State-of-Knowledge Assessment of Standards and Certification (2012) Toward sustainability: the roles and limitations of certification, Washington
Salmon G (2002) Round table on sustainable development voluntary sustainability standards and labels (VSSLs): the case for fostering them. Organisation for Economic Co-operation and Development, Paris
SAI – Social Accountability International (2012) About SAI. http://www.sa-intl.org/. Last accessed 13 Sept 2013
Sexsmith K, Potts J (2009) Voluntary sustainability standards and value chain governance: how sustainability standards affect the distribution of decision-making power in global value chains. International Institute for Sustainable Development (IISD), Winnipeg
UNFSS – United Nations Forum on Sustainability Standards (2012) United Nations forum on sustainability standards concept note. http://unctad.org/meetings/en/SessionalDocuments/ditc_tedb_ted0042_en.pdf. Last accessed 13 Sept 2013
UNIDO – United Nations Industrial Development Organization (2010) Making private standards work for you: a guide to private standards in the garments. Footwear and Furniture Sectors, Vienna
von Hagen O, Manning S, Reinecke J (2010) Sustainable sourcing in the food industry: global challenges and practices. Moderne Ernaehrung Heute 4:1–9

Part II
Formal and Private Standards: The Added Value

Having familiarised ourselves with the previous chapters as a foundation, we continue to build upon our understanding of voluntary standards in this section with the view of VSS from a legal perspective, directly comparing their use to traditionally understood formal legal standards.

Chapter 5 wrestles with the question of whether VSS offers competition to legal standards or has it added value to offer. Familiar legal standards regulate various aspects of human activity, developed over time, punctuated by precedents but notoriously slow to match the rapid pace at which human development tends to progress along. Voluntary standard systems act as an upstart to the accepted practice, proving more flexible and adaptable to the dynamics of human progress. We're introduced to both the basics and specificities of legal standards, in particular with detail upon sources, functions, limitations, and legal relevance. This continues with a discussion of voluntary standards under the same focus as legal standards, as well as the introduction of private standards into the law. How are legal standards included into private-law based systems? How do contracts reflect this? By clarifying such questions we hope to better understand the relationship between legal and voluntary standards, aided through determining their differences and respective advantages.

Chapter 6 follows on from this with an assessment of how useful private standards are at plugging any legal void, with specific reference to the global science community's recently emerged Wunderkind, the discipline of nanotechnology. The example of nanomaterials, presents a current real-world development showcasing the linkages between law and standardisation and the potential complementary overlaps which exist. Three areas are focussed upon, related to the standardisation of nanomaterials; encompassing social, economic and environmental potentials. Whilst the science is small, the scope for discussion is not. Nanotechnology branches into numerous and divergent topics, from water purification to electronics manufacture, and is defining the beginning of this twenty-first century. Where atomic technology and its misuse gave our predecessors pollution that was undeterminable to the naked eye, the potential of nano-scale pollution is arguably more sinister and requires robust controls and foresight.

Moving on from the sci-fi-esque realm of nanotechnology, Chap. 7 closes by exploring the concept of 'co-regulation', with specific reference to the EU Renewable Energy Directive (EU RED). This directive is at the crux of the debate concerning, effectively, European energy independence. As an attempt to kick-start the painful divorce from traditional fossil fuel sources and transition to realising the 'greener' promises of renewables, EU RED leads with its co-regulation approach. Based on a study conducted in 2012 by SQ Consult and commissioned by GIZ on behalf of the German Federal Ministry of Economic Cooperation and Development (BMZ), we are served details of the EU RED and its co-regulation approach as a starter, followed by analysis of the aforementioned co-regulations process. This is concluded with the results of stakeholder interviews and available documentation; digested down to a list of the ten main points which are of note.

Chapter 5
VSS and Legal Standards: Competition or an Added Value?

Eike Albrecht

5.1 Introduction

This chapter aims to discuss the relationship and interactions between Voluntary Sustainability Standards (VSS) and legal standards. The relationship between VSS and legal standards seems at first sight, relatively easy to describe. On the one hand, law is binding, created in a formalised procedure, following the regulations of constitutional and other law, and is enforced by authorities and courts. On the other hand, private law based voluntary provisions, are not binding (for everybody), created in partnership for regulating the private relations between contract partners, with no necessity to be enforced because all parties want to follow these standards.

But, the reality is not that simple: on one side, there is a tendency to privatise regulations and to shift it from state- or authority-based institutions to private bodies, but still to carry out state policy objectives (Reidt and Schiller 2012, no. 19). On the other side, the procedures for the creation of voluntary standards are increasingly institutionalised and are sometimes more formalised than legislative processes.

The number of law provisions and their relevance to our daily private and business dealings is increasing each year. In Germany alone (only at the Federal level), there are almost 250,000 provisions in 1,660 Federal Acts, 163,290 provisions in 2,661 Federal Ordinances, 83,654 provisions in 4,857 European Directives and Regulations (Hardinghaus et al. 2013, p. 52). The law provisions of the federal states and provisions of self governing bodies like communal bodies, other institutions like universities, public radio stations, social security insurances, trade chambers, etc., have to be added. The numerous private standards are relevant for the daily lives of citizens and businesses, in particular such provisions created by

E. Albrecht (✉)
Department for Civil and Public Law with References to the Law of Europe and the Environment, Brandenburg University of Technology (BTU), P.O. Box 101344, 03013 Cottbus, Germany
e-mail: albrecht@tu-cottbus.de

C. Schmitz-Hoffmann et al. (eds.), *Voluntary Standard Systems*, Natural Resource Management in Transition 1, DOI 10.1007/978-3-642-35716-9_5,

international standardisation committees such as the International Organization for Standardization (ISO) and the International Electronical Commission (IEC) (CIEL 2009, pp. 22ff.), European Standards (EN) and several national standards (for Germany for example DIN, VDI-standards, VDE-standards), as well as corporation related private standards, like Siemens norms (SN), also have to be considered.

The preparation of standards and thresholds requires enormous engagement of the various stakeholders, and the process of standardisation is costly. The costs for the experts in the private standardisation institutions alone (except state-organised working and expert groups) are estimated to be €650 million annually and another €90 million annually only for the budget of the German Institute for Standardization (Bahke 2006, p. 29).

The main focus of this chapter is, as already mentioned, to assess the relationship between legal and voluntary standards, clarify their differences and the specific advantages and disadvantages and finally to formulate an answer if private and legal standards are competing or are complementary. This chapter also doubles as an introduction to the next chapter (Chap. 6), which assesses the use and usefulness of private standards in the field of nanotechnology. Nanotechnology, as a relatively new technology, is so far not subject to major detailed regulation and Chap. 6 tries to find out if the absence of legal standards may be compensated by private standards. Following this introduction, Sect. 5.2 introduces the basics and specificities of legal standards, in particular discussion on sources, functions, limitations, and legal relevance. Section 5.3 discusses voluntary standards under the same focus as legal standards. Section 5.4 discusses the introduction of private standards into the law, Sect. 5.5 sheds light on how legal standards are included into private law based systems, in particular contracts. Section 5.6, the conclusion, tries to find an answer to the question, "are voluntary and legal standards competing, or do they have a complementary nature?".

5.2 Legal Standards

Legal standards are norms which are set by the state. This means the standards are developed, formulated, amended and finally publicised by the state. In respect to the contents of a standard, for example an action or trigger value of the Federal Soil Protection Act,[1] regulated in detail in the Federal Soil Protection Ordinance,[2] the source may be a local, regional, federal state, national, European or even an international one. The author or origin may be a communal or regional authority,

[1] Federal Soil Protection Act of 17.03.1998, Fed. Law Gazette I, p. 502, last amended by Art. 5 para 30 of the Act of 24.02.2012, Fed. Law Gazette I, p. 212; a translation is available in Mulloy et al. (2001), pp. 254ff.

[2] Federal Soil Protection Ordinance of 12.07.1999, Fed. Law Gazette I, p. 1554, last amended by Art. 5 para 31 of the Act of 24.02.2012, Fed. Law Gazette I, p. 212.

a federal state's ministry or other body, a national ministry or scientific institution or a combined working group of federal states and federal representatives, like the German working groups of the federation and the federal states on soil (Bund/Länder-Arbeitsgemeinschaft Boden—LABO), on water (Bund/Länder-Arbeitsgemeinschaft Wasser—LAWA) or on waste (Bund/Länder-Arbeitsgemeinschaft Abfall—LAGA), a European committee, like the European IPPC Bureau (EIPPCB), responsible for the Sevilla process for the development of the Best Available Technology—BAT, relevant for technical standards in licensing procedures of industrial installations of a certain size,[3] or an international institution like the World Health Organization (WHO) recommending certain drinking water standards (WHO 2011). But all these standards are at first sight nothing more than recommendations, as long as they are not binding or otherwise transferred to the legal reality following a certain transposition systematic which follows—of course—national principles and regulations.

5.2.1 Sources of Standards

This system to transpose standards to the respective relevant legal system includes standards set by formal parliamentary law, e.g. in an act or developed in a procedure derived from a formal parliamentary law, in particular in ordinances and administrative circulaires. The quality of legal standards is that they are finally legitimised by the representation of the people. This is clear for such standards, regulated by formal parliamentary acts. The members of the parliament are (in particular in a democratic system) elected by the people and the legitimation chain is clear and visible, regardless that the practice of legislation in reality may be questionable, in particular because the parliament is increasingly dependent upon the expertise of the government, in particular in urgent and complex matters (such as the regulation of nanotechnology or the finance sector in times of the recent Euro-crisis). On the federal level, the executive power which is responsible for the development and enactment of ordinances, and even more of administrative circulaires, is legitimised via the chancellor who is elected by the parliament. Thus, the legitimation chain is not as strong as for formal parliamentary laws, but is there (for details see Albrecht and Küchenhoff 2011, p. 47). Additionally, the competence for the development and enactment of ordinances must be regulated in a formal parliamentary law to a certain extent, as it is regulated within the German Constitution by Article 80 of the Basic Law.[4] Thus, the Parliament has the final responsibility and observance on

[3] See Art. 13 para 1 and 14 para 3 of the IE Directive [Directive 2010/75/EU of the European Parliament and of the Council of 24 November 2010 on industrial emissions (integrated pollution prevention and control)], Official Journal L 334, p. 17.

[4] Basic Law (Constitution of the Federal Republic of Germany) of 23.05.1949, last amended by Art. 1 of the Act of 11.07.2012, Federal Law Gazette I, p. 1478; a translation is available in Mulloy

these standards, following the principle of democratic legitimation by the people (Albrecht 2008b). On the European level there is a different standard setting procedure, but it also follows specific formalised rules.

5.2.2 Function of Legal Standards

Legal standards are necessary for the execution of law. This is particularly true for such fields of law where the execution of laws requires concrete standards and thresholds, for example environmental law. On one side, the general prohibition of the use of the environment (as a whole) is not possible, even though necessary. For example, in respect to climate change, it is expected that mankind has already taken the step to irreversible changes of the climate system (Latif 2006), and therefore the emission of greenhouse gases should be banned. But besides this of course absolutely relevant field, in most of the cases, an absolute and general prohibition of the use of environmental resources is not necessary. Several environmental sectors are able to remedy or repair themselves (Albrecht 2008a, p. 20) over time: over-exploited fish stocks can respawn and contaminated water can be cleaned by natural processes. The same is true for soil in so-called natural attenuation (NA) processes (Hennecke et al. 2008, pp. 46ff.), contaminants in the air may be washed out by the rain, and even ozone depleting substances are transformed finally into inert non-toxic substances (Beyerlin 2000, p. 153). But even though the environment has generally the ability to repair and remediate, human activities tend to overstress these self-repairing abilities of nature (von Weizsäcker et al. 1997, pp. 244f.). Thus, a general allowance of unlimited use of the environment is not possible and a system of standards and thresholds must be found to define the acceptable amount and limits of the use of the environment (Albrecht 2008a, p. 20).

5.2.3 Limitations of Legal Standards

Legal standards are necessary for the execution of law. This is true for most of the law fields to a varying degree. Legal standards are well known in building and construction law (e.g. minimum height of rooms, distances between buildings, stability parameters, etc., see Albrecht and Weiß 2008), in consumer protection and product liability law (maximum contents of substances in products, security standards, etc., see Albrecht 2008c), also social law (maximum floor space area for a state-paid apartment, etc.), and, to finish this enumeration, in financial law (International Financial Reporting Standards—IFRS of the International Accounting Standards Board—IASB, etc.). And it is particularly true for environmental law.

et al. (2001), pp. 1ff., and more actual under http://www.gesetze-im-internet.de/englisch_gg/index.html, last accessed 21.03.2013.

The more detailed such a system is, the more likely the defined level of acceptable environmental pollution and degradation is met.

On the other side a very detailed system entails complications and difficulties for the addressee of such regulations, like individuals, businesses, but also the state itself and the addressee of a provision or an administrative order based on such provisions does not know what is expected from them in detail (Albrecht 2008a, p. 20). For example, the order to remediate the soil of a contaminated piece of land is not precise enough to create a legal binding. At least certain minimum requirements and remediation targets have to be formulated (Knopp and Albrecht 1998, p. 32), for example thresholds for acceptable contamination of the ground water in case of securing a contaminated site (Albrecht 2003, p. 117). For the creator of standards and thresholds the difficulty is, to find the right balance between the optimum for the environmental protection (zero emission or zero contamination) and what is practical, accepted by the population and finally economically affordable (Albrecht 2008a, p. 20), which may be defined as the social optimum.

One major disadvantage in the development of legal standards is that the state loses more and more the competence to develop legal standards through their own capacities or expertise inside the state administrative and scientific institutions (Feldhaus 2000, p. 171). In these cases, there is a practical need to refer to voluntary standard systems.

5.3 Voluntary Standard Systems

The opposite of legal standards are not, as maybe expected, illegal or non-legal standards. Private standards are acknowledged by the law, often taken into consideration and in numerous cases laws refer to them. Such private standards may be defined in accordance to the standard DIN EN 45020, a consensually developed document, created by a (generally) accepted institution and which sets rules, guidelines or properties for activities or their results for general and repeated use, and which aspires to an optimal degree of order within a given content (DIN 2007).

Private standards are characterised as having the character of recommendations for correct technical or process-related functioning. Thus, private standards are relevant benchmarks in the legal system for the decision if a product, a procedure or a service has the characteristics which are expected by the interested stakeholders of the respected field.

But, because private standards are still recommendations, the use of standards is, in principal, free. Standards may be used but, generally speaking, there is no obligation to follow them. This is clarified by a central decision of the German Federal Civil Court on a case with relevance to German Industrial Norms (DIN), which stated that "private norms and standards are not legal provisions, but private technical regulations with recommendation character. They may reflect the generally accepted codes of practice, but they can also stay behind them".[5]

[5] Federal German Civil Court, decision of 14.05.1998, Neue Juristische Wochenschrift (NJW) (new juridical weekly journal) 1998, pp. 2814f. and 2815.

In the context of development and legitimation, voluntary standards follow different rules and they have in general, a different function. Voluntary standard systems are, from a legal perspective, agreements on the basis of voluntary participation developed in a more or less organised procedure and depend upon what kind of standard is in discussion. The participating parties, businesses, individuals and also authorities accept these standards as a basis for mainly contractual relations. For example, in a contract on soil sample taking, the parties can regulate the requirements, rights and obligations, and also the modus of how the samples have to taken in detail in the contract. Or they can refer in the specification of services, usually added as an annex to the contract, to a standard, e.g. for the definition of the concentration of copper in eluates and leachate the German Industrial Norm DIN 38406-7: 09.91, of arsenic in eluates and leachate the combined German, European and International Standard DIN EN ISO 11969: 11.96, or for the assessment of soil air the Association of German Engineers VDI-guideline 3865 sheets 2 and 3. The reference to such norms and standards ensures a certain standardised quality which makes results comparable. Also contracts are shorter and more practicable if standardised specifications could be shifted to annexes, referring to a private standard.

5.3.1 *Sources of VSS*

Several thousand standards and thresholds are known in the environmental sector in Germany and most of them are not legal standards. They have been developed by different institutions.

The best known German standards are probably the following:

- German Industrial Norm—DIN;
- Association of German Engineers—VDI standards;
- the Global Standards One—GS1;
- Association for Electrical, Electronic & Information Technologies—VDE standards.

To give an example on the background and functioning of private norming and standardisation institutions, the German Institute for Standardization (DIN) is used as an example. The German Institute for Standardization is a registered society on the basis of civil law (Bayerlein 2008, p. 52) and has its seat in Berlin since 1917. The German Institute for Standardization is the national institution representing Germany in European and international standardisation activities. It offers the interested stakeholders (producers, traders, industry, science, consumers, testing institutes and authorities) a forum to develop consensual norms and standards.

Consumers are represented by the consumer board of the German Institute for Standardization which has in total five members representing relevant consumer protection and scientific institutions (DIN 1994).

The principles of the German Institute for Standardization are regulated in the DIN 820 and are based on voluntariness, publicity, issue-relatedness, stakeholder participation, orientation on common welfare, internationality, transparency, consensuality, cartel-law related innocuousness, uniformity, user-friendliness, state-of-the-art of science and technology, economicality, market-orientation and use for the general public (DIN 1994).

These principles and other relevant provisions are regulated in several chapters or parts of the DIN 820. The most relevant parts for this chapter are part 1, part 4 (see Sect. 5.3.2), part 13 and part 15.

Part 1 regulates the general principles of a German Industrial Norm. Such a standard is for the general public and shall not lead to economical advantages of individuals (natural persons or businesses). The objective of standardisation is the improvement of quality in all areas of life. Norms and standards are a benchmark for correct technical characteristics and represent the so called "generally accepted codes of practice" (DIN 1994; Albrecht and Weiß 2008).

Part 13 regulates the transposition of European documents of CEN,[6] CENELEC[7] and ETSI[8] (DIN 2004), part 15 the transposition of international documents of ISO[9] and IEC[10] (DIN 2010).

5.3.2 Procedures to Develop VSS

Part 4 of DIN 820 regulates the quite formalised procedure for the development of a standard in the German Institute for Standardization between the application for a standard and the publication of it. Principally everybody can suggest the development of a standard. This suggestion will be discussed in an expert circle and, after a process of public participation, accepted as a standard and publicised (DIN 1998).

Thus, the procedure for standard development is usually a stepwise procedure. At first, the subject of standardisation has to be determined and it must be clarified if other similar subjects should be included into the procedure or excluded by comprehensible classification.

For the concrete work, a board of experts has to be established with participants from the relevant stakeholders (producers, user groups, scientists, representatives of

[6] European Committee for Standardization.

[7] European Committee for Electrotechnical Standardization.

[8] European Telecommunications Standards Institute.

[9] International Organization for Standardization.

[10] International Electrotechnical Commission.

authorities, consumers, etc.). A broad participation of the stakeholders secures the acceptance and applicability of the standard.

The next step is the preparation of a first draft which is opened for public discussion and includes also the possibility to raise objections. Opinions and objections are assessed and—if accepted by the board—included into the second draft which is again opened for public discussion. This step may be repeated several times, until a satisfactory result is reached and no relevant objections are raised. Then the final draft is prepared and in the usual way published.

To summarise this, standardisation procedures have in common (1) a stepwise procedure, (2) with the participation of stakeholders, (3) where in a consensual way (4) a standard is developed (5) by experts (and not by the standardisation organisation), and (6) is finally published.

This is of course true for procedures at the basis of DIN 820. Other standardisation organisations may follow other procedures and the principles may differ slightly, but they all have in common the aim that the standardisation is done for the benefit of the public and not for individuals.

5.3.3 Corporation Standards

Beside these common welfare-oriented standardisation institutions, there are also corporation based standards. Every business or corporation has certain rules, norms and standards for uniform treatment of specific cases, but in most of the cases, there is no complete and consistent standardisation system established. But in particular, such systems exist in large corporations, for example at German Rail (Deutsche Bahn AG) or Siemens AG. In respect to sustainability Siemens has established a standard for the environmental friendliness of products and installations in 1993 (Pfeiffer 2009, p. 140), the Siemens-Norm SN 36350, which is meanwhile transferred into an Environmental Protection (EP)-standard (Siemens 2013).

5.3.4 Limitations of VSS

Voluntary standard systems serve rationalisation, communication, securing of usability and quality, compatibility, convertibility, health protection and security, consumer and environmental protection. Standards have the aim to at least serve the 'generally accepted codes of practice' and they consider (and sometimes represent) the 'best available technology'. But, standards have also some limitations. In fast developing technical and technological fields, the procedure for setting standards could be (like the setting of legal standards) not fast enough to prepare standards, fully serving the objectives presented before. Second, private standards have not the same binding value as legal standards. And finally, national (private and legal) standards may be useless in an international or transboundary context. This is obvious in licensing and planning procedures (Helbron et al. 2006, p. 114), but also in private contracts, even though here the ruling of a defined legal system and specific standards is easier, possibly because the parties of a private contract are usually free in what they are agreeing to be part of in the contract, the so-called principle of private autonomy (Jauernig 2009, before Art. 104, no. 1).

5.4 VSS in Law

Private standards in voluntary standards systems are highly linked to legal standards. Often included in contracts, they serve as a source for interpretation of legal terms and are increasingly referred to directly by legal provisions. Even though private standards are agreements between private parties and participants, this may be in general acceptable, if certain requirements are taken into consideration. Thus, private standards are acknowledged by the law, often taken into consideration and in numerous cases laws refer to them. Private standards contain (usually) clear and detailed regulations. Thus, the reference to private standards within law could help to reduce legal insecurity in laws and in contractual affairs. The reference to private standards could relieve the state, individuals and business from too detailed legal provisions. But, there are some limitations to be observed; in particular it has to be assessed carefully, if the private standard is regulating what is intended by the private parties or the state.

5.4.1 Private Standards in Contracts

Voluntary standard systems are, from a legal perspective, agreements at the basis of voluntary participation, developed in a more or less organised procedure, depending on what kind of standard is in discussion. The participating parties, businesses, individuals, also authorities accept these standards as a basis for mainly contractual relations. For example, in a contract on soil sample taking, the parties can regulate the requirements, rights and obligations, and also the modus of how the samples have to be taken in detail. Or they can refer in the specification of services, usually added as an annex to the contract, to a standard, e.g. for the definition of the concentration of copper in eluates and leachate the German Industrial Norm DIN 38406-7: 09.91, of arsenic in eluates and leachate the combined German, European and International Standard DIN EN ISO 11969: 11.96, or for the assessment of soil air the Association of German Engineers VDI-guideline 3865 sheet 2 and 3 (see above Sect. 5.3). The reference to such norms and standards ensures a certain standardised quality which makes results comparable. Also contracts are shorter and more practicable.

5.4.2 VSS for the Interpretation of Undefined Legal Terms

Another function of standards is to support the interpretation of undefined legal terms. Such standards are not directly legally binding, but in many cases, these norms can be used in addition to legal terms and to interpret legal terms, in particular so called "undefined legal terms" (Albrecht 2008a, p. 24). To give an example: the question, "if a product is without material defect in respect to Art. 434 of the German Civil Code?" may be answered by interpretation of private standards. Concerning a product which does not fulfil the requirement of a certain German Industrial Norm which represents what is expected by the market participants, there is the indice that it has a defect in respect to Art. 434 of the German Civil Code[11] which addresses certain consumer rights, based on Art. 437 of the German Civil Code to the buyer of the product (Berger 2009, Art. 434, no. 14). Principally, with the compliance to a generally accepted standard like a DIN-norm it is assumed that the product, service or procedure complies with the "generally accepted codes of practice", as decided by court decision.[12] This assumption can be refuted, for example if the standard is in an amendment procedure or by an expert opinion, for example in a court process (Schulze-Hagen 2004, p. 5).

5.4.3 References in the Law to VSS

In cases where a system of private standards exist, and no legal standard system is established yet, but needed, the law itself may refer to private standards. Examples are the German soil protection law with numerous references to German, European and international standards, as regulated particularly in Annex 1 of the Federal Soil Protection Ordinance,[13] or noise regulation with more than 100 guidelines on noise reduction methods and noise values (Feldhaus 2000; details in Albrecht 2010).

To come back to the example from Sect. 5.3.2, the same standards which may be used for defining the content of a contract, are referred to in the Federal Soil Protection Ordinance to identify the requirement on specific investigation methods (location, duration, analysis methods, etc.) which may be regulated in an investigation and/or remediation order on base of Art. 9 and Art. 10 of the Federal Soil Protection Act (Knopp and Albrecht 1998, pp. 27ff. and 38). Thus, the addressee of

---

[11] German Civil Code in the version of the official publication of 02.01.2002, Fed. Law Gazette I, pp. 42 and 2909 and Fed. Law Gazette I 2003, p. 738, last amended by Art. 1 of the Act of 11.03.2013, Fed. Law Gazette I, p. 434. A translation of the German Civil Code is available in the internet: http://www.gesetze-im-internet.de/englisch_bgb/index.html, last accessed 30.08.2013.

[12] Higher Regional Court Munich, Neue Juristische Wochenschrift—Rechtsprechungsreport (NJW-RR) (*new juridical weekly journal—court decisions report*) NJW-RR 1992, pp. 1523f and 1524.

[13] Annex 1, Fed. Law Gazette I 1999, p. 1561.

an investigation order—if not otherwise regulated in the order—has to follow in respect to the definition of the concentration of copper in eluates and leachate the German Industrial Norm DIN 38406-7: 09.91, of arsenic in eluates and leachate the combined German, European and International Standard DIN EN ISO 11969: 11.96, or for the assessment of soil air the Association of German Engineers VDI-guideline 3865 sheet 2 and 3 (see Sect. 5.3.2).

In European law there is an even stronger tendency to shift the work for specification of requirements on product safety and quality to private bodies but with a different systematic (Fuchs 2005, pp. 2f.).

But, private standards are developed by private bodies. The disadvantage of private standards, in particular industry-centred standards, may be that the development could be driven by economic interests. Thus, these standards are at risk of being negatively influenced by interest groups in this process (SRU 2011). Thus, they are not democratically legitimised like legal standards (Bahke 2006, p. 23; see also Sect. 5.2.1) and could infringe against Art. 20 para 3 of the Basic Law, the principle of legal certainty (Albrecht and Küchenhoff 2011, p. 66; Feldhaus 2000, p. 182). Furthermore, if an act or an ordinance refers to a private norm, the content of the private norm may be amended or changed following the usual amendment procedures and intervals. Then the question is, if the reference in the act or the ordinance could be dynamic in a way that they refer to the respective actual version of the private standard.

The advantage would be that the law would follow the current state of discussion and regulation in the respective community. But on the other side, the norm-setting would be given to private norming organisations and it is not guaranteed that the democratically legitimised body would accept the norm content. This is generally regarded as incompatible with the principle of legal certainty, even in cases where federal law refers to federal state's law.[14] If a dynamic reference from one state law to another state law of a different level, but both democratically legitimised, is regarded as violation of the constitution, it is even more a breach of the constitutional principle of legal certainty, if the norm in question to be referred to, is a private (and not democratically legitimised) provision.

A reference to a static (private) standard is usually regarded as unproblematic in respect to constitutional concerns,[15] at least if the legislative organ has looked into the material content of the standard (Albrecht 2008a, p. 20).

Beside this more formal argument, material objections could also be raised: If a dynamic reference would be allowed, this would be nothing else than an anticipated acceptance of all future standards of the private norming organisation.

Thus, in the juridical literature the following requirements must be fulfilled:

---

[14] Federal Constitutional Court, decision of 01.03.1978, Neue Juristische Wochenschrift (NJW) (*new juridical weekly journal*) 1978, pp. 1475ff. and 1477.

[15] Federal Constitutional Court, decision of 26.01.2007, Beck-RS 2009, 31386, no. 12; see also Federal Constitutional Court, decision of 23.03.1982, Neue Juristische Wochenschrift (NJW) (*new juridical weekly journal*) 1982, pp. 2859ff. and 2860.

1. The members of the standardisation body must have expertise in the field of the standard (Sobczak 2002, p. 65),
2. The standardisation body must be balanced in a way that the different stakeholders are represented in a reasonable way in the standardisation body (Sobczak 2002, pp. 66ff.),
3. Participation of the public must be assured (Sobczak 2002, pp. 71f.),
4. Revision and publication must be possible in a structured and acceptable procedure. The addressees of the norm must have reliable and reasonable access to the norm content (Sobczak 2002, pp. 72f.).

In particular, the requirement to have access to the wording of a private standard which is referred to in a legal provision was subject to several court decisions[16] and also subject to decisions of the Federal Constitutional Court.[17]

5.5 Legal Standards in Private Law Relations

Without any limitations the agreement of compliance to legal standards in private law related cases, in particular in contracts, is possible. Even more, if a contractual provision violates legal requirements, the contract is (legally) regarded as invalid on the basis of Art. 134 of the German Civil Code.

Furthermore, contracts where the products to be sold, the services to be delivered or the procedures to be carried out violate legal standards, they are in principal suffering from a defect.

The question is which provisions have priority: the provisions on invalidity or the provisions on consumer rights. If the legal standard is meant to avert directly such contracts or contractual provisions, then the contract is invalid (Jauernig 2009, Art. 134, no. 1). This could be true for working contracts with immanent violation of employer safety rules. If contracts in question are generally accepted by the legal system, like this is generally the case for purchase agreements for products. If a product does not fulfil legal standards, then it has a material defect (Art. 434 German Civil Code). Thus the legal consequences are even stricter than in case of non-compliance with a private standard like a DIN-norm. In the second case, there is (only) an indice of a material defect (see Sect. 5.4.1).

Furthermore, Art. 906 of the German Civil Code regulate which immissions are to be borne by the owner of a piece of land. This provision states in conjunction

[16] Federal Administrative Court, decision from 29.07.2010, Neue Zeitschrift für Verwaltungsrecht (NVwZ) (*new journal for administrative law*) 2010, pp. 1567ff.; Higher Administrative Court Koblenz, decision from 26.03.2009, Neue Zeitschrift für Verwaltungsrecht—Rechtsprechungsreport (NVwZ-RR) (*new journal for administrative law—court decisions report*) 2009, p. 673; more generous the Higher Administrative Court Schleswig, decision from 11.08.2011, BeckRS 2011, 56394.

[17] Federal Constitutional Court, decision from 26.01.2007, Beck-RS 2009, 31386, no. 12.

with Art. 1004 of the German Civil Code that the owner has no right to stop immissions to his piece of land as long as legal immission levels are not exceeded.

To summarise, legal standards are influencing private law relations to a high extent. Legal standards are taken directly or indirectly as reference level for defect or not, to be accepted or not, and in some cases, valid or not.

5.6 Conclusion: Competition or Added Value?

To conclude this contribution, it is clear that both legal and private standards are of high relevance for our daily lives. VSS are becoming more and more important, in particular due to the speed of technological progress, scientific perception and globalisation, combined with a tendency to privatise state control and supervision and shift it to enterprises, NGOs and individuals. This makes it difficult for the legislative powers to regulate the relevant fields of environmental protection, consumer protection, labour and product safety and other fields in the necessary pace.

Therefore, the reference in law and contracts to private standards is useful and necessary. Of course, some constitutional barriers for a complete shifting of legislative powers to private standardisation organisations still exist, at least in Germany. But, if these requirements set by Art. 20 para 3 of the Basic Law, in particular the interpretation of the Federal Constitutional Court, are fulfilled; the meaning of VSS in the law will increase.

References

Albrecht E (2003) Die Kostenkonzeption des Bundes-Bodenschutzgesetzes (Cost conception of the Federal Soil Protection Act). Peter Lang Publishers, Frankfurt am Main

Albrecht E (2008a) Standards and thresholds in German environmental law. In: Schmidt M, Glasson J, Emmelin L, Helbron H (eds) Standards and thresholds for impact assessment. Springer, Heidelberg, pp 19–31

Albrecht E (2008b) Gesetzgebungsverfahren in Europa und in der Bundesrepublik (Legislative procedure in Europe and in the Federal Republic of Germany), Chapter 7.1. In: Ertel J, Clesle F-D, Bauer J (eds) Umweltkonforme Produktgestaltung (Sustainable product design). Publicis Publishers, Erlangen, pp 217–227

Albrecht E (2008c) Produkthaftung (Product liability), Chapter 7.4. In: Ertel J, Clesle F-D, Bauer J (eds) Umweltkonforme Produktgestaltung (Sustainable product design). Publicis Publishers, Erlangen, pp 239–243

Albrecht E (2010) The regulation of environmental noise in Germany. J Eur Environ Plan Law 7 (1):37–55

Albrecht E, Küchenhoff B (2011) Staatsrecht (Constitutional law), 2nd edn. Erich Schmidt Publishers, Berlin

Albrecht E, Weiß F (2008) Die Einhaltung der anerkannten Regeln der Technik in der Altbausanierung (The compliance with the generally accepted codes of practice in the rehabilitation of old houses). In: Zeitschrift für das Recht der Bauinvestoren, Bauträger und Projektentwickler (BTR) (J Law Investors Project Promoters Project Dev), pp 61–65

Bahke TH (2006) Technische Regelsetzung auf nationaler, europäischer und internationaler Ebene: Organisation, Aufgaben, Entwicklungsperspektiven (Technical regulation on the national, European and international level: organisation, tasks and development perspectives). In: Hendler R, Marburger P, Reinhardt M, Schröder M (eds) Technische Regeln im Umwelt- und Technikrecht (Technical rules in environmental and technical law). Erich Schmidt Publishers, Berlin, pp 13–30

Bayerlein W (2008) Zur rechtlichen Bedeutung von technischen Normen (On legal relevance of technical standards). Der Sachverständige – DS (Expert J), pp 49–53

Berger CHR (2009) Commentary on Art. 434. In: Jauernig O (ed) Bürgerliches Gesetzbuch (German civil code), 13th edn. Beck Publishers, Munich

Beyerlin U (2000) Umweltvölkerrecht (International environmental law). Beck Publishers, Munich

CIEL – Center for International Environmental Law (2009) Addressing nanomaterials as an issue of global concern. www.ciel.org/Publications/CIEL_NanoStudy_May09.pdf. Accessed 30 Aug 2013

DIN (1994) German Institute for Standardization, DIN 820, part 1: principles, 4th edn. Beuth Publishers, Berlin

DIN (1998) German Institute for Standardization, DIN 820, part 4: procedure, 2001, 1st edn. Beuth Publishers, Berlin

DIN (2004) German Institute for Standardization, DIN 820, part 13: transposition of European documents of CEN, CENELEC and ETSI, 10th edn. Beuth Publishers, Berlin

DIN (2007) German Institute for Standardization, DIN EN 45020: standardisation and related activities – general terms, 3rd edn. Beuth Publishers, Berlin

DIN (2010) German Institute for Standardization, DIN 820, part 15: transposition of international documents of ISO and IEC, 7th edn. Beuth Publishers, Berlin

Feldhaus G (2000) Umweltschutz und technische Normung (Protection of the environment and technical norms). In: Hendler R, Marburger P, Reinhard M, Schröder M (eds) Jahrbuch des Umwelt- und Technikrechts 2000 (Yearbook of environmental and technical law 2000). Erich Schmidt Publishers, Berlin, pp 169–189

Fuchs T (2005) Die Konkretisierung rechtlicher Anforderungen durch technische Regeln (The concretisation of legal requirements by technical standards), delegibus 9/2005. http://delegibus.com/2005,9.pdf. Accessed 30 Aug 2013

Hardinghaus B, Kuntz K, Neufeld D (2013) Mutter Staat (Mother state). In: DER SPIEGEL, vol 12. pp 52–58, 18 March 2013

Helbron H, Bölitz D, Schmidt M (2006) Erfahrungen aus dem Interreg-III-A-Projekt zur strategische Umweltprüfung für Sachsen, Polen und Tschechien (Experiences from the Interreg-III-A-project on strategic environmental assessment for Saxony, Poland and Czech Republic). In: Albrecht E, Nowacki K (eds) Die grenzüberschreitende Beteiligung der Öffentlichkeit und von Behörden in Deutschland und Polen (Transboundary participation of the public and of authorities in Germany and Poland). Lexxion Publishers, Berlin, pp 101–130

Hennecke D, Tränckner S, Joos A (2008) Wechselwirkungen der NA-Prozesse (The reciprocity of natural attenuation processes). In: Joos A, Knackmuss HJ, Spyra W (eds) Natürliche Schadstoffminderungsprozesse bei sprengstofftypischen Verbindungen (Natural attenuation processes in chemical explosive compounds). IABG, Berlin, pp 46–50

Jauernig O (2009) Commentary on Art. 104 and on Art. 134. In: Jauernig O (ed) Bürgerliches Gesetzbuch (German civil code), 13th edn. Beck Publishers, Munich

Knopp L, Albrecht E (1998) Altlastenrecht in der Praxis (The law of contaminated sites in practice), 2nd edn. Verlag für die Rechts- und Anwaltspraxis Publishers, Herne/Berlin

Latif M (2006) Globale und langfristige Strategie gegen den Klimawandel erforderlich (Global and long term strategies necessary for climate change). Zeitschrift für Europäisches Umwelt- und Planungsrecht (EurUP) (J Eur Environ Plan Law) 2006:267–270

Mulloy M, Albrecht E, Häntsch T (eds) (2001) German environmental law. Erich Schmidt Publishers, Berlin

Pfeiffer U-M (2009) Der Weg to Eco-Excellence (The way to eco-excellence). Publicis Publishers, Erlangen

Reidt O, Schiller G (2012) Commentary on Art. 2 of the Environmental Information Act. In: Landmann/Rohmer, Umweltrecht (Environmental law), commentary, supplementary sheets no. 66 (June 2012)

Schulze-Hagen A (2004) Die Bindungswirkung technischer Normen und der Anscheinsbeweis im Baurechtsprozess (The binding effect of technical standards and the prima facie evidence in building law cases), delegibus 10/2004. http://delegibus.com/2004,10.pdf. Accessed 30 Aug 2013

Siemens (2013) Standard für umweltverträgliche Produkte und Anlagen (Standard for environmental products and installations). Siemens AG, Berlin

Sobczak C (2002) Normung und Umweltschutz im Europäischen Gemeinschaftsrecht (Standardisation and protection of the environment in European Community law). Erich Schmidt, Berlin

SRU (2011) Vorsorgestrategien für Nanomaterialien (Precautionary strategies for nanomaterials). Sachverständigenrat für Umweltfragen (SRU) (German Advisory Council on the Environment), Berlin

von Weizsäcker E-U, Lovins AB, Lovins LH (1997) Faktor 4 (Factor 4). Droemer Knaur Publishers, Munich

WHO – World Health Organization (2011) Guidelines for drinking-water quality, 4th edn. WHO Press, Geneva

Chapter 6
VSS Where Formal Regulations Are Missing: Potential Study on Example of Nanotechnologies

Joel Goebelbecker and Eike Albrecht

6.1 Introduction

This chapter aims to discuss the potential of VSS where formal regulations are missing, in this case on the example of nanotechnologies. According to the US National Nanotechnology Initiative, nanotechnologies are "... science, engineering, and technology conducted at the nanoscale, which is about 1 to 100 nanometres. Nanoscience and nanotechnology are the study and application of extremely small things and can be used across all the other science fields, such as chemistry, biology, physics, materials science, and engineering". Nanotechnology is not just a new field of science and engineering, but a new way of looking at and studying (National Nanotechnology Initiative 2012). At nanoscale, the physical, chemical and biological properties of materials may differ in essential ways from the presently known properties of the same substance(s) of macroscopic size; mostly these changes are due to the increased relative surface area or quantum effects.

Besides the remarkable and promising opportunities of nanotechnologies (e.g. potential to solve global and future key issues, such as coverage of energy supplies, conservation of natural resources and comprehensive preventive and curative medical care) they have also substantial uncertainties regarding their possible risks; nanoparticles may pose a threat due to their currently unknown properties. Hence, it seems important to standardise their effects so as to legalise them more strictly in the future. At the moment, very few rules exist for the regulation of nanotechnologies directly. For example, the provisions on fine dust or haze in European law and their transposition into the national legal systems, such

J. Goebelbecker (✉) • E. Albrecht
Department for Civil and Public Law with References to the Law of Europe and the Environment, Brandenburg University of Technology (BTU), P.O. Box 101344, 03013 Cottbus, Germany
e-mail: joel.goebelbecker@tu-cottbus.de; albrecht@tu-cottbus.de

C. Schmitz-Hoffmann et al. (eds.), *Voluntary Standard Systems*, Natural Resource Management in Transition 1, DOI 10.1007/978-3-642-35716-9_6,

as the "Ordinance on Ambient Air Quality and Cleaner Air for Europe"[1] → 35th BImSchV (*Feinstaubverordnung*).[2]

In the EU regulation on chemicals REACH (Registration, Evaluation, Authorisation and Restriction of Chemicals),[3] carbon and graphite were excluded from Annex IV (substances that are considered to cause minimum risk) because of their nanoform usage possibilities. However, besides these rare examples, no direct regulation mechanisms are observed [at least at EU level (Lohse 2011, p. 44)]. Thus, the general provisions are applicable and specific risks may be answered via voluntary regulations for now, created by the actors in the field of nanotechnologies themselves. Beyond that, international standardisation committees such as the International Organization for Standardization (ISO) and the International Electronical Commission (IEC) at present develop the basis for a standardised nomenclature and standardisation of nanoscaled objects and procedures to work towards internationally coordinated efforts and definitions in the field of nanotechnologies (CIEL 2009).

The main focus of this chapter is to reflect on the main reasons for, and benefits from, implementation of VSS as an instrument aiding sustainable development of nanotechnologies which is highly linked with the question of precautionary assessment of risks to human health and the environment. In this context, this chapter supplements and continues the previous chapter (Chap. 5), giving a practical example on the connection points between law and standardisation, showing the possibilities of complementing one another in practise. The chapter starts following this introduction by exploring the technical potential of nanotechnologies themselves and discussing their need for standardisation (Sect. 6.2). The following Sect. 6.3 describes what voluntary standards might do better than compulsory regulation and Sect. 6.4 highlights the potentials of standardising nanomaterials in three different ways: First the social benefits will be discussed, secondly an economic outlook will be developed and thirdly it will be shown which potential for the environment can be expected in the sector of standards of nanotechnologies.

---

[1] Directive 2008/50/EC of the European Parliament and of the Council of 21 May 2008 on Ambient Air Quality and Cleaner Air for Europe (OJ L 152, p. 1).

[2] Ordinance on marking vehicles with low share of the pollutant load (Verordnung zur Kennzeichnung der Kraftfahrzeuge mit geringem Beitrag zur Schadstoffbelastung) of 10 October 2006 (Fed. Law Gazette I p. 2218), last amended 05.12.2007 (Fed. Law Gazette I p. 2793).

[3] Regulation (EC) No. 1907/2006 of the European Parliament and of the Council of 18 December 2006 concerning the Registration, Evaluation, Authorisation and Restriction of Chemicals (REACH), establishing a European Chemicals Agency amending Directive 1999/45/EC and repealing Council Regulation (EEC) No. 793/93 and Commission Regulation (EC) No. 1488/94 as well as Council Directive 76/769/EEC and Commission Directives 91/155/EEC, 93/67/EEC, 93/105/EC and 2000/21/EC (OJ L 396, p. 1) last amended by Regulation (EC) No. 1272/2008 of the European Parliament and of the Council of 16 December 2008 on classification, labelling and packaging of substances and mixtures, amending and repealing Directives 67/548/EEC and 1999/45/EC, and amending Regulation (EC) No. 1907/2006 (OJ L 353, p. 1).

In addition, based on an example of the ISO 31000:2009,[4] a risk management system able to handle nanotechnologies will be addressed in Sect. 6.5. Section 6.6 gives some recommendations on the standardisation process of nanotechnologies highlighting similarly which objectives standardisation cannot deliver. In this context, an answer to the question of the need for nano-specific laws will be approached. The chapter ends with conclusions in Sect. 6.7.

6.2 Current Trends

The experience with previous emerging technologies has prompted a growing demand for an approach to governance where the technological innovation has to be part of a unique process aiming to benefit society. Hence, sustainable growth has become a vital objective for many governments globally. However, the ethical, legal and societal aspects (ELSA) potentially connected to nanotechnologies are becoming ever more relevant and will progressively affect their governance approach (Mantovani et al. 2011).

At the same time, the technological development in the case of nanotechnologies is evolving rapidly in various directions. The following two sections consider the importance of nanotechnologies and their need for standardisation.

6.2.1 Why Is Nano Important?

Nanotechnologies as 'enabling technology' apply early on in the value chain, being used to design smaller, lighter, more durable and smarter materials resulting in products with significantly improved and in some cases entirely new functionalities. Yet, products and materials based on nanotechnologies are available to consumers in some countries already, and many more additional products and applications are currently in the research and development stage.

The 'new' properties of current and future applications of nanotechnologies are seen to have the potential to improve greatly the quality of life in nearly every sector and it is reasonable to predict that nanotechnologies will be the next disruptive technology because of the projected ability to impact and change so many areas of materials, applications and sciences. The innovation potential of nanotechnologies is still reaching much further ahead: Thus important contributions to solve global and future key issues (Federal Ministry of Education and Research 2009) such as medical care, coverage of energy supplies, and the

[4] ISO 31000:2009, Risk management—Principles and guidelines, ed. 1, published 15.11.2009, ISO copyright office, Geneva, Switzerland.

conservation of natural resources (resource savings) through the application of nanotechnological discoveries are expected (Tucker 2009).

In the field of nanotechnologies, both large corporations and small businesses are (and will be) involved (Federal Ministry of Education and Research 2010). Beyond that, many applications affect not only the industrial use, but especially contribute to the everyday life of consumers. In view of such progress, it is predictable that products derived out of nanotechnologies will be increasingly available to consumers worldwide in the coming years (Luther and Malanowski 2004). However, already today, products that can be realised only with the help of nanotechnologies have made significant sales. The global market for nanotechnologies (e.g. used in sun cream, colouring, even in food, as antibacterial coverage or medicine) was valued at nearly \$20.1 billion in 2011 and should reach \$20.7 billion in 2012 (BCC Research 2012). These numbers will after the increasing economic breakthrough even rise strongly. Total sales are expected to reach \$48.9 billion in 2017 after increasing at a 5-year compound annual growth rate (CAGR) of 18.7 % (BCC Research 2012).

6.2.2 *Why Standardisation of Nanotechnologies?*

Although, nanoparticles have been present for a long time naturally in the environment (e.g. volcanic eruptions, fires, and sea salt aerosols) or produced anthropogenically (e.g. burning wood or petrol, welding), it probably becomes generally problematic that the environment and those inhabiting it are faced with an unprecedented and ever-growing volume and diversity of nanoparticles (Mae-Wan 2010; Mantovani et al. 2010). So far, little is known about the exposure of nanoparticles with respect to human health and environment and their potential impact on them. However, concrete evidence is available, that there are interactions of nanoparticles with biological systems (Monica et al. 2006). In a recent study, researchers examined whether gold nanorods could readily pass from water to the marine food web. Their findings suggest that nanoparticles move easily into the marine food cycle and are absorbed in marsh grasses, trapped in biofilms and consumed by filter feeders, such as clams (Ferry et al. 2009). Moreover, a number of publications show that nanoparticles may pose special risks because of their unique properties. In terms of small size, it is important to note that the tiny nanoparticles are able to overcome especially those (biological and physical) barriers that usually remain unconquerable for larger particles (Führ et al. 2006).

Due to this exceptional nature of nanomaterials, the current methodologies employed to conduct risk assessments, toxicological assessments and life cycle analysis of products containing or consist of nanotechnologies may be ineffective or may not currently exist. There are presently almost no standard test methods for measurement of human or environmental exposure to nanoparticles (Hatto 2007). In further consequence, the effects of many nanomaterials are not yet sufficiently evaluated. Initial investigations show that the environmental risks should receive special attention; the studies have speculated that an increased hazard can at least not be excluded (NanoKommission der deutschen Bundesregierung 2008).

To solve these problems, the use of specific hard regulation is advocated by some parties, but so far, the strategies from authorities worldwide have been essentially on probing the extendibility of existing regulatory schemes for nanotechnologies. In the last few years, voluntary measures have been endorsed by public bodies and industry to build confidence and trust, promote safety or gather data. To support the regulatory efforts, an intense activity to increase the knowledge base and to develop standards, methods and protocols is also going on (formally since 2005) involving acknowledged bodies, such as International Organization for Standardization (ISO), European Committee for Standardization (CEN), Organization for Economic Co-operation and Development (OECD) and, recently, World Health Organization (WHO) (Mantovani et al. 2011).

As progress accelerates in the manufacture and characterisation of nanoscale materials and nano-enabled products, it will become increasingly important to researchers, manufacturers, regulators, and other stakeholders to have agreed upon nano standards. Such standards will include definitions with which to communicate; testing and characterisation methods to compare results; and materials properties to facilitate commercialisation of the many and varied applications and uses of nanomaterials (Secretariat of CEN/TC 352 2007).

6.3 Voluntary Nano Standards

The OECD and the ISO have set up special committee groups on nanotechnologies to monitor and address their challenges (IRGC 2009). These organisations are currently working on the standardisation of methods to identify and measure potential risks derived from nanomaterials and their applications and have already published guidelines on health and safety practices for nanomaterials in the workplace, and terminology used for nanotechnologies and nanosciences. They are currently developing standards on a range of other nano-related topics, such as nanoparticle measurement methods, and the safe handling and disposal of nanomaterials. In addition, several nano-specific risk strategies have also been designed to help companies assess, monitor and manage the possible impacts of nano-based products and processes (CENELEC 2012). What these (and other) voluntary standards can deliver is outlined in the following sections.

6.3.1 Stricter than Law?

Private standards have a much larger role in human society than just agreed measures. Put simply, a standard is an agreed, repeatable way of doing something (BSI 2012). However, in the standard-developing process, many stakeholders have to be heard and included, which might lead to the consequence that a 'middle way' will be developed, 'more-or-less' satisfying all attendees; by comparison a legislator would not have these problems. Alternatively, legal standards are created usually in a formalised procedure which is time consuming and in particular in such fields where the innovation speed is high, not fast enough to keep up pace with the scientific progress. However, standard initiatives usually aim to complement existing regulation (or prepare the ground for new ones), in this case, helping to gather detailed information on the introduction and use of nanomaterials and nano-related products to the market. However, their voluntary nature has some drawbacks, when endorsed by public/government bodies they received a moderate response, so that it was suggested, for example in the case of reporting schemes, to make them mandatory. On the other hand, when promoted by private companies, these measures are treated by some stakeholders with suspicion and of little value in their opinion (Mantovani et al. 2011). Nevertheless, even with their relative lack of force when compared to legal standards, voluntary standards can play an important, constructive role in the present state of nano-specific regulation, to build a knowledge base to support policy and regulatory decisions (Mantovani et al. 2011). They might also be used by companies as a strategic tool to reduce their regulatory burden, when handling nanomaterials.

To summarise, a private standard usually should (at minimum) respect the law, and even be tighter (e.g. more specific) but there may be cases, in which there are sometimes stricter laws than what is agreed internationally as a standard. Indeed, private standards are usually voluntary; however, they can become obligatory if they progress to becoming legally-binding (e.g. by contract) or their thresholds are used as guidance values, e.g. for undefined legal terms (Albrecht 2008).

6.3.2 Faster than Law?

In the case of nanotechnologies the above question can clearly be answered with a 'yes', as till now only very few laws try to address nanotechnologies. For example, in the EU there will be, among others, labelling requirements for cosmetics[5] (perhaps soon: novel foods[6]) and the obligation to carry out studies[7] for food

[5] Regulation 1223/09/EC of the European Parliament and of the Council of 30 November 2009 on cosmetic products (OJ L 342, p. 59).

[6] See Regulation 258/97/EC of the European Parliament and of the Council of 27 January 1997 concerning novel foods and novel food ingredients (OJ L 043, p. 1) last amendment Regulation 596/2009 of 18 June 2007 (OJ L 188, p. 14) and the Commission staff working document—Accompanying document to the Regulation of the European Parliament and of the Council on novel foods and amending Regulation (EC) No. xxx/xxxx [common procedure]—Summary of the impact assessment [COM(2007) 872 final] [SEC(2008) 12] (SEC/2008/0013 final, 14.1.2008).

[7] Art. 4 and Art. 6 of Regulation 258/97/EC of the European Parliament and of the Council of 27 January 1997 concerning novel foods and novel food ingredients (OJ L 043, p. 1) last amendment Regulation 596/2009 of 18 June 2007 (OJ L 188, p. 14).

additives.[8] In contrast, the process to standardise these technologies seems to be proliferating and indeed availability of appropriate standards seems to be pivotal to implementing an appropriate regulation for nano-related products (Mantovani et al. 2010). The law in this case (maybe true for every innovative technology) has a problem with 'knowing' and 'defining' new technologies and procedures. So it might be right to say, that the standardisation-committees have and had the advantage of (broader) knowledge regarding nanotechnologies.

Added to this, most of the international standard organisations indeed have become very efficient in coordinating the associated consensus processes in such matters. Thus, they gather information relatively quickly and are able to come to an inclusive agreement within a short timeframe.

However, the speed of the process of standardisation cannot move quicker than the information that can be generated out of the research and development and in many cases to come to consensus, cultural changes often are needed in some sectors. The pace of standardisation will always be dependent on the acceptance and pace of implementation of the policies which the standards support. However, there may be some different redundant standards with the same regulative topic. Hence, there might be a time following the publication of (a) standard(s), in which a leading standard (adopted by the majority of involved stakeholders) will have to win through, and such a process could take a long time. The lawmaker again does not have such 'problems'. Hence, at least in theory, the law could be faster than the standard-maker(s), because here only one party within a formalised procedure can decide which way to go. Indeed, in this case, the process of standardisation is clearly leading the legislative one.

6.3.3 *Laws Following Standardisation?*

As addressed in the previous sections, there are efforts underway to elaborate a regulatory framework to address many of the aspects related to the use of nanotechnologies, but it is largely acknowledged that there is the need to improve technical guidance documents used for the application and implementation of existing regulatory frameworks, as well as to develop new ones. The availability of appropriate standards is pivotal to implementing an applicable regulation for nano-related products (Mantovani et al. 2011).

Until now, the standardisation-initiative's aim has been to complement existing regulation, helping to gather detailed information on the introduction and use of nanomaterials and nano-related products on the market (e.g. type, use, quantity, and safety aspects of the material or related product). Thus, voluntary measures can play an important, constructive role in the present state of regulation: For

[8] Regulation 1333/08/EC of 16 December 2008 of the European Parliament and of the Council on food additives (OJ L 354, p. 16).

nanotechnologies it seems special, that for the more risk attached to the issue, the more government involvement is likely (Lohse 2011, pp. 58f.). Currently, it seems that most governments have a preference for the possibility of self-regulation of the industry. However, with the further development of nanotechnologies, it is likely that with its expanding technical possibilities, the risks of the applications will rise, and would make a governmental legal approach more likely; which then will possibly follow, or at least take into account some approaches of standards.

The legislature is able to take over private standards, indeed: Like in almost all fields of the German environmental law the use of standards and thresholds is commonly practiced and is necessary for its systematic and reasonable execution, to make it applicable and functional by defining legal terms or giving thresholds to users. However, the German Federal Constitutional Court has set some requirements to allow the takeover of private standards (BVerfGE 49, 89—Kalkar I).[9]

6.4 Potentials on Standardising Nanomaterials

Nanotechnologies encompass different research fields and find their way into a large variety of sectors and markets. However, that makes a standard based and uniform definition complex and difficult. Nevertheless, standardisation-processes play an important role in the short and medium term in dealing especially with the current uncertainties about the regulatory situation of nanotechnologies. Standards can support disclosure and sharing of information, definition and dissemination of guidelines and best practices, provide common principles and values and facilitate trust between different current and potential stakeholders. Thus, they do not primarily intend to replace regulation or any other legislative requirement but instead aim to help complement those (e.g. definitions or thresholds) or help during the redefinition of existing hard regulation (Mantovani et al. 2010).

Current focus (Secretariat of CEN/TC 352 2007) of standardisation efforts of nanotechnologies is centred in the four broad areas of:

- Terminology and nomenclature (providing a common framework for communications about nanotechnologies for commercial, scientific, and legal purposes);
- Nanomaterials (characterising physical and chemical properties of nanomaterials for various applications);
- Safety and risk assessment (developing evaluation methods to prove suitability, toxicity, health and potential environment effects on human body);
- Nanometrology (developing methods, equipment and systems to measure basic characteristics of nanoproducts).

[9] Federal Constitutional Court's Decisions (BVerfGE) Vol. 49, p. 89—Kalkar I, decision 2 BvL no. 8/77 from 08.08.1978.

6.4.1 Social Potential

Standards and compliance are the keys ensuring the quality and consistency of physical, chemical and biological measurements throughout society (BSI 2012). Standards exist at different levels and with different scopes: National standards such as ANSI (American National Standards Institute) in the United States or DIN (German Institute for Standardization) in Germany, regional standards such as the standards set by the Pan American Standards Commission or the EN standards in the European Union, and international standards such as the IEC or ISO that are recognised in most of the states worldwide.

Standards generally create above all comparability. For nanotechnologies this is applicable, however, very detailed chemical, physical, pharmaceutical, technical or biological information may not be understandable in detail, at least by the (private) end users and therefore might be more beneficial for the business to business (B2B) communication (e.g. producer to processor). Here, standards for nanotechnologies can provide the essential framework for industries and governments to maintain domestic and foreign confidence in goods and services and are also the key to enhancing global competitiveness, attracting investment and encouraging and supporting innovation, benefiting from committees of manufacturers, users, research organisations, government departments and consumers working together to meet the demands of society and technology (Standards Australia 2012).

On 08.04.2006, an article published by the Washington Post entitled "Nanotech Raises Worker Safety Questions", lamented that no state or federal occupational safety regulations relate to the specific risks of nanomaterials, even though many laboratory and animal studies have shown that nanoparticles are or at least some could be problematic for health (of workers) and environment (Weiss 2006). Additionally, downstream users in the supply chain need security and so, in the matter of social recognition, one facet of standardisation might become vital: The labelling of nanoproducts to protect consumer health and ensuring fair practices. Consequently, future standards or labels should give end users confidence that products are safe and reliable, and that they will perform as they are intended. Here, standards could establish consistent expectations and help generally ensure those expected properties or features are met by the products.

For end-users, a label refers to mainly product features and also serves declaration and security purposes, in justified cases it also includes information on safe handling and disposal of products, and hence, a nano label seems appropriate when the consumer should be informed in regard to a product on the inherent quality or environmental, health and safety properties. Thus, labelling is a key management tool in risk regulation, meeting generally different objectives: On the one hand it marks and enables the mature consumers purchasing decisions and protects them from misleading information, on the other hand it should enable and promote innovative product development. Consumers are thus included in the risk management of various product groups. Nano-specific labelling requirements are for example increasingly used in EU law, initially in the areas of cosmetics, foods and biocidal[10] products.

As opposed to this, voluntary labelling could not yet penetrate the market significantly (Mantovani et al. 2012). Moreover, it would be most beneficial, that information about the nature of the processing and use of nanomaterials would also get back to their manufacturers and suppliers, as a bi-directional transfer of information allows on each stage of the supply chain the optimal estimation of potential risks, thus helping to use the whole potential of nanotechnologies and cutting their risks to the lowest possible level. Art. 34/38 of the European REACH Regulation already demands such a procedure, which establishes the flow of information between manufacturers and users; but this is up to now mainly linked to chemical, not nanotechnological (e.g. quantum physical) effects.

6.4.2 Economic Potential

The economic potential of standardisation of nanotechnologies is enormous. Not only can trade barriers be reduced; standards as mentioned also create a common language that manufacturers and end users can utilise to communicate on issues like quality and safety. Thus, standards help in promoting product compatibility and interoperability, overcoming trade barriers for global markets and fostering the diffusion and adoption of new technologies in general. In addition, they give participants of the development process (e.g. scientist, producers, traders, authorities or consumer protectors) early access to technological knowhow. Moreover, the participants may be able to influence how certain test or measurement guidelines are documented, thereby affecting the content of the standard, in the case of a pending or an already developed standard.

International standardisation is a way to overcome technical barriers of inter-local or inter-regional commerce caused by differences among technical regulations and standards developed independently. These technical barriers mostly arise when different groups come together, each with a large user base, doing some well-established practice that between them is mutually incompatible. Establishing standards, preferably at the earliest opportunity, is one way of preventing or overcoming this problem. However, typically for any new dynamic area at the beginning is that there is a mixture of vocabulary and terminology causing confusion and retarding the adoption of new developments. The early publication of standards provides a relatively consistent set of terms that will address these issues.

Furthermore, standards, particularly open standards, contribute to the standardisation of interfaces and products, leading to larger markets due to lower market segmentation. Larger markets induce more competition between suppliers.

[10] Regulation 528/12/EC of the European Parliament and of the Council of 22 May 2012 concerning the making available on the market and use of biocidal products (OJ L 167, p. 1).

This in turn causes falling prices, higher unit sales of products, more research and development and more technical progress for a better balance of supply and demand (e.g. weaker fluctuations in the price fluctuations of supply or demand), and lower transaction costs by simplified contract negotiations and contracts (Smith 1776; Buxmann and König 1998; Morasch 2006). But moreover, the cost of standardisation will enter into decisions about when, where, and if product or process standards are used (David and Thompson 2008). Here, the standpoints could be one of both; that process standards are harder to monitor and would therefore be more costly, especially in a third party auditing situation, or that they are an investment where at least the economic benefits will outweigh the expenses.

It is likely that standards will vary according to the specific nanotechnology in question. However, global integration will require cooperation among competing institutions. But typically, the tension that results from competition limits cooperation on regulation. Additionally, who integrates with whom becomes a point of contention (IFAS 2007). For some enterprises, the use of standards is a strategic tool to raise competitiveness; others might see standardisation only as an added cost of doing business.

In recent years 'nano' has often been used as an effective sales slogan, presumably for conventional products that have nothing to do with nanotechnologies (Eisenberger et al. 2012). This is not only unpleasant for consumers but also for producers of actual nanoproducts, as they invest considerable research and development work in their products. Therefore, there were isolated cases in several countries of voluntary labelling applied to nanotechnology in the form of a so-called private label and seal, but which has not yet significantly penetrated the market. To date, there is no established negative labelling in the form of special 'nano free' labels, but in the future enterprises may try occasionally to inform consumers about products that contain no nanoparticles (Eisenberger et al. 2012).

In 2004, the reinsurer Swiss Re expressed among other concerns that nanotubes could have similar effects on human health, such as in the case of asbestos, and therefore recommended insurers to limit the liability for nanotechnologies (Swiss Re 2004). Likewise, the insurer Allianz sees conceivable risks that could have not only health related, but also far-reaching economic consequences if not handled professionally (Allianz SE 2005). Regarding this, for any assurance-seeking company it should be conclusive to gain an advantage, if it has a standardised risk management system implemented. Beyond that, a compliance with standards could be a reason for an insurer to make a contract with an enterprise handling nanomaterials; at least it is very likely, that a company without standards and risk management would not find insurance, or get relatively hard contracts in any case. This might predominantly be true for the matter of environmental harms, especially harms threatening biodiversity (Knopp 1995).

And one question remains to be explored: How does the risk profile of a company change, if it works with nanomaterials? Possibly a standardised risk management system is required which takes into account the specific characteristics of nanotechnologies. This can ultimately affect the overall assessment of the value of a company. Here it will be interesting to watch whether future nanotechnologies receive good valuations from society, or such as genetic engineering and nuclear technology have slipped into the negative, which then could be fatal for due diligence.

6.4.3 Ecological Potential

As already mentioned, the labelling could play an increasing role for the risk-and technology-regulation where traditional instruments are limited. As a result, the states and the authorities may observe voluntary labelling by the industry carefully and will force it into compulsory labelling when the voluntary approach fails. The Royal Society and The Royal Academy of Engineering (UK) already recommended that given the emerging evidence of serious toxicity risks, nano-ingredients should be subject to new safety assessments and face mandatory product labelling (RS and RAE 2004).

Besides, unlabelled and unstandardised nanomaterials might be very risky when in the processing, use or disposal of any sanitary or environmentally hazardous substance is handled unknowingly. Hence, by passing information down the value chain by using standardised labels, sustainability is highly promoted by standardisation. In the subject of ‘best practices’ and similar matters it helps to bring all the developers, manufacturers, distributors, users, and firms on the reuse or disposal side to a table and discuss an integrated view. Standardisation brought to end-users could also help to strengthen their involvement in sustainable development of nanoproducts, by enabling the users to compare the products.

By harmonising standards at a global level, there seems to be the agreement that the main focus needs to be on public health and environmental impacts (IFAS 2007), and if nano standards evolve from current standards, there will be a combination of national and international standards. It might be possible to begin by agreeing on principles for standards rather than on specifics. Indeed, international standards have more potential to become politicised while national standards can be developed in a manner that is relevant to local conditions (IFAS 2007). For nano applications, if there are environmental consequences, they must be related to local and national situations. However, nanotechnologies exhibit unique features and do not have national boundaries. Some nanoproducts, if persistent, e.g. some inorganic or carbon nanoparticles (Reijnders 2012), could have international implications if they are released into the atmosphere or water cycle. Therefore, the question of the right of a country to refuse to be in contact with the product needs to be addressed. There also is the issue of the right of a government to reject exposure of its citizens to certain materials (IFAS 2007).

One of the other many challenges that must be overcome is how to prioritise which standards to develop next, based on measurement best practices and characterisation processes. It has to become clear to understand whether the measurement tools available today are the right tools from an international perspective, taking into account current technical developments and those of the foreseeable future. However, standards could provide clear guidance regarding the currently questionable disposal of manufactured nanomaterials and could support manufacturers and others in making decisions as to the most appropriate way to dispose of their process waste. As increasing numbers of products incorporating nanomaterials are made, the need for manufacturers to safely dispose of the process waste also increases. This will not only be useful to manufacturers, but also to those involved in waste disposal, research and development on nanomaterials and the regulation or monitoring of waste and waste disposal.

Deliberately manufactured nanoparticles are important technological materials with many benefits but also attendant risks and hazards; certain standards should also help in their assessment and management.

6.5 ISO 31000:2009 – A Brief Introduction

In the capital market, for example, risk management is known as an obligation due to changes in the German Stock Corporation Act[11] since 1998. There is a worldwide standard on risk management: The international standard ISO 31000:2009.[12] In conjunction with the revised ISO IEC Guide 73:2009[13] "Risk management – Vocabulary" the documents were published in late 2009.

ISO 31000:2009 provides principles and general guidelines on risk management (risk being defined here as the "effect of uncertainty on objectives") and is not specific to any industry or sector. The design and implementation of risk management plans and frameworks will need to take into account the varying needs of a specific organisation, its particular objectives, context, structure, operations, processes, functions, projects, products, services, or assets and specific practices employed (ISO s. a.). The familiar 'top-down' approach in the standard offers generally a basis to deal with emerging risks, such as those associated with nanotechnologies or related processes and is above that able to take into account all the different risk conditions in an organisation. However, it will not automatically deliver thresholds or values to deal with nano-related risks. A schematic view of the standards framework is shown in Fig. 6.1 below.

As depicted, ISO 31000:2009 offers continuous stages: Establishing context is about setting the parameters or boundaries around the organisations risk appetite and risk management activities. It requires consideration of the external factors and the alignment with internal factors such as strategy, resources and capabilities (AIRMIC et al. 2010). It involves defining the location and extent of the system

[11] Art. 91 para 2 Stock Corporation Act (Aktiengesetz) of 06.09.1965, Fed. Law Gazette I, p. 1089, last amended 20.12.2012, Fed. Law Gazette I, p. 2751.

[12] ISO 31000:2009, Risk management—Principles and guidelines, ed. 1, published 15.11.2009, ISO copyright office, Geneva, Switzerland.

[13] ISO IEC Guide 73:2009, Risk management—Vocabulary, ed. 1, published 2009, ISO copyright office, Geneva, Switzerland.

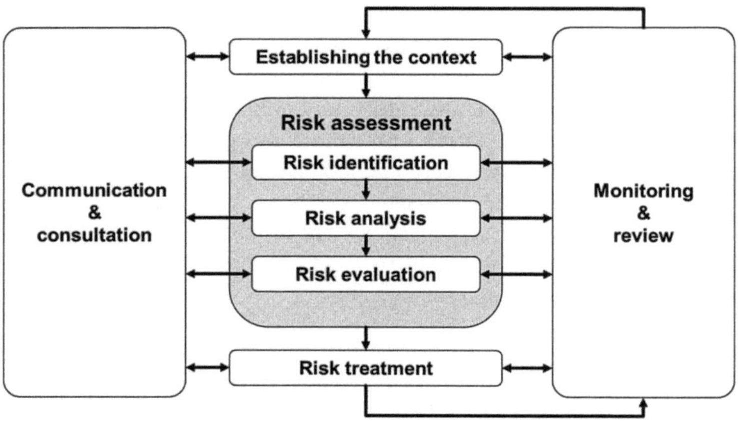

Fig. 6.1 Risk management process of ISO (adapted from ISO 31000:2009, Clause 5)

and the processes operating in its area that may generate risks. It is important to decide which subgroups (e.g. site producing nanomaterials) the risk management plan shall address. The subsequent risk assessment aims to explore (1) the potential impact (i.e. high level of damage) on (2) a particular value (e.g. environment) from (3) a hazardous process (e.g. production of nanoparticles). Thus, as part of establishing the context, the economic, social, political and environmental values where the plan applies should also be described. In addition, there must be defined risk criteria for the risk assessment, including the preparation of likelihood and consequence scales and their combination into a risk matrix, to be able to determine the level of risk. It is also important to define the level at which a single risk is considered acceptable, tolerable or intolerable; here, it is wise to modify the acceptance level for local conditions in consultation with all stakeholders (e.g. providing relevant data and research findings on nanomaterials). The importance of the process of establishing context must not be underestimated. Setting the wrong context is a risk in itself, because all of the steps in the subsequent process of the standard are dependent upon it (Krause and Borens 2009).

Risk assessment: Comprises the single processes of identifying, analysing and evaluating risks. Concerning nanotechnologies it is expected that there will be risks mostly in the product and its processes, but as well as in an uncertain legal environment or standard which is prone to development and change. Hence, an operator should utilise a range of risk identification techniques, e.g. set up a process of how scientific studies on effects of nanomaterials may be followed.

At this point, the ISO/IEC 31010 provides further guidance on how to select and apply systematic methods for risk assessment. As far as nanotechnologies are concerned, it must be assumed that there will be scarce available data to estimate a reliable level of risk. However, the risk analysis considers possible causes, sources, likelihood and consequences to establish the inherent risk. Existing management controls should also be identified and effectiveness assessed to determine the level of residual risk (AIRMIC et al. 2010). The risk assessment process inherently requires that uncertainty is transparently described, but also, provides for a scale of likelihood or consequence to be ascribed to what may possibly occur. Finally, risk evaluation, as defined in ISO 31000:2009 involves comparing the results of the risk analysis with risk criteria, to determine whether the level of risk is acceptable, tolerable or intolerable. Concluding the three steps of risk assessment in a short overview (Krause 2009):

1. Risk identification, e.g. emission of a substance; short and long term exposure;
2. Risk analysis, e.g. likelihood and the consequence associated with each risk; finally the overall level of risk (e.g. high, medium, low);
3. Risk evaluation, e.g. the intolerable and tolerable level of risk and residual risk; execution and effect of controls or mitigating actions.

Next step, risk treatment: The risk owner in general is able to treat risks by avoiding them completely, modifying their likelihood or influencing the extent of their consequences. First and foremost the process of developing management options as part of the risk management plan should aim to reduce, avoid or eliminate intolerable risks as a first priority. Management options considering nanotechnologies could be designed to reduce the likelihood of their risks (e.g. implement work practise guidelines to reduce the probability of an emission of nanoparticles) or their consequences (e.g. implement an emergency management plan to reduce the result of possible emission), or both. To decide which of the management options to choose from, a cost benefit analysis could determine which of the possible risk treatments will provide the best benefit, relative to cost; however treating the highest risks first should always take priority (Krause and Borens 2009).

Monitoring and review: This process enables tracking of all risks, to ensure they remain within an acceptable range. The monitoring and review process is interwoven throughout the entire risk management procedure proposal of the ISO 31000:2009 and could be particularly beneficial if the changing environment (e.g. social or political, legal and regulatory climate) of nanotechnologies is taken into consideration. Any modification here should be a trigger for the user of the standard to review the risks in light of those changes. Alternatively as part of the monitoring and review, if the risk profile of a certain indefinable or uncertain risk source has, as under some circumstances single nanotechnologies or materials, not changed, it may be wise to extend delaying the handling (e.g. of unknown nanomaterials) until such time as the likelihood and consequences of the distribution risk can be better defined. On the matter of some nanotechnologies, it seems this could be especially appropriate at the present stage of development and knowledge level (Krause and Borens 2009).

6.6 The V in VSS

As shown, standards are powerful instruments to support the development of new technologies like nanotechnologies and help to make them sustainable in many ways. However, standards *per se* are not legally binding, but they can become that by laws and regulations of the legislature or by contracts in which compliance is agreed to be binding. Here standards are often being used to fill undefined legal terms, for example, the term 'state-of-the-art' or 'best available technology' used for instance in the Integrated Pollution Prevention And Control Directive (IPPC),[14] and retrieving legal significance.

However, a few functions may not be deliverable by standards: In this section the need for laws should become clear. Today, several regulatory agencies worldwide focus essentially on the following actions (Mantovani et al. 2010):

1. Provide or improve technical guidelines and procedures to support safety assessment for specific types of nanomaterials or nano-related products.
2. Adapt or strengthen pre-market notification procedures to ensure nanomaterials are reviewed before entering the market, including options for mandatory reporting schemes.
3. Introduce amendments and changes into existing legislation to ensure inclusion of nanomaterials and nano-related products (e.g. specific definitions, risk management procedures, labelling, restrictions, or the exclusion of carbon and graphite of the Annex IV of the REACH regulation, etc.).

The availability of suitable standards therefore is pivotal to implement an appropriate regulation for nano-related products (Mantovani et al. 2011). However, due to the innovative production processes enabled by nanotechnologies and the peculiar behaviour of the matter at the nanoscale, the system of written and physical standards established for the macroscopic and microscopic world, cannot easily be scaled down to the nanoscopic world (Mantovani et al. 2010).

6.6.1 Standards Are Not Laws

Standards in general, especially voluntary ones, all share a weakness—obvious as it might be: As long as they are not agreed on the basis of private law agreements, e.g. B2B-contracts, they are voluntary! Hence, whenever it becomes too difficult for joining enterprises, it might be unsurprising that the participant simply withdraws from the standard. Indeed, standards are in general lacking the power of force to sanction violations. However, if a voluntary standard (or a fragment of it) becomes part of an agreement (e.g. as described above) with sanctions included, it may lead to a different outcome.

Nevertheless, a future evolution of nanotechnologies regulation(s) could influence the path of the entire development of nano-related products and processes. However, even if an enterprise would comply with all standards, especially international ones, this would still not be a guarantee of legality within single states of its

[14] Directive 2008/1/EC of the European Parliament and of the Council of 15 January 2008 concerning integrated pollution prevention and control (OJ L 24, p. 8).

processes or products. In addition, the existence of a published standard does not imply that it is always useful or correct. For example, if an item complies with a certain standard, there is no assurance at all that it is fit for any particular use; therefore validation of suitability thus becomes necessary.

Certain countries around the world are making it a primary part of their own research plans to guide or fertilise the process of standardisation to benefit from the following development of the standardisation. For instance, the Chinese Ministry of Science and Technology has made the drafting of nanotechnologies research standards part of its national basic research plan (MOST s. a.).

Other countries (e.g. Canada, Japan, and USA) are striving for leadership positions within standard organisations (ANSI s. a.), too, so that they can help shape the standards to which everyone must adhere later on. It so happens that numerous different standard setting organisations globally are highly active in defining standards (Tucker 2009). So the main question might be: Which one will manage to become dominant (e.g. most common)?

6.6.2 Limitations of Standards: Need for Laws?

As shown, standards in general help make life simpler and increase the reliability and the effectiveness of many goods, services, and processes. They are intended to be aspirational—a summary of good and best practice rather than general practice. And standards are designed for voluntary use and do not impose any regulations (BSI 2012). However, private standards are one tool in the regulatory spectrum of the legislator to provide a solution to a problem (possible risks of nanotechnologies). The disadvantage of an industry standard is that the establishment and development generally is driven by economic interests and hence the published standards may be controlled, or at least be influenced, by interest groups along this process (SRU 2011). Here, there is also a high potential for laws and legislation to handle the risks of nanotechnologies and to assure sustainable development in every way by selecting the correct standard to be adopted or enforced (SRU 2011).

In terms of nanomaterials, as a special form of substances, their properties and effects still leave many knowledge gaps in the analysis of the regulatory framework, which makes a continuous precaution-oriented handling of those materials impossible. These shortcomings are partly due to the peculiarities of nanomaterials (SRU 2011). Accordingly, the need for nano laws is in demand. Though, the above mentioned shows that a proper regulation might not be possible without the utilisation of standardisation: In the first stage, it should be build knowledge about regulatory procedures and gaps and in parallel develop standards for self-regulation. Then enforced self-regulation in the medium term should be made possible followed finally by strict legislation in the long term. Here, there even might be an independent ‘nano-law’ possible (Mantovani et al. 2010).

6.7 Conclusions and Recommendations

Self-regulation initiatives, such as standards, play an important role in the short and medium term to deal with the current uncertainties and ambiguity about the regulatory situation for nanotechnologies. They can support disclosure and sharing of information, definition and dissemination of guidelines and best practices, provide common principles and values and facilitate trust between different current and potential stakeholders. As clearly stated in the general objectives of most of these initiatives, their aim is not to replace regulation or any other legislative requirement but instead to help complement those (Mantovani et al. 2010).

Private standards offer the possibility to regulate necessary issues where the state is not able to regulate or to execute. For example, Peine (2011) stated on the example of the Equipment Safety Act[15] which serves as transposition of the European Directive on General Product Safety[16] transformed into German law, that difficulty, complexity, and dynamics (Breuer 1976) of technology makes a reference to technical regulations necessary and legitimate to gain control over the complexity of the future. Here, the German Constitution is the framework for political action which does not omit the technological future (Peine 2011).

Indeed, in the case of nanotechnologies, at least, law and private standardisation could and should go well together. Both take into account human and cultural factors, and undeniably, nano risks are eventually managed by people, not processes or tools. There will be the need to respect different perceptions, but also different settings and positions: There might be no 'one-size-fits-all' approach; the law as well as those responsible for crafting standards should respect that.

However, finding the suitable standards or laws, trying to understand them and to obey them, one could be forgiven for becoming lost; nonetheless there is still one more item to consider: There might be one thing that strict legislation cannot force and voluntary standards cannot deliver either. The best way to be truly "sustainable" is to form individual opinions, run research independently, collect expertise and finally be transparent and open: Inform stakeholders and decision makers—even if the message is not a good one; create a forum for communication, e.g. as it is regulated in the European REACH-approach. This would potentially be more appropriate instead of uncritically investing "only" on private standards and laws and hoping everything will work out well. Similarly the risk management, at least, must stay dynamic, iterative and responsive to change. Likewise this is true for respective standards and laws and might especially be true on the matter in question of nanotechnologies. Nevertheless it is also true for every other possible issue.

---

[15] Art. 1 of the German Equipment Safety Act (*Gesetz zur Neuordnung der Sicherheit von technischen Arbeitsmitteln und Verbraucherprodukten*) Fed. Law Gazette I pp. 2 and 219, last amendment on 07.07.2005, Fed. Law Gazette I p. 1970.

[16] Directive 2001/95/EC of the European Parliament and of the Council of 3 December 2001 on general product safety (OJ L 11, p. 4).

Hence, it is imperative to get away from mere blind compliance with mandatory or voluntary rules (passive risk mitigation) and come to a lively integration, following the depicted change of mindset to active and preventive risk defence.

References

AIRMIC, Alarm and IRM – The Association of Insurance and Risk Managers, The Public Risk Management Association, The Institute of Risk Management (2010) A structured approach to Enterprise Risk Management (ERM) and the requirements of ISO 31000. http://theirm.org/documents/SARM_FINAL.pdf. Accessed 29 Aug 2013

Albrecht E (2008) Standards and thresholds in German environmental law. In: Schmidt M, Glasson J, Emmelin L, Helbron H (eds) Standards and thresholds for impact assessment. Springer, Berlin-Heidelberg, pp 19–31

Allianz SE (2005) Opportunities and risks of nanotechnologies. Report in cooperation with the OECD International Futures Programme. Allianz SE, Munich

ANSI – American National Standards Institute. ANSI-Accredited U.S. TAG to ISO/TC 229 Nanotechnologies. http://www.ansi.org/standards_activities/standards_boards_panels/tc229.aspx. Accessed 29 Aug 2013

BCC Research (2012) Nanotechnology: a realistic market assessment. Report highlights. Report code: NAN031E. http://www.bccresearch.com/report/nanotechnology-market-applications-products-nan031e.html. Accessed 29 Aug 2013

Breuer R (1976) Direkte und indirekte Rezeption technischer Regeln durch die Rechtsordnung (Direct and indirect reception of technical rules by law). Archiv des öffentlichen Rechts (Arch Public Law) (AöR) 101(1976):46–88

BSI – British Standards Institution (2012) What is a standard? http://www.bsigroup.com/en/Standards-and-Publications/About-standards/What-is-a-standard/. Accessed 29 Aug 2013

Buxmann P, König W (1998) Das Standardisierungsproblem (The standardisation problem): Zur ökonomischen Auswahl von Standards in Informationssystemen (The economic choice of standards in information systems). In: Wirtschaftsinformatik (Business informatics) 40

CENELEC – European Committee for Electrotechnical Standardization (2012) European mandates. http://www.cenelec.eu/aboutcenelec/whatwestandfor/supportlegislation/europeanmandates.html. Accessed 29 Aug 2013

CIEL – Center for International Environmental Law (2009) Addressing nanomaterials as an issue of global concern. www.ciel.org/Publications/CIEL_NanoStudy_May09.pdf. Accessed 29 Aug 2013

Commission staff working document – Accompanying document to the Regulation of the European Parliament and of the Council on novel foods and amending Regulation (EC) No. xxx/xxxx [common procedure] – Summary of the impact assessment [COM(2007) 872 final] [SEC(2008) 12] (SEC/2008/0013 final, 14.1.2008)

David K, Thompson PB (2008) What can nanotechnology learn from biotechnology? Social and ethical lessons for nanoscience from the debate over agrifood biotechnology and GMOs. Elsevier, Burlington

Eisenberger I, Greßler S, Nentwich M (2012) Zur freiwilligen und verpflichtenden Nano-Kennzeichnung (Voluntary and mandatory labelling of nano). NanoTrust-Dossier Nr. 031, March 2012: http://epub.oeaw.ac.at/ita/nanotrust-dossiers/dossier031.pdf

Federal Ministry of Education and Research (2009) Nano.DE-Report 2009. Status Quo der Nanotechnologie in Deutschland (Status quo of nanotechnology in Germany). Bundesministerium für Bildung und Forschung (BMBF) (Federal Ministry of Education and Research (BMBF)). Department "Nanomaterialien; Neue Werkstoffe" (Unit "nanomaterials; new materials"). Berlin, Bonn

Federal Ministry of Education and Research (2010) Technologieanalyse. Potentiale der Nanotechnologie (Technology analysis. Potentials of nanotechnology). http://www.bmbf.de/de/4877.php. Accessed 10 July 2010

Ferry JL, Craig P, Hexel C, Sisco P, Frey R, Pennington PL, Fulton MH, Scott GI, Decho AW, Kashiwada S, Murphy CJ, Shaw TJ (2009) Transfer of gold nanoparticles from the water column to the estuarine food web. Nat Nanotechnol 4:441–444. doi:10.1038/nnano.2009.157 [published online: 21 June 2009]

Führ M, Hermann A, Merenyi S, Moch K, Möller M, Kleihauer S, Steffensen B (2006) Rechtsgutachten Nano-Technologien (Legal expertise of nano-technologies) – ReNaTe. Sonderforschungsgruppe Institutionenanalyse und des Öko-Instituts (Society for Institutional Analysis and the Institute for Applied Ecology) e.V., Darmstadt, FKZ 363 01 108

Hatto P (2007) Chairman UK NTI/1 and ISO TC 229 nanotechnologies standardisation committees: an introduction to standards and standardisation for nanotechnologies. http://shop.bsigroup.com/upload/Standards%20&%20Publications/Nanotechnologies/Nano_Presentation.ppt. Accessed 29 Aug 2013

IFAS – Institute for Food and Agricultural Standards (2007) An issues landscape for nanotechnology standards: report of a workshop. Cowles House. Institute for Food and Agricultural Standards, Michigan State University Lansing, East Lansing

IRGC – International Risk Governance Council (2009) Policy brief: appropriate risk governance strategies for nanotechnology applications in food and cosmetics. International Risk Governance Council, Geneva

ISO 31000:2009 (2009) Risk management – principles and guidelines, Ed. 1, published 15.11.2009, ISO copyright office, Geneva Switzerland

ISO IEC Guide 73:2009, Risk management – vocabulary, Ed. 1, published 2009, ISO copyright office, Geneva Switzerland

Knopp L (1995) (Umwelthaftung) Betriebliche Umwelthaftung [(Environmental liability) Company environmental liability], Bonn

Krause L (2009) Das Risiko und Restrisiko im Gefahrstoffrecht (The risk and residual risk in the hazardous substances law). NVwZ 8:496ff

Krause L, Borens D (2009) Das strategische Risikomanagement der ISO 31000 (The strategic risk management of ISO 31000), two-section, Section 1: ZRFC 4/2009, pp 180ff; Section 2: ZRFC 5/2009, pp 227ff

Lohse EJ (2011) (Noch) Ungewisse Risiken: gesetzgeberische Handlungsmöglichkeiten und -pflichten am Beispiel der Nanotechnologien [(Still) Uncertain risks: possibilities and obligations of the legislative power based on the example of nanotechnologies]. In: Scharrer J, Dalibor M, Rodi K, Fröhlich K, Schächterle P (eds) Risiko im Recht – Recht im Risiko (Risk in the law – law in the risk). Nomos, Baden-Baden, pp 37–60

Luther W, Malanowski N (2004) Das wirtschaftliche Potenzial der Nanotechnologie. Technikfolgenabschätzung – Theorie und Praxis (The economic potential of nanotechnology. Technology assessment – theory and practice): Nr. 2, 13. Jahrgang – Juni 2004 (No. 2, 13. Year – June 2004). http://www.itas.fzk.de/tatup/042/tatup042.pdf

Mae-Wan H (2010) ISIS Report 10/03/10. Nanotoxicity in Regulatory Vacuum. http://www.i-sis.org.uk/nanotoxicityInRegulatoryVacuum.php. Accessed 29 Aug 2013

Mantovani E, Porcari A, Morrison MJ, Geertsma RE (2010) Developments in Nanotechnologies Regulation and Standards 2010 – Report of the observatory Nano. June 2010. www.observatorynano.eu

Mantovani E, Porcari A, Morrison MD, Geertsma RE (2011) Developments in Nanotechnologies Regulation and Standards 2011 – Report of the Observatory Nano. July 2011. www.observatorynano.eu

Mantovani E, Porcari A, Morrison MD, Geertsma RE (2012) Developments in Nanotechnologies Regulation and Standards 2012 – Report of the Observatory Nano. April 2012. www.observatorynano.eu

Monica Jr J, Lewis PT, Monica JC (2006) Preparing for the future health litigation: the application of products liability law to nanotechnology, 3 NANOTECH. L. & BUS. 54
Morasch K (2006) Ökonomie der Informationsgesellschaft (Economics of the Information Society). https://dokumente.unibw.de/pub/bscw.cgi/d4804801/OekonInf2006-Folien34u35.pdf. Accessed 29 Aug 2013
MOST – Ministry of Science and Technology of the People's Republic of China. S&T Programmes. National Basic Research Program of China (973 Program). http://www.most.gov.cn/eng/programmes1/200610/t20061009_36223.htm. Accessed 29 Aug 2013
NanoKommission der deutschen Bundesregierung (Nano Commission of the German Federal Government) (2008) Verantwortlicher Umgang mit Nanotechnologien (Responsible use of nanotechnologies). Bundesumweltministerium (Federal Environment Ministry), Berlin. Bericht und Empfehlungen der Nanokommission der deutschen Bundesregierung (Report and Recommendations of the Nano Commission of the German Federal Government)
National Nanotechnology Initiative (2012) What is nanotechnology? http://www.nano.gov/nanotech-101/what/definition. Accessed 29 Aug 2013
Peine FJ (2011) Gerätesicherheitsrecht (Equipment safety law). In: Schulte M, Schröder R (eds) Handbuch des Technikrechts: Allgemeine Grundlagen Umweltrecht – Gentechnikrecht – Energierecht – Telekommunikations- und Medienrecht – Patentrecht – Computerrecht (Handbook of technology law: general principles environmental law – biotechnology law – energy law – telecommunications and media law – patent law – computer law), 2nd edn. Springer, Heidelberg
Reijnders L (2012) Human health hazards of persistent inorganic and carbon nanoparticles. J Mater Sci 47(13):5061–5073
RS and RAE – Royal Society and Royal Academy of Engineering (2004) Nanoscience and nanotechnologies: opportunities and uncertainties. http://www.raeng.org.uk/news/publications/list/reports/nanoscience_nanotechnologies.pdf. Accessed 29 Aug 2013
Secretariat of CEN/TC 352 (2007) International Standards in Nanotechnology – a report made available by Nanoposts.com, an information exchange website of the Technology Transfer Centre (IoN). Document CEN/TC 352 N 68
Smith A (1776) An inquiry into the nature and causes of the wealth of nations, Book I, Chapter II, 4th edn 1786, in: The Works of Adam Smith, vol II, pp 1811–1812 [reprint Aalen 1963]
SRU – Sachverständigenrat für Umweltfragen (2011) Vorsorgestrategien für Nanomaterialien (Precautionary strategies for nanomaterials). Sachverständigenrat für Umweltfragen (German Advisory Council on the Environment) (SRU), Berlin
Standards Australia (2012) Benefits of standards. http://www.standards.org.au/StandardsDevelopment/What_is_a_Standard/Pages/Benefits-of-Standards.aspx. Accessed 29 Aug 2013
Swiss Re (2004) Nanotechnologie. Kleine Teile – grosse Zukunft? (Nanotechnology. Small parts – great future?). Swiss Reinsurance Company Ltd, Zurich
Tucker JL (2009) Standards will help ensure order in nano-enabled industries. Keithley Instruments, Inc., U.S.A. No. 3051
Weiss R (2006) Nanotech raises worker-safety questions. Washington Post, 8 April 2006. http://www.washingtonpost.com/wp-dyn/content/article/2006/04/07/AR2006040701725.html. Accessed 29 Aug 2013

Chapter 7
Recognition of Private Sustainability Certification Systems for Public Regulation (Co-Regulation): Lessons Learned from the EU Renewable Energy Directive

Martina Gaebler

7.1 Introduction

In a globally acting economy with a growing demand for sustainable products and services there is an increasing interest to include sustainability aspects in state regulation. In doing so, state regulation can make use of private regulatory systems, or sustainability standards systems. In fact governments increasingly use private certification systems to implement their political interests. The use of private regulation systems in public regulation is called 'co-regulation'. Co-regulation can have several benefits if implemented in an efficient and effective way. By using a co-regulation approach at national (or supra-national) level, governments are able to promote sustainable economic activities in globalised supply chains. Instead of establishing and maintaining costly control systems at state level, governments make use of already existing private control mechanisms that regulate and manage global supply chains. There are some examples how co-regulation can work in practice. The timber procurement regulations in the Netherlands, the UK and Germany are prominent examples. However, there is little academic literature available. Lister (2011, p. 29) analyses co-regulation from a governance perspective: "the mixing and temporal sequencing of various public, private, and co-regulatory instruments at the different stages of the policy cycle constitute a co-regulatory governance system". In the context of environmental policy Gunningham and Grabosky (1998, p. 15 in Lister 2011, p. 29) explain that "recruiting a range of regulatory actors to implement complementary combinations of policy instruments tailored to specific environmental goals and circumstances, will produce more effective and efficient policy outcomes".

One example, where private certification schemes are used on a supra-national and state level, is the European Union Renewable Energy Directive (EU RED). It

---

M. Gaebler (✉)
Programme for Social and Environmental Standards, Deutsche Gesellschaft für Internationale Zusammenarbeit (GIZ) GmbH, Eschborn, Germany
e-mail: martina.gaebler@giz.de

C. Schmitz-Hoffmann et al. (eds.), *Voluntary Standard Systems*, Natural Resource Management in Transition 1, DOI 10.1007/978-3-642-35716-9_7,

introduces sustainability requirements for liquid biofuels that are counted towards the national renewable energy targets and are eligible for financial support (Directive 2009/28/EC). Companies can prove compliance via private certification schemes that are recognised by the member states or the European Commission. Similar forms of co-regulation are also relevant for a number of other sectors, for example, in forestry (EU Timber Regulation), public procurement policies, and possible extension of RED to solid and gaseous biomass. The recognition of private sustainability certification schemes for public regulation involves a number of actors and tasks. It is a complex process that needs constant revision. It is therefore time to evaluate the experiences made in the EU RED process in order to identify lessons learned for future co-regulation processes.

The chapter starts by exploring the concept of co-regulation, where Sect. 7.2 will give an introduction to the concept of co-regulation, the EU RED and its co-regulation approach. Section 7.3 presents the results of the analysis of the EU RED co-regulations process and Sect. 7.4 concludes the chapter by drawing together the ten main lessons to be learned. The work is mainly based on a study conducted in 2012 by SQ Consult commissioned by GIZ on behalf of the German Federal Ministry of Economic Cooperation and Development (BMZ), as well as on GIZ briefing papers.

7.2 Introduction to the Concept of Co-Regulation and the Renewable Energy Directive

The term co-regulation is used to explain the combination of private and public regulation. Governments increasingly make use of private regulatory instruments to implement policies. We can distinguish three pathways of co-regulation:

1. Governments set binding goals which they enforce by officially recognising private compliance schemes. Details on implementation and verification of compliance are left to the discretion of a private scheme (this is the approach the EU has taken with its regulation for sustainable liquid biofuels).
2. Governments can adopt private regulations into national laws.
3. Governments may support private schemes without legislation or adoption, e.g. by creating conducive legal and regulatory frameworks (national accreditation), support private party implementation directly (e.g. by providing loans) or to participate in the development of private schemes.

Co-regulation policy combines the advantages of private and public regulatory spheres while avoiding the disadvantages. Strengths from governmental regulation include the democratic legitimacy, applicability to all firms within state jurisdiction and the enforceability through state supervisory agencies. Weaknesses of governmental regulation include the often slow development, no applicability outside state jurisdiction and high implementation costs for private sector parties. On the other hand, private regulation is often flexible, quick and innovative in nature, while being international in terms of focus and applicability. Governmental regulatory strengths can in turn compensate the weaknesses of private regulation schemes such as their lack of government legitimacy, voluntary nature and limited sanctions.

Especially concerning globalisation and the assurance of sustainability, co-regulation appears to be highly relevant. With co-regulation, national governments have developed a new approach to help them step-up regulation efforts and benefit from globalisation by promoting sustainable economic activity for production processes beyond their own judicial boundaries.

Co-regulation also has inherent risks, e.g., different and perhaps contradictory demands from governments on private systems may increase implementation costs. Also, governments may view private regulatory initiatives as competitive, and may act accordingly. Furthermore, when governments do not fully understand the logic and functioning of the private schemes, co-regulation may not be efficiently implemented or used at all. Finally, for protectionist purposes, governments may also misuse sustainability regulations, thereby endangering the neutrality and credibility of such schemes.

Private regulatory initiatives need to complement government regulation. They can work by supporting each other, but private schemes cannot substitute for government regulation. In co-regulation, private initiatives are there to enhance governmental (sustainability) policy implementation and to create a more efficient regulatory environment.

7.2.1 Introduction to the EU Renewable Energy Directive (RED)

Introduced in 2009, the EU RED recognition of sustainability certification schemes was one of the first examples of co-regulation in the area of sustainability criteria. Despite all criticism, it was possible within two to 3 years to establish a co-regulation system that allows bringing certified sustainable biomass onto the EU market.

In the EU RED, the European Commission has set the ambitious target of increasing the proportion of renewables in the EU's energy consumption to 20 % by 2020, including 10 % in transport (Directive 2009/28/EC). EU RED requires biofuels and bioliquids[1] that are brought onto the EU market to meet certain sustainability criteria. Companies selling biofuels in the EU Member States

[1] According to EU RED, 'bioliquids' means liquid fuel produced from biomass for energy purposes other than transport, including electricity and heating and cooling, 'biofuels' means liquid or gaseous transport fuel produced from biomass. For the purpose of simplicity we will refer only to biofuels, which should be taken to include biofuels and bioliquids according to the EU definition.

('obligated companies') must prove to the responsible Member State authority that the biofuel they brought onto the Member State's market complies with the criteria mentioned below. In order to do so, the obligated companies must use independent auditing. If obligated companies cannot prove compliance with the EU RED criteria, their biofuels cannot be counted towards the national renewable energy targets and cannot receive financial support (e.g. tax relief). The sustainability criteria apply both to internal EU production and to imports of biofuels and biomass from third party countries.

Mandatory sustainability criteria of the EU RED include:

- Greenhouse gas emissions saving of at least 35 % (50 % from 2017 and 60 % from 2018);
- No feedstocks to be derived from land with high biodiversity value;
- No feedstocks derived from land with high carbon stock;
- Use of a chain-of-custody system (mass balance) to trace sustainable products.

The EU RED also contains reporting requirements on additional issues (e.g. soil, water and air protection; social sustainability; etc.). Economic operators have to *report* their actions relating to these additional issues but do *not* have to *comply* with certain requirements. These reporting requirements have not yet been further defined by the European Commission.

Each EU Member State must provide a regulatory framework for companies to report on the RED-compliance of their biofuels. In that framework they must specify which reporting and certification rules companies must follow. Member States can do this by establishing their own certification scheme or by recognising voluntary certification schemes. The EU Commission also recognises certification schemes. A scheme that is recognised by the EU Commission must automatically be recognised by all EU Member States.

Thus obligated companies and their corresponding supply chain operators have two options for demonstrating compliance with sustainability criteria:

a. Using a certification scheme recognised by the EU Commission (EU-wide recognition). As of August 2013, the EU Commission accepted 14 schemes and another batch of schemes is in the pipeline[2];

[2] Systems that have been recognised (EC 2013): (1) International Carbon and Sustainability Certification (ISCC), (2) Roundtable on Sustainable Biofuels (RSB), (3) Roundtable on Responsible Soy Association (RTRS), (4) Bonsucro/Better Sugarcane Initiative, (5) Greenergy—Brazilian bioethanol verification programme, (6) Abengoa RED Bioenergy Sustainability Assurance Scheme (RBSA), (7) Biomass Biofuels voluntary scheme (2BSvs)/France, (8) NTA 8080 certification scheme, (9) REDcert (German industry scheme), (10) SQC (Scottish Quality Farm Assured Combinable Crops (SQC) scheme), (11) Red Tractor (Red Tractor Farm Assurance Combinable Crops & Sugar Beet Scheme), (12) Ensus (voluntary scheme under RED for Ensus bioethanol production), (13) Roundtable on Sustainable Palm Oil (RSPO), (14) BioGrace (GHG calculation tool). Schemes awaiting recognition include: CARBIO (Argentinean soy scheme) and Neste Oil Scheme.

b. Using a certification scheme recognised by an EU member state (MS-wide recognition). Currently only Germany and the Netherlands have recognised voluntary certification schemes for their respective markets.

7.2.2 Introduction to the RED Co-Regulation Approach

The Directorate General for Energy (DG ENER) is in charge of the recognition procedure at the European Commission. The technical assessment of the voluntary schemes is outsourced to a contractor. The technical assessment performed by the contractor is an iterative process in which applicants, i.e. the certification schemes, are requested to solve the issues found during the evaluation. If the scheme passes the technical assessment, DG ENER begins an Inter Service consultation with other Directorate Generals for their co-approval of the assessment. Once the Directorate Generals have approved the technical assessment, DG ENER starts the commitology process with the Member States' Advisory Committee. This Advisory Committee comprises representatives from Member States. The Advisory Committee votes its approval, though the result of this voting is not binding for the EC. DG ENER then makes its recommendation to the EC for the adoption of a formal decision for the recognition of the voluntary scheme. The EC Decision is valid for 5 years. Private schemes may present modifications after formal recognition. In such cases, DG ENER decides whether the initial recognition is affected. If it is affected, a new assessment would be required though it is not clear at this moment if and how the full process is applied.

Based on the information collected from 23 key informant interviews and available legal documents, the EC recognition procedure is analysed in terms of its efficiency and effectiveness. The following aspects are considered in the analysis:

- Availability and clarity of administrative procedure (e.g. clear responsibilities and description of each administrative step, public availability of procedures, timelines for administrative steps, and length of procedure).
- Transparency and confidentiality (e.g. management of procedure with transparency towards the applicant and other market stakeholders, confidentiality of scheme documents).
- Technical assessment framework (e.g. quality of assessment framework and involvement of stakeholders/experts in the establishment of the framework).
- Cross acceptance rules (e.g. existence of rules governing the cross acceptance of certificates by different recognised schemes, existence; and clarity of rules).
- Parallel recognition procedures in Member States (e.g. Member States allowing the recognition of other private voluntary schemes for their markets, relationship and similarity to EU recognition rules).

7.3 Analysis of RED Co-Regulation Approach

7.3.1 Administrative Procedures

Administrative procedures are available for processing applications from first reception until the final Commission decision. However, they are not pro-actively communicated and applicants mainly receive information upon direct request. There are no clear administrative procedures available in case a system undergoes changes after official recognition and in case the system suffers serious problems or fraud. This lack of clear communication and lack of clarity regarding administrative procedures contributed to a strong perception of lacking transparency and unequal treatment of schemes.

There is no binding timeline for the Commission to accept schemes (German law grants the responsible authority 6 months). The general length of the recognition procedures are approximately 12 months. This is due to various consultation processes at EU and Member State level (see Fig. 7.1), as well as the sheer amount of schemes that applied for recognition (more than 20). In addition, the lack of clarity in the assessment framework concerning some essential aspects contributed to delays in the process. Overall, applicants felt that the process was too lengthy. Not all schemes were accepted at the same time and so there was a risk that market distortions occurred due to delayed recognitions.

Another aspect identified in the context of process duration is the personnel capacity at Commission level. At DG ENER, one officer is responsible for the entire recognition procedure who runs this process amongst a number of other tasks. Although the assessment of scheme documentation was outsourced to a contractor, there is a strong perception amongst stakeholders that the personnel capacities were not sufficient to effectively deal with the amount of applications in a timely manner.

In some aspects the procedures were adapted over time in order to increase efficiencies. In-person meetings of Member States were, for example, changed to written procedures. While this shows flexibility in a learning-by-doing environment, it also had some (unintended) consequences as the information flow between Member States was reduced.

7.3.2 Transparency and Confidentiality

Overall, there is a strong perception that the recognition process is in transparent. This perception seems to result from a lack of pro-active communication from the recognising authority as well as to the continuous clarification process of the assessment framework when the recognition process had already started. In addition, the Commission applied rather strict confidentiality rules concerning the documentation of schemes that was sent to the stakeholders of the advisory processes (DGs and Member States) and later published on the Commission

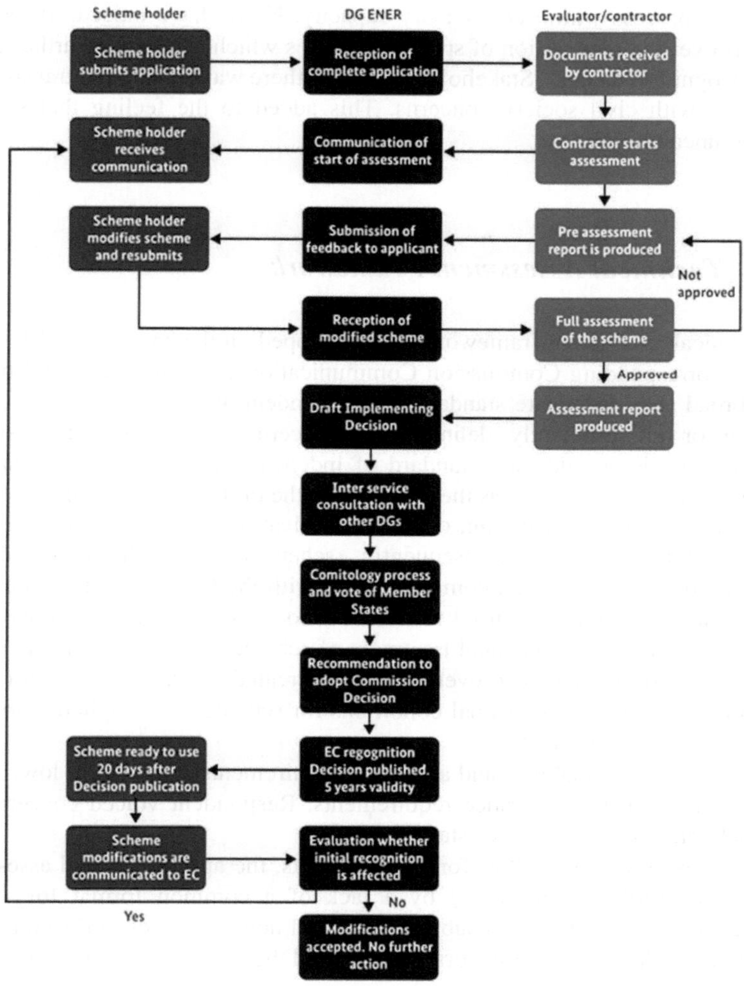

Fig. 7.1 Overview of the administrative steps in the EU renewable energy co-regulation process (*Source*: BMZ 2013)

website. The assessment framework was not publicly available until the first official recognitions were published on the Commission's webpages. Schemes were free to blacken information in their documentation that they considered commercially sensitive. As some schemes made extensive use of this possibility by blackening entire documents (e.g. their assurance system), the possibility to carry out effective advisory services was limited.

Participation of civil society is not officially foreseen in the process but it has become clear that there is a strong interest from the public to at least be well informed about sustainability schemes, if not to be involved in the recognition and monitoring process within an advisory capacity. NGOs have indeed voiced their concerns over the recognition of specific schemes which resulted in further delays of the recognition process. Stakeholders felt that there was not enough transparency in dealing with civil society concerns. This added to the feeling that schemes received unequal scrutiny.

7.3.3 Technical Assessment Framework

The technical assessment framework was developed on the basis of the EU Directive and corresponding Commission Communications. Because this legal basis is rather broad (e.g. 'adequate standard of independent auditing'), essential criteria were not, or not sufficiently, defined. This concerns, for example, the assurance systems (what is an adequate standard of independent auditing?) and chain-of-custody requirements as well as the criterion on the protection of highly biodiverse grasslands. The time-consuming, ongoing clarification process presented a major challenge for applicants. Consequently, schemes found different solutions depending on their individual communication with the Commission, which led to inconsistencies across recognised schemes. Also, it nurtured the perception of a lack of transparency and unequal treatment of schemes. In some areas the assessment framework was adjusted over time. This created improved clarity for future applicants but also led to unequal conditions for schemes that applied before and after the changes were made.

The level of sustainability and assurance requirements is relatively low. This is particular true for the assurance requirements. Respondent voiced concerns that many schemes are not fraud resistant enough.

For the assessors, as well as for the applicants, the application and assessment process was further complicated by a lack of a common format for scheme documentation. Some schemes submitted several hundred pages of documentation which made it difficult for assessors to easily find the relevant elements and to cope with such an excess of information.

7.3.4 Cross-Recognition of Schemes

The RED does not contain any rules on cross acceptance of schemes. There are also no guidelines from the EC side. However, all stakeholders consider this a relevant and necessary issue. Schemes cannot automatically accept material certified under other EC approved schemes, unless such rules are part of their scheme documentation. If a scheme would like to introduce such rules after its official recognition, the Commission would review the respective clause and decide whether the change is acceptable.

While many stakeholders see cross acceptance as a useful instrument to handle the quantity of schemes on the market, there is also a concern regarding the potential to 'greenwash'. This is because recognised schemes differ considerably in their level of content and assurance requirements. As a result of this, a scheme with a relatively narrow scope (e.g. covering only mandatory criteria) and a low assurance level, accepts certificates from schemes that also cover non-mandatory criteria with stricter assurance systems and vice versa. Companies who do want a choice of certification schemes in order to show their commitment to sustainability, are prevented from doing so. Certified material might be passed on in the supply chain with the claim of the more robust system although it has actually not been produced under that standard.

A concern related to cross acceptance expressed by many stakeholders is that there are no overarching control of trade with certified biomass ('clearinghouse') at the EU level. Selling the same amount of 'certified' biomass into several schemes and reporting those within Member State reporting is a real danger, no matter if intentional or in good faith. Cross acceptance, especially without clear rules, raises the risk level even further.

7.3.5 Parallel Recognition Structures

In addition to the EU Commission's recognition procedure, EU Member States have the possibility to approve private certification schemes for their own market. The United Kingdom, Germany and later the Netherlands have used this option. The recognition of schemes was useful as an alternative to the creation of national certification schemes and it also served to bridge the time until the Commission officially recognised schemes in July 2011. These procedures at member State level also allow for flexibility for accepting a scheme especially designed for local/ regional characteristics.

However, the assessment frameworks at Member States level and Commission level were not exactly the same. Therefore, schemes that had already been accepted at Member State level were asked to change their documentation. This resulted in two different standards versions under one scheme.

Market actors feel that these double structures are confusing and costly, and, at least in the case of internationally operating schemes, needless as soon as schemes are EC recognised. In some context, such measure might however make sense to create local solutions for local peculiarities.

7.4 Conclusions: Lessons from Renewable Energy Directive Co-Regulation

The following lessons are drawn from the analysis of stakeholder interviews and available documentation. They do not necessarily reflect the opinion of all interviewees.

7.4.1 Lesson One: Communicate Pro-Actively

The analysis shows that the communication strategy shapes the perception of the process credibility and also efficiency. While it is important that respective contact persons at the recognising institution are available for questions from applicants and other relevant parties (e.g. Member States), pro-active communication on clear administrative procedures, status of applications and timelines, reasons for delays etc., towards all parties (and not on an individual basis) adds to transparency and credibility of the process. This could be done via the agencies websites, newsletters and/or mailing lists. In particular, efficiency can be increased if solutions to open issues in the assessment framework are communicated openly and not on an individual scheme basis. In case of consultation processes with several actors (e.g. the Member States Advisory Committee), comments and concerns regarding the technical assessment results should be shared amongst stakeholders in order to enable cross learning and knowledge exchange, making the process more effective. Also, to increase efficiency, actors that have an advisory or decision-making role should be informed in a timely manner about the list of applicants in the pipelines and the status assessments.

7.4.2 Lesson Two: Have a Clear and Complete Assessment Framework

The assessment framework needs to be complete, containing clear criteria and guidance. There are clear benefits in allowing diversity in the schemes, i.e. allowing for nationally appropriate or crop-specific solutions. However, there should be a highest common denominator in all schemes, which assures an adequate level of credibility and performance. Stakeholder consultations on recognition criteria can be a useful tool for including expert and civil society inputs and reaching broad-based acceptance of the framework.

7.4.3 Lesson Three: Set Robust Criteria for Verification Requirements

An important problem related to the effectiveness and reliability of co-regulation under RED for biofuels sustainability lies in the risk of recognising schemes with a low level of assurance. The current differences in assurance among recognised schemes have impacts on the quality of audits, and therefore on the certification costs. A recognition system should include clear and internationally accepted criteria for third party verification of sustainable production practices. This includes the standards such as ISO 17065 or 17021 and the principles of the ISEAL Alliance Code of Good Practice for Assuring Conformance with Social and Environmental Standards.

7.4.4 Lesson Four: Have Clear and Transparent Administrative Procedures

Formal guidelines on administrative steps and accompanying (indicative) timelines help foster clarity and reliability in the process. This includes timelines not only for the internal and/or external advisory and consultation processes, but also for the performance of the entity managing the recognition. Submission of scheme documentation and the assessment can be facilitated on both sides by setting a common format for documentation. Also, public consultation periods are a useful instrument for addressing the concerns of third parties and finding a solution together with the involved stakeholders within the assessment process. Public consultations can also be an instrument for monitoring the post-recognition performance of schemes. It is a way to gather information and process it in order to effectively perform the necessary oversight. Making the procedure publicly available, along with clear guidelines and expectations for applicants, responsibilities and contact persons, helps to create trust and avoid misunderstandings. The procedure should be open to adaptations, especially in a learning-by-doing environment; however, any changes should be openly communicated.

7.4.5 Lesson Five: Limit Confidentiality to a Minimum

All documents relevant for the assessment of a scheme should be made available to assessors and to the public. Their identity should be disclosed. All relevant stakeholders must be able to make an informed opinion and/or perform their regulatory duties by reviewing and assessing all relevant application documents. Certification schemes often follow logics of varying complexity and can only be fully assessed when all necessary information is made available. The assurance system is the basis of a scheme's credibility and should especially be available. The information provided to advisory or consultative groups should therefore be adequate and not limited to elements copied and pasted into the assessment framework. The public has an interest in transparency whenever taxpayer money is used to support sustainable production and should therefore be able to examine recognised schemes. This does not prevent the scheme from protecting documents with copyrights and does not mean that the scheme has to display its entire business model.

7.4.6 Lesson Six: Allocate Sufficient Human and Financial Resources and Capacities

Recognition of private certification schemes involves a number of tasks: the technical development of the assessment framework, the technical assessment of the applicants' documentation, the administrative steps in establishing the procedures, acceptance and processing of applications, communication with all related parties, collaboration with second- and/or third-party advisors or decision-making structures, as well as the establishment and operation of a monitoring system. To run these processes smoothly and in a timely manner, co-regulators need the specific technical expertise, personnel capacities and sufficient funds. The case-specific needs should be carefully analysed and allocated. Expertise on specific technical issues should be gained through the use of independent experts. Applicants should not underestimate the resources needed for developing a new scheme or adapting an existing one and the approval process. Especially for schemes that rely on multiple stakeholders with often limited capacities, such a process can absorb capacities of several (staff) members and/or additional financial funds may be needed to contract external support.

7.4.7 Lesson Seven: Establish Clear Rules for Changes in Schemes and Scheme Failure

Co-regulation does not stop with the official recognition of schemes. The public authority recognising schemes should retain the capacity to monitor the performance of such schemes—otherwise the effectiveness of co-regulation might be reduced. If states attach financial or other support to proving sustainability (as in the case of the RED), monitoring systems must also aim to avoid refunding financial support paid in the case of system failure. A co-regulation system should also maintain its flexibility to allow schemes to adapt or improve their systems and should therefore provide a clearly defined mechanism for scheme changes.

7.4.8 Lesson Eight: Establish a System to Control Volumes and Reporting of Certified Materials Across Schemes

Produced and traded certified volumes should be controlled across the system boundaries of one certification scheme. This is important because a given scheme can only control what happens within its system. At the same time it is interlinked with other systems because economic operators often use several certification schemes (hold several certificates) and the systems themselves might accept other certificates into their supply chains (see Lesson Nine). A control mechanism cutting across all schemes at EU level (or national level if applicable) can reduce the risk of (involuntary) double-counting or fraud.

7.4.9 Lesson Nine: Establish Rules for Cross Acceptance of Schemes

Regulators should establish rules for the cross-acceptance of certificates between schemes. These rules should consider the criteria covered by schemes as well as their assurance mechanism, including the chain-of-custody models in particular. All schemes that are recognised under one co-regulation mechanism should be able to accept each other's certificates into the scheme's own supply chain. This enables economic operators to purchase and sell materials independently from a specific scheme while ensuring that the requirements are met. Cross-acceptance of schemes is however, not trivial. A particular challenge arises when schemes significantly differ in criteria coverage and assurance systems. Market players might not be able to differentiate certificates from their preferred schemes any longer and there is a potential for (involuntary) greenwashing.

7.4.10 Lesson Ten: Avoid Duplication of Efforts at Member State Level and EU Level

Having double recognition structures allows for the management of different administrative speeds accounting for specific local/regional characteristics. The assessment frameworks at national and EU-level should, however, not contradict each other; this would result in different versions of schemes for one scheme holder. If a scheme holder applies first at national level and then at EU level (e.g. to take advantage of different administrative speeds in the implementation of the recognition system), national recognition should be phased out—regardless of the version of scheme approved—as soon as the scheme is recognised at EU level.

References

BMZ – Bundesministerium für wirtschaftliche Zusammenarbeit und Entwicklung (German Federal Ministry for Economic Cooperation and Development) (2013) Recognition of private certification schemes for public regulation. Lessons learned from the Renewable Energy Directive, Study by GIZ and SQ Consult, GIZ, Eschborn

Directive 2009/28/EC of the European Parliament and of the Council of 23 April 2009 on the promotion of the use of energy from renewable sources. Official Journal of the European Union L 140/16 of 05.06.2009

EC – European Commission (2013) Renewable energy. Biofuels – sustainability schemes. http://ec.europa.eu/energy/renewables/biofuels/sustainability_schemes_en.htm. Last accessed 30 Aug 2013

Lister J (2011) Corporate social responsibility and the state. International approaches to forest co-regulation. UBC, Vancouver

Part III
Voluntary Standard Systems: The Development Dimension

Having been introduced to the basics and general legal situation surrounding voluntary standards, Part III is characterised by studies in standard development. With a definitive European perspective, contributors consider: Germany's national position on voluntary standards development; focus on impacts of VSS and the 'how-and-why' of their measurement; the synergy between life cycle assessment tools and voluntary standards in consideration of wood products within the built environment being regarded as carbon stores; and VSS certification influencing market strategies in former Soviet states.

The scope for Chap. 8 begins on the German national level in determining the use of voluntary standards in German development policy. Such policy considerations encompass support for sustainable production and trade conditions, and are essential in encouraging the implementation and development of legal and voluntary standards. We uncover the situating of voluntary standards within the overarching aspirations of German development policy, along with recounting previous development experience within the field of standards. Voluntary standard development showcases multi-actor involvement as an advantageous attribute, and as such their development features in international discussions and forums as we see in this chapter. From this follows how standards are acknowledged within trade and investment agreements, at both intra- and international level. With the associated confusion caused by the proliferation and multiplicity of standards, there is a description on the database and internet platform tools available in enhancing transparency and the exchange of information on standards.

A discussion and analysis of the impacts of voluntary sustainability standards is set out in Chap. 8. The development of standards, in relation to their contributory factors e.g. multi-stakeholder initiatives, claims and impacts, legitimacy, etc., along with the motivating reasons for measuring impacts are included. Methodological approaches applied in helping untangle the problems with impact measurement are described, combined with the importance for standardised language for comparison and collaboration. Economic, environmental and social impacts are defined and elucidated, with comment on how VSS can contribute to sustainable development closing this chapter.

Chapter 10 explores the use of standards in respect of wood products for use in the built environment, and how these are applied in identifying and measuring the respective environmental impacts and benefits. We are introduced to the discussion on how sustainably sourced timber products can be viewed as a form of carbon storage, with case studies of carbon valuation highlighted. The background of 'Life Cycle Assessment' methodologies (LCA) is outlined, along with discussion on pending European legislation regarding environmental product declarations. These prepare the reader for the proposed potential relationship between timber certification and LCA tools to produce environmental data of greater accuracy than is available currently.

Chapter 11 gives insight into the penetration of voluntary standards into former Soviet states, and how producers have reacted to the potential competitive edge offered through environmental standard certification. Whilst developed predominantly in western nations, VSS is a relatively new concept to former soviet industries which were familiar only with former 'command-and-control' principles of administration. This chapter presents some of the obstacles and potential market advantages involved when states experiencing a 'transition economy' attempt to develop and implement market strategies based on voluntary environmental standards.

Chapter 8
The Role of Voluntary Standards in German Development Policy

Evita Schmieg

8.1 Introduction

This chapter will elaborate on the role of voluntary standards in German development policy. Today, social and ecological conditions of production are at the heart of intense discussions on globalisation. Of particular concern are production conditions in developing countries, where lack of capacity to implement international agreements is often observed. Private ecological and social standards play an important role in the efforts of private companies to introduce more sustainability in their supply chains. Since these developments directly affect developing countries, private standards have increasingly been of concern in German development policy. In some cases this chapter discusses private voluntary standards together with legally binding requirements. There are several reasons for this. Firstly, private standards often rely on binding international conventions—e.g. the International Labour Organization (ILO) core labour standards are the basis of almost every private social standard. Additionally, from a government's point of view, both are ways to achieve the same objectives and some instruments—like modern trade and investment agreements—deal with both. Additionally, governments sometimes even rely on private standards when designing regulatory initiatives. For example, the Guatemalan government has made Forest Stewardship Council (FSC) certification mandatory for forestry firms operating in the 'Mayan Biosphere Reserve' (IAWG 2011, p. 24). Therefore, a discussion of the issues surrounding private standards is often linked to legal instruments.

German development policy (the Federal Ministry for Economic Co-operation and Development, BMZ) has been involved in voluntary standards since the mid-1990s. Bilateral development-cooperation in the traditional form of capacity building, improvement of competitiveness and institution building offers significant

E. Schmieg (✉)
Stiftung Wissenschaft und Politik (SWP), Ludwigkirchplatz 3-4, 10719 Berlin, Germany
e-mail: evita.schmieg@swp-berlin.org

C. Schmitz-Hoffmann et al. (eds.), *Voluntary Standard Systems*, Natural Resource Management in Transition 1, DOI 10.1007/978-3-642-35716-9_8,

support to standards. This chapter concentrates on newer forms of development policy which are more specific in the area of standards. The projects mentioned here are in no way exhaustive or representative, however, they give an overview on the policy developments of the previous years. The first typical projects dealing with standards issues were classical public–private partnerships where the government worked together with private enterprises on specific projects. Development policy today works increasingly on systemic issues linked to voluntary standards and in multi-stakeholder forums.

The chapter starts by situating voluntary standards into the broader picture of the goals of German development policy. Section 8.3 summarises some of the early experiences of German development policy in the area of standards. Section 8.4 describes a selection of international dialogue and discussion forums and how voluntary standards are dealt with at that level. Section 8.5 gives some indications on the reflection of standards in regional or international trade and investment agreements. Section 8.6 describes instruments which contribute to more transparency on the multiplicity of standards and a better exchange of information. Section 8.7 describes the entry points for development policy in the internal discussion in Germany surrounding the link between standards and development issues. Section 8.8 gives a final remark.

8.2 Objectives of Development Policy and the Role of the State

Today, the Millennium Development Goals (MDG's) state the international consensus on the objectives which all countries try to pursue, defining the important pillars of a development path for developing countries to follow, supported by donors. The overall objective of the MDG's is to fight poverty and improve the living conditions of people in developing countries. The conditions under which production in developing countries takes place is directly linked to this issue: Are children exploited or is forced labour involved in the production process? Are workers free to form associations to co-operate in following their interests? Are workers free from discrimination? All these questions refer to the core labour standards (or: Fundamental Principles and Rights at Work) as defined within the International Labour Organization (ILO). The link between these issues and the living conditions of people in developing countries is obvious. The link is just as obvious for environmental standards: Are preventive measures observed when pesticides are used for flower production? Is drinking water affected by residual water? The ecologic and social conditions, under which people work, are central for their living conditions and well-being. Since the late 1980s there have been international discussions on the role of environmental and social issues in globalisation and what a human face of globalisation means. Non-governmental organisations played a crucial role in underlining the importance of the issue and in enhancing consumer interest in sustainable goods. Companies have become increasingly aware of the importance of sustainability issues not only for ethical but also for economic reasons. All this contributed to the evolvement of a variety of standard systems, labels and company initiatives for increased sustainability.

These developments have been reflected in a range of international discussions: In 1998 the decision on the core labour standards has been taken within the ILO. At the same time the European Union proposed to negotiate social standards in the upcoming World Trade Negotiations—but this effort failed, mainly due to the strong opposition from developing countries who feared that these new concerns raised by industrialised countries would be no more than hidden protectionism. However, the issue gained ground in other forums: The United Nations Global Compact was created in 1999, the OECD decided on its Guidelines for Multinational Enterprises in 2000 (revision 2011), in Heiligendamm in 2007 the G8 decided that they would include social standards in Free Trade Agreements and in 2011, the UN decided on the Guiding Principles for Business and Human Rights 2011 . . . to name just some of the major developments.

All these developments are milestones of international discussions, but they only partially constitute binding international law. The ILO core labour standards, comprising eight individual conventions which are internationally recognised human rights, have been ratified by 135 member states of the ILO. However, even with regard to those, we observe great weaknesses concerning their implementation in a range of developing countries. To give just one example, the ILO estimates "that some 215 million children across the world are still trapped in child labour" (ILO 2012).

Against this background, voluntary standards since the 1990s have shown to be a very practical instrument to improve ecologic and social conditions on the ground. In the beginning, they evolved on a completely private basis. One could argue that the involvement of the state (via development co-operation or otherwise) is unnecessary. But in reality it is not as simple as that, there are good reasons for development policy to get involved and quite often the involvement of the state is even a necessary precondition for projects and programmes to take place. There are various starting points for development co-operation in the area of private voluntary standards:

- The interest on a policy level might go beyond the interest of standard systems or private enterprises. This is most obvious if one looks at the results of projects and programmes on meso and macro levels, i.e. institution building and raising awareness in the government of partner countries. Sustainable change is sometimes only possible if governmental institutions in partner countries are involved in projects. Development co-operation can thus play a facilitating and convening role.
- Often, the introduction of standards is only possible on the basis of certain preconditions—e.g. with regard to the health situation or the educational background of workers—which go beyond the reach or possibilities of individual firms. The co-operation of the public and the private sector can thus bring about

results each partner alone could not reach. Other stakeholders like trade unions, non-governmental organisations and researchers, also play an important role in that context by bringing in their own views and experiences. It has therefore been shown that multi-stakeholder initiatives are particularly successful in introducing and implementing standards.

- Development co-operation has not only long standing experience in working with different target groups in developing countries but also different methodologies in doing so. Cooperation with producers in developing countries, training tools and experience with the target group of poor people overall is the traditional comparative advantage of development cooperation, which governments are able to bring into the common approaches with private companies.
- When co-operating on standards issues, development policy can support broader target groups and can bring about transparency, knowledge transfer and dissemination of results to an extent that an individual company would not be interested in. Development policy can thus contribute to spreading positive external effects of knowledge.
- One of the major entry points for development co-operation is to support small scale producers to cope with the demands of standards systems and the multiplicity of standards.

Development co-operation ministries and agencies thus became natural partners to standards organisations as well as private enterprises in their efforts to introduce and implement private voluntary standards.

8.3 Some Experiences of German Development Policy in the Area of Standards

Development co-operation has supported voluntary standards in a range of ways during the last 15 years. It began mainly with public private partnership projects (PPPs), which were followed by multi-stakeholder initiatives aimed at supporting the elaboration or improvement of specific standards. In some cases, these initiatives started comprehensive processes which then became independent from public support through development co-operation and still contribute to the implementation of sustainability standards.

The Common Code for the Coffee Community (4C) provides a good example for this type of development. In 2003, the German Coffee Association (DKV), German development co-operation (BMZ and its implementing agency Gesellschaft für technische Zusammenarbeit, (GTZ), today Gesellschaft für internationale Zusammenarbeit, (GIZ), trade unions and nongovernmental actors agreed to establish a code of conduct for sustainable coffee production via a multi-stakeholder process. Soon after, about 80 % of exporting raw coffee producers of the world and about 70 % of producers of the finished product (among them Nestlé, Kraft Foods, Sara Lee, Tchibo and the Brazilian Association of roasters) were represented in the project. The Swiss development co-operation joined as an additional donor. The objective of the initiative has been to improve the living conditions of coffee producers, improve product quality and safeguard the environment by developing a basic sustainability concept with a code of conduct for the sustainable production, post-harvest processing and trading of green coffee. The project turned out to be very attractive for its members and on December 1, 2006, the 4C Association was officially founded, an independent association representing producers, trade and industry and civil society. As of July 2013, 4C counts 280 members from all over the world, uniting "all relevant coffee stakeholders in working towards the improvement of the economic, social and environmental conditions of coffee production and processing to build a thriving, suitable sector for generations to come" (4C Association 2013).

Another successful example is the cooperation between BMZ/GIZ and the Foreign Trade Association of German Retail Trade (AVE—Außenhandelsvereinigung des deutschen Einzelhandels). The AVE developed its social standard in 2002, which was then implemented on a pilot basis in 11 countries with the support of the BMZ. The projects combined an approach of audits with implementation plans and were complemented by multi-stakeholder round tables within the 11 countries, where the topic of social standards was discussed within the different country contexts. On the basis of the AVE the Business Social Compliance Initiative was founded in 2003. These two projects—4C and AVE—are very positive examples which show that cooperation between government and private initiatives can bring about long lasting and positive structural effects well beyond the scope of individual private projects.

A more recent project is COMPACI (Competitive African Cotton Initiative), a follow-up to CmiA (Cotton made in Africa), a co-operation of BMZ, DEG, the Bill and Melinda Gates Foundation and the Aid by Trade Foundation as the private partner, which aims to improve the living conditions of small cotton farmers and their families. About 300,000 farmers in Africa have been trained on good agricultural practices and implementation of sustainability standards. Overall, it was possible for farmers to increase their income by 60 % in the first phase between 2009 and 2012. These are remarkable results. The Aid by Trade Foundation had been founded in 2005 by Michael Otto, owner of the Otto Group. Otto made the clear decision not to rely on existing sustainability standards but rather to develop his own system—Cotton made in Africa. Tchibo and others joined in that approach, which also bears the challenge of establishing a new system in the market. Although progress has been made, the amounts of cotton that can now be produced under the CmiA system still exceed demand. It is thus a promising development that a benchmarking of CmiA (which is mainly known in the European market) with the Better Cotton Initiative (better known in the US market) has been carried out. This could provide the basis for better marketing possibilities for the participating farmers.

The experiences with multi-stakeholder dialogues over the years allowed some lessons to be drawn, which are presented below in Sects. 8.3.1–8.3.4.

8.3.1 Issues of Stakeholder Co-operation

Most standards initiatives are based on an alliance of partners who often sustained long-standing conflicts with each other in the past and their relationship was at best one of mutual mistrust. This is a difficult basis for discussions within a common initiative and can make it difficult to achieve common outcomes. There are some basic conditions which have to be met to ensure that this is possible:

- The stakeholders all need to be clear about their individual interests in the common approach. Usually there are different reasons behind the individual stakeholder's interests in participating in a common approach. But it is necessary that the interest is there to follow a *joint* approach, i.e. the stakeholders have to be aware that they can achieve their individual objectives only through interaction as a group.
- A neutral facilitator can then play the role of initiating discussions and organising the technical support to get the process going. This is where development co-operation can use its convening power to bring stakeholders together which had so far been unable to co-operate. Further, development co-operation can intervene to correct a kind of 'market failure': In the beginning of the process it is difficult for the participants to judge the overall benefit they can derive from it. Private enterprises might therefore be reluctant to finance such processes on their own and NGOs do not have the means to do it.
- During the process, a lot depends on the moderator who has the difficult task of bringing groups together who not only have different interests but also different communication cultures.

8.3.2 Credibility

Standards initiatives can achieve their objectives only if they are credible, because they need to compete in the market and therefore convince buyers/consumer of the value of their approach. The credibility of standards initiatives is very much dependant on the level of stakeholder participation. To bring a broader group into the process can firstly assure that all relevant information is fed into the process and more importantly, conflicts between different interests can be solved during the development process of the standard. With all their different elements, standards can only be sustainable and credible in the long run if the different groups of society which are interested in the issue support and do not challenge these standards. Coming to joint approaches requires time, but patience is necessary.

8.3.3 Involvement of Partner Governments

The relationship between standards initiatives and governments has to be looked at with care. Voluntary standards—since they are not regulations set by governments—usually have the advantage of not being regarded as establishing barriers to trade. Governments are not automatically a factor in these initiatives, as these are of a private nature. Governments at a national and also regional level do, however, play an important role and should not be neglected. As regulatory bodies they should use their technical responsibilities to ensure that these processes function in the long-term. The 4C initiative had been a learning process to all participants: It has proven to be necessary to explicitly inform and integrate the governments of a developing country in the process of developing the common code through specific workshops to ensure their continuing support to the process. Although the initiative had always informed governments via the bodies of the International Coffee Organisation (the ICO was an extraordinary member of the initiative from the start), this was not enough to ensure that the governments of developing countries were informed of all relevant parts of the project. During a period in 2005 some government representatives formulated opposition to the initiative within the ICO. The opposition was overcome, however, by informing governments in greater detail about the process, its participants and its objectives.

8.3.4 Power Imbalances

In practice, there might be power imbalances between different stakeholder groups—e.g. with regard to their ability to finance travelling or their capacity to take an active role within the decision-making structure of the initiative. These imbalances result from different negotiating skills and different capacities to organise themselves within the groups. In the case of the 4C project this, for example, led to the organisation of separate workshops for the constituency group of the producers to prepare their common position for the steering group negotiations. These power imbalances need to be addresses to ensure the final acceptance of the outcome. Development co-operation can be necessary to overcome these bottle-necks.

8.3.5 Contribution of German Development Policy Towards the Promotion of Voluntary Standards

The approaches of German development co-operation towards sustainability standards and what has been achieved was subject to a comprehensive independent evaluation that had been carried between 2006 and 2008 (Ramm et al. 2008). The evaluation showed an overall positive impact of cooperation in the area of voluntary sustainability standards. It was possible to contribute to raising incomes of the target groups, i.e. poor farmers and workers (economic sustainability). Further, it was possible to improve living and working conditions of the target groups. Farmers in standards related projects showed a higher degree of organisation, with cooperatives and rural communities showing improved social cohesion (social impact). Standards contributed to the protection of national resources, e.g. in the agricultural sector, the use of pesticides was reduced by up to 50 % and water and energy was saved (ecological sustainability). The evaluation also showed—as most evaluations in development co-operation do—that a multi-level approach proved to be efficient, i.e. assistance should not only be given to farmers (micro level), but the functioning of intermediary institutions has to be looked at, e.g. local training centres, local certification institutions (meso level), and projects and programmes prove to be successful in the long term if they are backed by government policies (macro level).

However, the evaluation also underlined some challenges, above all those stemming from the multiplicity/proliferation of standards (high costs for producers and confusion for consumers). Other recommendations referred to the need to upscale positive effects, to invest in capacity building in the target groups, to establish and strengthen local certification institutions and national accreditation systems and to connect activities more effectively at different levels. The study pointed out that “On the whole, however, the topic of voluntary standards is not yet sufficiently embedded in policy advisory activities.” (Ramm et al. 2008, p. xiii).

Amongst other things, the evaluation highlighted the importance of capacity building in developing countries. To help improve capacity at all levels—the farmer or small scale producer, institutions and in partner country governments—is one of the major objectives of development policy. A whole range of projects touch on this issue—not just the examples of public private partnership projects mentioned above, but also purely bilateral projects of projects with multinational institutions like the ILO. German co-operation in the area of quality infrastructure deserves accentuation in this context. From the beginning, the Physikalisch-Technische Bundesanstalt (PTB), the German national metrology institute, has been an implementing agency for development co-operation. Its co-operation specialises in support to measurement, standardisation, testing and quality assurance and therefore directly helps to improve the preconditions for the implementation of mandatory and private quality standards, but often also sustainability standards. For example PTB has been supporting the accreditation bodies from Argentina, Brazil, Mexico Costa Rica and Ecuador to achieve international recognition. These five accreditation bodies later signed a memorandum of understanding with GLOBALG.A.P. to provide accreditation services. The international recognition has been a prerequisite here. Support to quality infrastructure, but also to capacity building in general, is and remains one of the major areas of support in German development policy.

However, in building on the experience so far and following upon the recommendation of the evaluation, the approach to supporting standards in development co-operation changed in several ways. Co-operation with individual standards—initiatives was from then on more strictly based on specific criteria, e.g. impact, interests of business in developing countries, relevance to poverty, and process of defining the standards (for details cf. GTZ 2006, Guiding Principles). Since then, important objectives of development co-operation activities in the standards area have been to contribute to the efforts of harmonisation and/or better co-ordination between standards initiatives. One important pillar of doing so is through supporting benchmarking exercises. Another way is through supporting national and international dialogue on the issues of private voluntary standards as well as transparency and harmonisation efforts.

8.4 International Discussions on Standards in the Development Context

Not so long ago, international dialogue on standards issues was very controversial. When the EU presented the idea to the members of the World Trade Organization (WTO) at the end of the 1990s, to include negotiations on social standards in the forthcoming WTO trade negotiations, there was tremendous opposition by developing countries and the idea never came to life. At that time, private standards were only beginning to emerge and what effect these private standards would have on international trade and developing countries exports in 10 years was as yet unknown: According to the Codex Alimentarius Commission, company-specific labels accounted for 14 % in 2000 and roughly 22 % of total retail food sales at global scale in 2010 (ITC 2010). Developing countries have problems with the multiplicity of standards, with a lack of transparency, problems of compliance with standards in order to stay competitive internationally, with the capacity to implement standards requirements and with costs of compliance.

Today, the discussion of the role of standards in trade takes place in different forums. It is thus not astonishing that developing countries brought up the discussion in the WTO SPS (sanitary and phytosanitary standards) committee, that private standards would be a trade barrier. In 2011, agreement had been reached on five actions on how members might deal with private sector standards for food safety and animal and plant health (WTO 2011). The actions cover defining private standards, sharing information, and cooperation between the WTO's SPS Committee and other organisations. No agreement could be reached on other actions such as developing guidelines and codes of conduct and clarifying governments' legal obligations under the SPS Agreement. Although this agreement will already contribute to improved transparency, it is the contested action, however, which would really bring life into the standards world e.g. internationally agreed guidelines for private standards could provide transparency and clarity about the contents of standards and how to evaluate those. Such a major step forward, however, is currently impossible at an international level. It is another question, whether the SPS committee is the right place for such a discussion.

Another important international dialogue took place during 2011 G20 discussion of private voluntary standards and responsible investment in value chains. This discussion was based on the report “Promoting standards for responsible investment in value chains” of the Inter-Agency Working Group (IAWG 2011). The United Nations Conference on Trade and Development (UNCTAD) was the co-ordinator; UNDP, ILO, OECD and the World Bank were members of the IAWG. Germany (with the federal ministry for economic cooperation and development BMZ in the lead) and Saudi Arabia have been co-facilitators to the G20 Working Group on the private investment and job creation pillar of the G20 multi-year action plan on development. The IAWG report to the high-level development working group gives an overview on where the international discussion on standards has progressed to. It classifies standards as: (a) Intergovernmental organisation standards, which could be (1) normative instruments (like the ILO Tripartite Declaration of Principles concerning Multinational Enterprises and Social Policy and the OECD Guidelines for Multination Enterprises) or (2) international initiatives like the UN Global Compact; (b) Private standards, which are (1) Multi-stakeholder initiative standards, (2) industry association codes and (3) individual company codes. The report goes on to identify key policy issues like differences, overlaps and inconsistencies, the relationship between voluntary CSR standards and national legislation, and the possible danger of standards acting as trade and investment barriers. The report draws lessons from standard setting and implementation and outlines policy approaches for governments to take with regard to private standards. These comprise government purchasing policies, capacity building, regulation or incorporation of CSR in international trade and investment agreements. The report also contains guiding questions which governments should consider when dealing with private standards, which can serve as a guideline for governments to use when checking if private standards will have a positive developmental impact.

The *first* remarkable issue about that process is that it was agreed by the G20 to establish such a working group at all. Standards had always been suspected (whether it be true or false) to be used for protectionist purposes. Since private standards increased in importance, participating governments have now grown an interest in receiving a more detailed understanding on the role of private standards in responsible investment. This led to the G20’s demand to the IAWG. Before, it was shown to be difficult to start discussions with the more advanced developing countries on CSR issues. The so-called Heiligendamm process with the BRIC (Brazil, Russia, India, China) nations dealing with, amongst other subjects, investment including CSR issues, which started after the G8 meeting in 2007, showed more differences than commonalities (G8 2009).

The *second* remarkable issue is that there was a common report, prepared by five international organisations on the issue of private standards. Some years ago, this seemed out of reach, because at least some of the organisations at that time were not willing to deal with private standards in-depth. The ILO, for example, had been very reluctant with regard to a working group on ‘Decent Trade’ (a reference to the decent work concept of the ILO) that had been convened by BMZ in 2008. The working group had prepared an input into the Geneva Trade and Development Forum 2008, a conference on issues of trade and development in Crans Montana, Switzerland. The reluctance of the ILO may have resulted from a different appraisal of the importance of private voluntary standards and their role in international trade at that time and what the relationship between the ILO and this issue would be.

However, even though the IAWG came up with a substantial report which is really suited to give guidance to governments about what private voluntary standards are, how they are or can be related to national and international law and initiatives and what benefits and challenges can be derived for national governments from the existence of such private standards, the support to this report by the G20 itself is rather vague. The G20 development working group in its own report (G20 2011) summarises the IAWG report's contents and highlights that standards should not be abused for protectionist purposes. The latter remark shows a serious concern of some G20 countries which played a very important role within the discussions. Stronger words of support of the G20 to the IAWG report could have been formulated, e.g. that the G20 would invite countries to think about implementing some of the policy approaches in their own countries. But stronger reference to the content of the report was not possible. The G20 discussion—although a major step forward—thus underlines the need to further strengthen dialogue with developing country governments at an international level.

It is the objective of the United Nations Forum on Sustainability Standards (UNFSS), which came into being in 2012, to provide a neutral discussion ground for an intense international dialogue on the role and effects of private sustainability standards. The UNCTAD formulated in the Information Note that the "Forum will be a platform for providing information, analysis, and capacity-building assistance on these standards, with a particular focus on their potential value as tools for developing countries to achieve their sustainable development goals and boost production and exports of sustainably produced products" (UNCTAD 2012a). The UNFSS started with the first grant from the Swedish government in 2012 and Germany decided to support the UNFSS through a secondment of a part-time technical expert working in the GIZ Programme Office for Social and Ecological Standards. Time will show whether the forum will have the convening power and the impetus to start an intense dialogue amongst the private sector, governments of industrial and developing countries, non-governmental organisations, international organisations, trade unions and the global scientific community. But there are numerous issues that deserve further discussion at an international level: The impact of standards, the relationship between voluntary standards and legally binding requirements, the possibility for governments to use standards for their policy objectives, how to cope with the multiplicity of standards and the effects on costs especially for small-scale producers and so on. The work programme for the next years is substantial and hopefully the UNFSS will be able to contribute in shedding further light onto some of these issues and thereby contribute to a better understanding of sustainability issues.

8.5 Standards in International Trade and Investment Agreements from a Development Perspective

An important entry point for underlining the importance of standards issues, and broadening their impact, are negotiations on free trade areas and comprehensive trade and investment agreements. However, trade and investment agreements can only contain what all sides have agreed upon. This is of course also true for substantive clauses on sustainability standards. To introduce such clauses thus assumes some common understanding of the importance of sustainability issues on the sides of the respective negotiating partners. The EU now brings up the so-called sustainability issues (social and environmental concerns) in all negotiations. One example where the issue has been successfully included is the Economic Partnership Agreement of the EU with the Caribbean States (CARIFORUM). Reference to core labour standards is included in different chapters. Art. 72 of the agreement demands that, "Investors act in accordance with core labour standards as required by the International Labour Organization (ILO)" (EU 2008). Additionally, the agreement demands parties to ensure that foreign direct investment is not encouraged by lowering domestic standards (Art. 73). But the agreement also contains an individual chapter (Chap. 5) on social aspects, which makes special reference to the core labour standards and decent work. The clauses refer to the necessity of implementing the core labour standards of the ILO and the importance of the core labour standards and decent work in general. Parties agree on the importance of monitoring and assessing the operation of the agreement on decent work and foresee consultations on the issue. If consultations do not lead to satisfactory results, "any Party may request that a Committee of Experts be convened to examine such a matter." (EU 2008, Art. 195).

Although critics may argue that these clauses are not strong enough, because they do not foresee economic sanctions, in comparison with other agreements they are already far reaching, since they substantially exceed best endeavour clauses. The UNCTAD, in its World Investment Report 2012, points out that increasingly International Investment Agreements include reference to sustainable development issues, instead of concentrating solely on the protection of investor interests (UNCTAD 2012b, p. 89). A range of agreements concluded in 2011 contains reference to the protection of health and safety, labour rights, environment or sustainable development within the treaty preamble. Although a reference in the preamble is of course weaker than an article in the substantial provisions, such clauses may be important in the case of disputes, because they point out considerations which are to be taken into account by arbitration panels. A substantial number of agreements explicitly recognise that parties should not relax health, safety or environmental standards to attract investment.

A reference to sustainability issues in international agreements is possible in different ways. Above, possibilities discussed refer to international legal requirements. But agreements can also refer to private standards and call for increased corporate social responsibility of investors. The number of agreements with such clauses is still small, but growing (IAWG 2011, p. 27). CSR specific provisions are mentioned in the preamble, but also in substantive provisions requiring foreign investors to respect international CSR standards (e.g. in the Free Trade Agreement Canada—Columbia). However, it is unclear what impact the introduction of clauses, referring vaguely to private investor behaviour, can have. The European Parliament reiterates "with regard to the investment chapters in wider FTAs, it calls for a corporate social responsibility clause and effective social and environmental clauses to be included in every FTA the EU signs" (European Parliament 2011). The Parliament thus wants to strengthen private action by referring to CSR clauses, but also demands legally binding clauses asking for government action.

However, in trade instruments, there is also the possibility to provide incentives to improve compliance with private sustainability standards. Such an incentive could be introduced, e.g. into the Generalised Scheme of Preferences (GSP). Within these preferential trading schemes, industrialised countries grant preferential treatment to imports from developing countries—usually in form of a reduction on import tariffs. The EU foresees an even more preferential treatment for developing countries' exports if the partner countries have ratified and implemented a range of international conventions. This is meant as an "incentive for the respect of core human and labour rights, environmental and good governance standards through the GSP + scheme" (EC 2013). These criteria—which are of course non-discriminatory and transparent in order to comply with WTO law—do thus refer to international law. It would, however, also be possible to introduce private sustainability standards into such a scheme and thereby support the sustainability efforts of individual companies. To that end, additional preferences could be granted to those products exported to the EU which are proven to comply with sustainability criteria. The EU would have to define these criteria for sustainability standards in a non-discriminatory way. Standard initiatives would then have to prove that they comply with these criteria. Enterprises could then, when applying these standards, register for preferential import treatment. Recently, BMZ has conducted a study on the possibility of introducing such an idea into the GSP. It shows that a range of practical problems still have to be discussed but that this would be a possible way forward and could provide an incentive for private enterprises to comply with sustainability standards. However, there is still enough time to discuss and improve the idea and to find support from other stakeholders, as the GSP of the EU has just undergone a revision and is now settled for the next years.

Besides its participation in the EU decisions on trade and investment agreements, German development co-operation has been actively contributing to the German position towards discussions within the ILO and the revision of the OECD guidelines on Multinational Enterprises. BMZ is actively participating in the discussions within the German OECD national contact point, when cases are concerning companies which are active in developing countries. Germany has also supported the secretariat of the United Nations Global Compact.

8.6 Information and Transparency

As has been mentioned several times, the multiplicity of standards is problematic for small scale farmers, industrial producers searching for sustainable inputs and consumers alike. In this situation, German development policy aims at contributing to increased transparency in standards issues. One way to do so has been German (together with Swiss) support of the Trade for Sustainable Development (T4SD)-project of the International Trade Centre (ITC). Germany's decision to support was instrumental in setting up the T4SD database, which aims at providing, "comprehensive, comparable and verified information on voluntary sustainability standards… The Standards Database covers all facets of standards systems, such as their contextual background, product and geographic scope, provisions and requirements … governance structure, stakeholders' engagement, implementation mechanisms and verification systems…" (ITC 2012). The idea was to create a comprehensive system that would allow consumers, manufacturers, retailers, institutions and all those interested in standards issues to have easy access to information about individual standard systems and to compare their content. This project provides a good example of a fruitful mix of bilateral and multilateral instruments, in that the ITC T4SD project has not simply been supported financially. Additionally, BMZ had asked the Programme Office for Social and Ecological Standards (GIZ) to closely work together with ITC in order to bring in the long standing experience of German development policy in the field of standards.

Based on the T4SD project, Germany and Switzerland together decided to put up 'Kompass Nachhaltigkeit', an internet platform which aims at making the T4SD data on sustainability standards more easily accessible and usable for purchasing decisions of official purchasing entities as well as small- and medium-sized enterprises. In 2009, development policy was actively engaged in the adaptation of the German Competition Law ("Gesetz gegen Wettbewerbsbeschränkungen") in order to allow for the integration of social considerations in government purchasing decisions. A government's decision to purchase in a sustainable manner can have a large impact on demand for sustainable products, since government procurement is a very large market. Additionally, such a government decision can serve as an example to private enterprises and other organisations and underlines the will of the government to anchor sustainability issues in its policy more effectively. For issues of policy coherence it was therefore sensible for German development policy to engage in the opening of competition law towards social sustainability issues. This then led to an increased demand of public procurement officials and NGOs for solid information on products which do comply with environmental and social standards. The Kompass Nachhaltigkeit was meant as an instrument to supply this demand.

However, increasing numbers of consumers, producers and companies see a need for an instrument that not only makes available comprehensive information on existing sustainability standards, but also for a means with which to assess the performance of different standard systems in order to make more meaningful comparisons. Since 2011, the Sustainable Standards Transparency Initiative

(SSTI), whose members are GIZ—on behalf of BMZ, ISEAL and ITC, has been working towards this more comprehensive benchmarking objective. The idea is to compare not only the content of standards but also their implementation systems, e.g. in how far they are able to monitor non-compliance with the standard requirements and formulate an appropriate response. Currently, the initiative is in the process of collecting criteria for the assessments, in a next phase, an IT tool shall be programmed to make the information ready for use. The more interest and support the tool receives from companies, consumers and other users, the more it is able to become comprehensive and to be improved. It is therefore desirable that other actors and donors join the SSTI as well as the T4SD project in the near future.

8.7 The Link to German Stakeholders and to the Public

Development policy and development objectives have to be anchored in the society to ensure long lasting support for this policy field. Discussions and projects on sustainability standards are an ideal way of making development objectives comprehensible for consumers and the general public. The Fair Trade movement and its actors have been working with this concept for decades. It has always been the objective of Fair Trade to directly contribute to the improvement of the living conditions of small-scale producers in developing countries, but at the same time to create awareness about trading conditions and the situation of producers to make the trading system, in itself, 'fairer'. The link with the product becomes a responsibility of the consumer at a personal level, with the living conditions of the producer in a developing country thus becoming very direct and comprehensible. Fair trade thereby contributes to improve the knowledge of the general public about development issues. Therefore Fair Trade has, for a long time, been a close partner for development policy. BMZ traditionally financed projects of German Fair Trade actors and supported the implementation of the major information campaign 'Fair Feels Good' (2003–2007). The doubling of Fair Trade sales in the last 3 years to 477 million Euro in 2011 (Forum Fairer Handel 2012) shows the increasing interest of German consumers in Fair Trade and standards issues. The support to Fair Trade through development co-operation is thus also a possible way of increasing awareness in German society about development issues in general. However, currently an effort is being made to co-operate to an even greater extent with Fair Trade approaches in the classical field of bilateral development co-operation. A typical approach for development policy is, for example, to support small scale producers in their efforts to comply with the Fair Trade Standard. Although this kind of project has also been carried through in the past, this might be an area for a further strengthening of co-operation in the future.

Close co-operation of development policy also exists with other German actors. At a very early stage, BMZ had already launched a German multi-stakeholder process on social standards: The Round Table Codes of Conduct (Runder Tisch Verhaltenskodizes) had started in 2001 as a forum for exchange and dialogue on the introduction and implementation of social standards in the supply chain. The round table is moderated by BMZ and the secretariat provided by GIZ on behalf of BMZ. Members of the roundtable are from private companies, company associations, trade unions, non-governmental organisations and the German ILO office. The roundtable has always been an instrument which was linked to concrete practical questions of relevance to the stakeholders. Common projects—like the above mentioned AVE project—have been issues of discussion and points of reference. Changes in the discussions at the roundtable are exemplary for the change in perception of standards issues in the society. While the issue of living wages was too controversial to be sincerely dealt with at the roundtable 10 years ago, in 2011 the roundtable members agreed on an action programme for a living wage (Round Table 2011). This paper reflects a common analysis on the issue of a living wage and outlines ideas on methods to progress towards wages in developing countries which allow for a decent standard of living for workers. The issues discussed and identified at the roundtable and experiences exchanged have often helped to improve and further refine approaches towards social standards of companies, but also in development co-operation projects. The Round Table Code of Conduct is also linked with the broader CSR discussion within the German CSR Forum.

A more recent stakeholder approach is the Forum Nachhaltiger Kakao (Sustainable Cocoa) which was launched in 2012 by the government (BMZ and Ministry of Agriculture), industry and trade associations, private companies in the confectionary sector, as well as non-governmental organisations and other interested actors. The objective is to exchange information on good practice and to commonly elaborate further approaches which can be taken towards increased sustainability in the cocoa sector in cooperation with producer countries.

8.8 Conclusions

Sustainable production and trade conditions, respecting all pillars of sustainability, are a major objective of development policy. Therefore, development policy has been playing an active role in supporting the development and implementation of legally binding, as well as voluntary standards. Lastly, efforts were more directed towards contributing to transparency, harmonisation and benchmarking. However, one major natural area of activity for development co-operation is support for small scale producers, as well as workers, in developing the necessary capacity for implementing standards.

The standards world is in a continuous process of change. After the emergence of more and more sustainability standards, there is a change in direction towards increased transparency and the possibility to compare standards. Also, there is a tendency for countries to rather regulate ecological and social conditions; an example is the recent regulation of India on biologically produced cotton. Another important development is the policy of large companies to choose development of their own sustainability concepts rather than rely on individual standard systems (which are also not always able to provide the necessary large amounts).

Development policy can, in that context, bring in its comparative advantages at several levels and users can continue to build upon their cumulative experience of its use.

References

4C Association (2013) Mission and vision of the 4C Association. http://www.4c-coffeeassociation.org/about-us/mission-and-vision.html. Last accessed 04 July 2013

EC – European Commission (2013) Generalised scheme of preferences. http://ec.europa.eu/trade/policy/countries-and-regions/development/generalised-scheme-of-preferences/index_en.htm. Last accessed 04 July 2013

European Parliament (2011) Report on the future international investment policy (2010/2203 (INI)). http://www.europarl.europa.eu/sides/getDoc.do?type=REPORT&reference=A7-2011-0070&language=EN. Last accessed 04 July 2013

EU – European Union (2008) Economic partnership agreement between the CARIFORUM states, of the one part, and the European Community and its Member States, of the other part. Official Journal of the European Union L 289/I/3

Forum Fairer Handel (2012) Wachstum trotz Krise: Fairer Handel in Deutschland boomt – doch Rohstoff-Spekulationen gefährden den Erfolg. Pressemitteilung, 10.08.2012. http://www.forum-fairer-handel.de/#fairerhandel_pressemitteilungen. Last accessed 04 July 2013

G8 (2009) Concluding Report on the Heiligendamm process, L'Aquila. http://www.g8italia2009.it/static/G8_Allegato/06_Annex_1__HDP_Concluding.pdf. Last accessed 04 July 2013

G20 (2011) 2011 report of the development working group. http://www.g20.utoronto.ca/2011/2011-cannes-dwg-111028-en.pdf. Last accessed 04 July 2013

GTZ – Deutsche Gesellschaft für Technische Zusammenarbeit GmbH (now GIZ) (2006) Guiding principles on the provision of support by German Development Cooperation for voluntary social and ecological standards initiatives, commissioned by Federal Ministry for Economic Cooperation and Development, Eschborn

IAWG – Inter-Agency Working Group on the Private Investment and Job Creation Pillar of the G20 Multi-Year Action Plan on Development (2011) Promoting standards for responsible investment in value chains: report to the high-level development working group, September 2011. http://unctad.org/sections/dite_dir/docs/diae_G20_CSR_Standards_Report_en.pdf. Last accessed 04 July 2013

ILO – International Labour Organization (2012) Tackling child labour: from commitment to action. ILO/International Programme on the Elimination of Child Labour (IPEC), Geneva

ITC – International Trade Centre (2010) Voluntary standards in developing countries: the potential of voluntary standards and their role in international trade. International Trade Forum 3/2010

ITC – International Trade Centre (2012) Voluntary standards. http://www.intracen.org/trade-support/voluntary-standards/. Last accessed 04 July 2013

Ramm G, Fleischer C, Künkel P, Fricke V (2008) Introduction of voluntary social and ecological standards in developing countries: synthesis report. Evaluation Reports 43. Federal Ministry for Economic Cooperation and Development, Germany (BMZ), September 2008

Round Table – Round Table Codes of Conduct (2011) Round Table (RT) Codes of Conduct action programme for a living wage. http://www.coc-runder-tisch.de/index.php/en/ueber-den-runden-tisch-en-2/publikationen-runder-tisch-en. Last accessed 04 July 2013

UNCTAD – United Nations Conference on Trade and Development (2012a) Information note: UN forum on sustainability standards to be launched soon. http://unctad.org/en/pages/InformationNoteDetails.aspx?OriginalVersionID=22. Last accessed 04 July 2013

UNCTAD – United Nations Conference on Trade and Development (2012b) World Investment Report – Towards a new generation of investment policies, Geneva
WTO – World Trade Organization (2011) Sanitary and phytosanitary measures: Formal meeting. http://www.wto.org/english/news_e/news11_e/sps_30mar11_e.htm. Last accessed 04 July 2013

Chapter 9
Voluntary Sustainability Standards: Measuring Their Impact

Carsten Schmitz-Hoffmann, Berthold Hansmann, and Sophie Klose

9.1 Introduction

Standards are an instrument to translate the vision of sustainable development into concrete and practicable steps, whose impacts can be measured and aid further development. Voluntary sustainability standards (VSS) are one part of the answer to the call for a socially and ecologically compatible form of globalisation. As they have been shown to improve worker living conditions and protect natural resources in developing countries, the German Government regards these standard systems as an important tool in combating poverty. The German Government therefore actively supports the application of VSS as one instrument in attaining the Millennium Development Goals (MDGs). The consolidation of voluntary sustainability standards contributes to the achievement of several priority tasks of the MDGs by 2015, such as:

- halving extreme poverty (MDG1)
- providing universal primary education (MDG2)
- promoting gender equality and empowering women (MDG3)
- ensuring environmental sustainability (MDG7)
- creating a global partnership for development (MDG8)

C. Schmitz-Hoffmann (✉) • B. Hansmann
Programme for Social and Environmental Standards, Deutsche Gesellschaft für Internationale Zusammenarbeit (GIZ) GmbH, Eschborn, Germany
e-mail: carsten.schmitz-hoffmann@giz.de; berthold.hansmann@giz.de

S. Klose
Programme for Social and Environmental Standards, Deutsche Gesellschaft für Internationale Zusammenarbeit (GIZ) GmbH, Eschborn, Germany

Department of Environmental Planning, Brandenburg University of Technology (BTU) Cottbus, P.O. Box 101344, 03013 Cottbus, Germany
e-mail: sophie.klose@tu-Cottbus.de

C. Schmitz-Hoffmann et al. (eds.), *Voluntary Standard Systems*, Natural Resource Management in Transition 1, DOI 10.1007/978-3-642-35716-9_9,

Voluntary initiatives support the consolidation of the MDGs through the development and implementation of a number of voluntary sustainability standards and codes of conduct. Those initiatives have covered, by now, almost all sectors. Their scope ranges from forestry and agriculture to textiles, natural stones and dams.

Over the past 10 years the German Federal Ministry for Economic Cooperation and Development (*BMZ – Bundesministerium für Wirtschaftliche Zusammenarbeit und Entwicklung*) and the German Society for International Cooperation (*GIZ – Deutsche Gesellschaft für Internationale Zusammenarbeit*) contributed by implementing projects, providing advisory services to existing initiatives, cooperating with governments of developing country, establishing dialogue platforms, conducting evaluations, providing financial funds and supporting the application of necessary frameworks within which standards initiatives can operate. This chapter discusses and analyses the social, economic and environmental impacts of voluntary sustainability standards.

9.2 Background: How Initiatives Have Developed—Contributing Factors

The development and implementation of voluntary sustainability standards emerged for two major reasons. Initially, risk assessment and management cannot be fully covered by national standards and law regulations due to their limited influential sphere and the complexity of international supply chains. Therefore voluntary standard systems are important tools as they implement the international conventions into their standard schemes and hence enable the measurement of compliance. The second major factor is based on market competition and reputation. The differentiation of products and services on the global market, through standard certification and labels in contrast to uncertified products, poses a unique selling point through which economic and marketing conditions can potentially be improved.

Sustainability standards are developed by industry, non-profit organisations, trade associations or others, committed to finding solutions to specific problems. Mostly, they are designed in so-called 'multi-stakeholder initiatives' which comprise actors from the private sector, public sector and civil society. Such initiatives may foster a results-orientated political climate and a willingness to move beyond the lowest common denominator. Thus, voluntary standard initiatives demonstrate the feasibility of sustainable business practice with the support from the relevant involved stakeholders.

The success and acceptance of standards initiatives are based upon their claims and impacts (social, environmental and economic), credibility and legitimacy. While credibility of voluntary sustainability standards directly affects consumers' and producers' acceptance, the right to implement measures to achieve legitimate policy objectives, such as the protection of human health and safety or the environment, is an important factor as long as the regulations and certification procedures do not create unnecessary obstacles to trade, i.e. technical barriers to trade (TBT), under international trade rules (see WTO 2013 for a discussion and resources on TBT).

Both the International Organization for Standardization (ISO) and the International Social and Environmental Accreditation and Labelling Alliance (ISEAL), the global association for voluntary sustainability standards, developed guidance for the development of standard schemes with emphasis on improving effectiveness, impacts and credibility of voluntary sustainability standards (Annex 1 to Chap. 1 of this volume provides an overview of the ISEAL Credibility Principles; see also Sect. 1.3 defining VSS credibility).

9.2.1 The Motivation for Measuring Impacts

Voluntary sustainability standards systems are designed to address the most pressing social and environmental challenges, accordingly, it has been assumed that sustainability standards that are credible and effective can bring significant positive social, environmental and economic impacts. However, voluntary standard initiatives are facing pressure from all sectors to make a stronger case for proving the positive impacts that their programs are having. To better analyse and understand impact, external as well as internal impacts are separately monitored. The following stakeholders have differentiated interests and motivations in impact assessments of standard initiatives:

- Governments are interested in ascertaining whether voluntary sustainability standards provide positive economic benefits and market opportunities for their industries or whether they act as a technical barrier to trade. In addition, governments want to know how far VSS can be used to prove compliance of regulative processes;
- Donors and development agencies want to know whether their investment in voluntary sustainability standards have been justified and to which extent they are promoting sustainable development and poverty alleviation;
- Industry is looking at voluntary sustainability standards to determine which one is most cost-effective in delivering the social and environmental impacts that their customers and risk management strategies are demanding;
- Consumer groups and NGOs want to see a level of guarantee that these voluntary sustainability standards are delivering on their stated social and environmental claims; and
- The producers themselves need to evaluate the benefits of incurring costs of coming into compliance with the voluntary sustainability standard requirements.

However the most benefit from impact assessment is gained by the standard systems themselves. In the course of exploring and measuring impacts, internal organisational learning and improvement can be an even stronger motivation than meeting external expectations. A robust Monitoring and Evaluation (M&E) programme will enable a VSS system to refine and improve both the content of the standard as well as the relevance and efficiency of the audit process. Information gathered by the audit process and other methods of impact analysis can be provided to external stakeholders for performance verification as well as being used for internal learning and development processes.

9.2.2 Methodological Approaches

For measuring impacts there is no one definite methodological approach that fits all requirements of assessment. Depending on which information is to be gathered and which aspect is to be concentrated on, different approaches can be appropriate. However, information gathered through altered approaches is often complementary and can be used to create a more complete picture of the impacts of VSS. By combining information from several sources, the overall picture is likely to be more complete.

Nevertheless, there are still gaps in the evidence base, making generalisations about impacts difficult. The most dominant challenges in designing an appropriate impact study are the following: the challenge of identifying an appropriate counterfactual—as a comparable scenario; challenges posed by evolving systems that continuously adjust and modify; and the limitation of observations from individual pairs or small sets of cases, mostly case studies. Moreover, social and ecological values, especially the prevention of their losses, are extremely important impacts although they are difficult to measure and to quantify.

Much of the evidence base is relatively new, and the impacts in many cases have not been studied over sufficiently long periods of time. Initially, impact assessments were not incorporated area-wide in standard initiatives. Qualitative, snapshot studies have predominated. Only later, when evidence for impacts of standards became more important, impact assessments were conducted on a more regular basis. Therefore the majority of the information revealed during these assessments is based on either medium-term studies (3–5 years), depicting anecdotal evidence with little detail, or broader studies analysing a wide spectrum but with less detailed information.

Many voluntary sustainability standard systems which were developed in recent years gradually employed more rigorous qualitative and quantitative methods. Methods and approaches have been professionalised and selectively included a monitoring and evaluation program to track impacts and keep records of changes in social and ecological sectors.

While it is unrealistic to assume a common methodology for all case studies, it could be useful to explore whether there are common elements of good practice. In agreeing on common core issues, impacts can be assessed in a more systematic manner that will provide more meaningful and comparable results.

9.2.3 A Common Language

In order to better understand what constitutes good practice in measuring impacts, it is most important to define a common language, including descriptions of data formats and definitions on basic terms that are addressed. This common language is a prerequisite not only for discussions among stakeholders but is also necessary for later comparisons of findings and methodologies.

One important step towards a common understanding has been introduced with a concept posed by the standard umbrella organisation ISEAL Alliance.[1] Its members agree that the most appropriate language to describe their work is the result-based language of Monitoring and Evaluation.

It is important to keep in mind that the challenge in measuring the impacts of voluntary standard systems is to highlight the causal link between the changes that are caused through the compliance with the requirements through the application of good practices and the long-term impacts on society and the environment.

9.2.4 Core Issues

At a high aggregation level, all stakeholders are likely to share the same interest in core issues, ranging from biodiversity to child labour impacts. On this basis, each assessment method defines its own criteria and definitions describing the precise area that is to be studied more specifically.

However, setting those common core issues is important in granting a certain degree of comparability. One of the weaknesses of most impact assessments to date is that many assessments take place in isolation, resulting in a lack of comparability and therefore not matching with the consumers' needs and requirements. The extent to which the impacts are being measured should be consistent across standards and production units, in order to directly influence the comparability and, thus, usefulness of the data.

9.3 The Impact of Voluntary Sustainability Standards

Impacts describe the changes in the quality and resilience of ecosystems, changes in resource efficiency, livelihoods, and changes in social welfare within the workplace and wider community (RESOLVE Inc. 2012).

---

[1] ISEAL was founded in 2002 and aims at setting basic rules and guidelines for standard initiatives in order to support standard implementation on the market and communities. Working together with standards in sectors like fishing, agriculture, forestry and many more, ISEAL also supports measuring their social and environmental impacts (ISEAL Alliance 2008).

Most VSS focus their impact assessment on the actual production processes of certain products (management unit) e.g. in forestry, however there are certain impacts on regional, national and even international level (beyond management unit) which need to be considered in order to capture the entire impact of the operation of the VSS.

Dependent on the scale applied, potential impacts and outcomes of standard systems are diverse and can be classified in several ways. Impacts on a small scale farmer unit may differ drastically compared to a whole landscape unit depending on the outreach of a standard system. Furthermore standard systems address a wide variety of sustainability aspects. Impact assessment therefore needs to take into consideration the diversity of intended achievements.

To assess voluntary sustainability standard systems on their impact on the aspect of sustainability, a number of frame conditions have to be set and defined. VSS systems can be distinguished according to their range, type of resources affected and the functional or structural impacts.

Functional impacts describe changes in the function or mode of utilisation of resources, whereas structural impacts describe the change in structure or composition of resources. Both are often closely interrelated as structural changes can create the preconditions for functional changes.

The assessment of VSS systems is generally based on the separate consideration of the three major pillars of sustainability (environment, economy and social conditions; see Table 9.1).

Environmental impacts usually cover three major fields of study: ecosystem integrity; biodiversity; and pollution and waste. Therein, ecosystem integrity is assessed towards the persistent supply of ecosystem services. Biodiversity is an indicator of the health of an ecosystem, based on data of different types of flora and fauna. The indicator 'pollution and waste' examines the effects and management of excess material like agrochemicals, entering the environmental system during production processes.

Economic impacts are also measured by different indicators: the net enterprise income, considering the changes in income of the farmers versus respective costs, as well as the business opportunities gained by, for example, opening global markets to farmers. In contrast to social impacts, information on economic impact is comparatively easy to measure—and therefore to collect. Hence, fairly good data is available to conduct analysis.

The social impacts of standard schemes have been studied historically less intensively than either ecologic or economic impact. Three major fields of social impacts are important: working and living conditions; rights and benefits; and relationships with the wider community. Through many initiatives, businesses are also encouraged to integrate further human rights principles and their impacts into their business operations, including fair working conditions, equality, dignity, health and security of workers, housing and standards of living, the integration of indigenous peoples, land and culture and human rights in the supply chain. On the other hand, these human rights are universal and should never remain criteria of voluntary sustainability standard schemes—but shall be covered by legal

Table 9.1 Overview of social, economic and environmental issues addressed by standards schemes (*Source*: RESOLVE Inc. 2012)

Environmental issues	Description
Loss of biodiversity	Both natural biodiversity and agro-biodiversity, including local varieties, geographic overfishing, and disease transfer to wild species
Conversion of natural ecosystems	Destruction of forests, primary tropical forests and peat land; closely related to biodiversity loss and climate change
Pollution/contamination of air, soils, and water	Leaching form pesticides, nitrates, and phosphates contaminating water, land, and air; waste treatment and disposal; water use
Soil degradation, erosion, and/or desertification	Intensification of production leading to poor soil quality and infertility
Climate change	Indirect contributions to greenhouse gas emissions through deforestation and energy use; direct contributions from cattle, manure, nitrogen in soils, etc.
Social issues	**Description**
Working and living conditions	Health and safety, housing, medical care
Rights and benefits	Freedom of association, working hours, discrimination
Community development	Land rights, food security, education
Economic issues	**Description**
Income profitability	Changes in income, price premiums
Business opportunities	Market access, access to credit, technical assistance

requirements. However, due to the fact that not all countries subscribe to the optimal notion of human rights, many standards have taken up universal human rights requirements within their criteria system.

Improvements in working and living conditions in standard initiatives can be displayed through improved welfare in developing countries such as school education or increasing income per household. Rights and benefits in most initiatives aim to empower (especially women and indigenous people) and enable participation in decision making processes of single farmers or farming communities. Several standard initiatives also include a tool for strengthening the community and its development. For example, by offering a 'social premium' the community earns extra money they can spend in construction of new community housing, purchasing new machinery for harvests or the improvement of infrastructure.

9.3.1 Impacts of Standards Initiatives on Three Examples

To illustrate impacts of voluntary standard initiatives, three practical examples for ecological, economic and social impacts are presented below.

Environmental Impacts

The Marine Stewardship Council (MSC) was launched in 1999. The certification programme aims to use its eco-label to reward fisheries with sustainable practices, to influence purchasing practices of people towards sustainable seafood and contribute to improving the conditions of the world's oceans. MSC examines impacts of its programme upon the environment on a regular basis. The last studies showed (MSC 2006).

In MSC certified fisheries the greatest quantified change observed was made in the stock status. In post-certification, 21 fisheries increased their stock, whereas 18 fisheries had to decrease their stock status in order to comply with new, updated requirements in the MSC certification. During the review of MSC in 2006 further improvements had been identified in bycatch, biomass target reference points, the status of endangered species and the related seabird mortality rate per year. In comparison to earlier studies, pressure exerted from bycatch decreased, with less endangered species being caught (seal or endangered seabird bycatch). In regard to impacts on the reduction of pollution and waste, no specific numbers are available yet; however a positive impact is expected on a larger scale.

Concerning impact on a wider scale through the certification against MSC standards, participants achieved significant improvements in management practices and in stakeholder engagement. A number of fishery managements also adapted a more holistic approach, focusing on wider environmental concerns.

Generally, fisheries, like any other sector, are making the largest improvements prior to certification through the application of good practices. After certification, fisheries slowly and continuously improve performance i.e. in stock number and health conditions, encouraged by the use of certification conditions such as management help with improved market access. Therefore, certification is rather a control process of improved conditions than a tool for significant improvements after having gained the certificate.

Additionally, analysis of the evidence and stakeholder views confirm that 'on the water' environmental improvements have occurred in MSC-certified fisheries and these improvements are incremental throughout a fishery's involvement with the program.

Economic Impacts

The Fair Trade Labeling Organization coordinates labelling of products like cocoa, coffee, sugar, wine, bananas, cotton, tea, honey and many other agricultural goods that are traded in 'fair' conditions for producers and consumers. The system aims at providing higher and minimum prices to ensure welfare benefits for producers of agricultural goods in developing countries.

There is substantial evidence in literature that the Fair Trade label has positive economic impacts mainly through improved productivity and higher prices, which leads to improved economic stability (e.g. CEval 2012). Nevertheless Fair Trade certified farmers state higher incomes mostly leading to improved welfare for smallholder producers. The minimum guaranteed prices and pre-financing by buyers of Fair Trade-certified farmers also works as a buffer against economic fluctuations on national markets.

Aside from enhanced access to markets, technical assistance and credit access is also provided. Within the Fair Trade certification farmers often get the ability to access loans that can either be used for the expansion of the farm or for reinvestment e.g. technical equipment. They can also access options to diversify income sources, or in contrast to specialise on one specific product but in a higher quality market. Reports also state that certification against Fair Trade standards may also positively affect business management and improve risk control for smallholder farmers. As a whole, the model of Fair Trade provides several benefits for participating farmers.

Social Impacts

Measuring, naming and quantifying social impacts are complex processes. The challenge is to distinguish the changes caused by applying standard systems themselves or from naturally occurring changes in social structures and livelihoods by external influences. Almost every voluntary standard initiative has a social component included, providing principles and criteria for social improvements. Impacts of standards on the social situations generally can be measured by assessing the changes in living and working conditions or the improvement of welfare for farmers and communities over a long period of time.

Impact assessments of the Forest Stewardship Council (FSC) and other standard initiatives studied impacts over a longer period of time and tracked patterns of changes that seem to be consistent throughout standard initiatives. Through the certification against FSC or others, indigenous groups and smallholder farmers were empowered to play a more significant role in decision making processes concerning issues such as forest concession. Especially programs for the empowerment of women and temporary workers were started to improve their access to e.g. labour rights. Improved hygiene and work safety on the ground are also important successes. The FSC also assessed that providing land tenure security has motivational influences upon farmers and workers.

A number of voluntary standard systems also offer a 'social premium' for the community on every kilo of agricultural commodity they produce. The extra money is distributed in community development or spent on implementing beneficial projects. A Social Premium Committee (mostly consisting of farmers) decides on how the money is to be spent, usually by financing local buildings, technical modernisation, infrastructure and many more potential improvements.

Generally, social impacts are difficult to measure. Most impacts assessed are strongly dependent on the certification scheme, location and time. Measuring direct impacts of voluntary standard systems on living conditions, welfare and social safeguards is highly difficult to distinguish from natural external processes and influences. A general assessment therefore is difficult and very system and place specific.

9.4 Voluntary Sustainability Standards: Contribution to Sustainable Development

Poverty, pollution and working conditions: the opportunities and challenges of today's economic globalisation are closely connected. Opportunities to improve living conditions, raise educational standards and establish better healthcare through participation in global supply chains are certainly visible. Yet at the same time, while we are all competing globally, there is an equal risk of joining a race to the bottom when it comes to the environment and social conditions. The work of German Development Policy is dedicated to supporting a just and socially equitable process of globalisation, which is based on the concept of sustainable development. Germany therefore actively supports the Millennium Development Goals, striving for a combination of economic success, social justice and peace, ecological balance and political stability. Although considerable progress has been made, there is still work to be done. A wider application of voluntary sustainability standards is one instrument to aid in reaching these goals, inextricably linked to sustainability. After more than 30 years of active German Development Policy in the field of standards, it is apparent that market-driven standards are not only able to contribute to relieving poverty, but are also able to support development towards social inclusion and democracy. Experience continues to teach us that the obstacles to development and freedom are best countered with instruments and arguments that are embedded in a market environment. The living conditions clearly change for the better when voluntary standards are adopted. It goes without saying that these standards cannot of course replace legislation and international agreements. Since such standards are market driven, they follow a different logic. Nevertheless, their economic success is an argument for further application: monitoring and enforcement are merely complementary elements of this.

Voluntary sustainability standard systems have introduced a new form of partnership between civil society organisations and businesses, shifting the landscape of sustainable production and consumption in important ways. They have opened avenues for public and stakeholder interests to participate in defining standards that have subsequently become societal and even regulatory norms. They have raised public awareness of, and demand for, more sustainable products. And, they have put the missing pieces of a sustainable supply chain in place, from technical assistance and extension to supply chain tracking systems. However, they should be limited in number if they are to serve as an instrument of market transparency to consumers on the demand side. Although, voluntary sustainability standards are market-driven instruments and as such they have to be consistent with business principles, one main obstacle for further promotion and scaling up of market penetration is the fact that there is a strong competition between the various standard initiatives and therefore and accompanying lack of cooperation.

Despite all the positive results and impacts mentioned above, the weakness of VSS continues to be the scale of market penetration. Most of the VSS have reached a market penetration of less than 15 %. In order to increase the proportion of people benefiting from VSS—both producers and consumers—new and better interlinked activities are needed and the existing systems need to undergo a structural reform to be able to expand and to scale up to reach 60–80 % market penetration.

References

CEval – Center for Evaluation of the Saarland University (2012) Assessing the Impact of Fairtrade on Poverty Reduction through Rural Development, Fairtrade Impact Study Final Report, Commissioned by TransFair Germany and Max Havelaar Foundation Switzerland. Saarbrücken, Germany. http://comunicarseweb.com.ar/download.php?tipo=acrobat&view=1&dato=1359385781_Fairtrade-Impact-Study.pdf. Last accessed 31 Aug 2013

ISEAL Alliance (2008) Measuring the impacts of certification – background paper, London

MSC – Marine Stewardship Council (2006) Environmental benefits resulting from certification against – MSC's Principles & Criteria for Sustainable Fishing – Final Report, London

RESOLVE Inc., Steering Committee of the State-of-Knowledge Assessment of Standards and Certification (2012) Toward sustainability: the roles and limitations of certification, Washington

WTO – World Trade Organisation (2013) Technical barriers to trade. http://www.wto.org/english/tratop_e/tbt_e/tbt_e.htm. Last accessed 31 Aug 2013

Chapter 10
Environmental Standards and Embedded Carbon in the Built Environment

Callum Hill and Andrew Norton

10.1 Introduction

As the awareness of the potential impacts of climate change increases, so will the imperative to reduce the level of anthropogenic atmospheric fossil carbon dioxide in the atmosphere. One very important consideration in this is the use of forests and forest products as long term atmospheric carbon stores. This chapter discusses the current situation relating to voluntary standards and how the environmental benefits associated with sequestered carbon in timber products may be measured and reported. When timber products from sustainably managed forests are utilised in the built environment this can have a positive environmental benefit (Hill 2011) but the question is how should this benefit be reported? With increasing emphasis being placed upon the environmental performance of goods and services it is essential that robust procedures are developed to validate claims. The importance of considering the built environment as an additional atmospheric carbon sink in long-life products is now receiving particular attention. Simultaneously, Europe is about to adopt legislation dealing with the certification of timber products. One reason for this is the failure to universally adopt voluntary certification schemes.

The purpose of this chapter is to explore in detail the current situation with respect to the standards and methodologies that are being used to determine the environmental impacts and benefits associated with wood products. The chapter

C. Hill (✉)
Norsk Institutt for Skog og Landskap, Ås, Norway

JCH Industrial Ecology Limited, Bangor Business Centre, 2 Farrar Road, Bangor, Gwynedd, Wales LL57 1LJ, UK

Renuables, Llanllechid, Gwynedd, Wales, UK
e-mail: enquiries@jchindustrial.co.uk

A. Norton
Renuables, Llanllechid, Gwynedd, Wales, UK
e-mail: a.norton@renuables.co.uk

C. Schmitz-Hoffmann et al. (eds.), *Voluntary Standard Systems*, Natural Resource Management in Transition 1, DOI 10.1007/978-3-642-35716-9_10,

goes on to consider impending future legislation in the European Union. There is now an opportunity to combine timber certification and life cycle assessment tools to produce much more useful environmental data. This initiative can be combined with an extended chain of custody of the timber products through the value chain to end-of life. The chapter starts by discussing why it is important to consider the use of timber from sustainable forestry in the built environment as a carbon store (Sect. 10.2). In Sect. 10.3 the chapter then continues with the description of the current situation that is developing in Europe with respect to Type III environmental declarations (also known as environmental product declarations) and in particular the standards that are emerging in this area. Some background to the methodology used in life cycle assessment (LCA) is given in Sect. 10.4. In Sect. 10.5 the scientific principles behind the concept of using timber in the built environment as a carbon store are discussed. The chapter then goes on to describe the current situation with respect to environmental standards and carbon storage in timber (Sect. 10.6). The issue of certification is covered in Sect. 10.7. Some case studies discussing the topic of carbon valuation and forestry are given in Sect. 10.8. Finally, in Sect. 10.9 conclusions are drawn and recommendations are made.

10.2 Background

We are in an unprecedented time in the history of humanity. The results of human activity are now having an impact on a global scale. Anthropogenic carbon dioxide emissions are seen as making a substantial contribution to climate change (Intergovernmental Panel on Climate Change—IPCC 2007). Since pre-industrial times (before 1750) the concentration of carbon dioxide in the atmosphere has risen from a baseline level of 280 ppm (parts per million) to about 380 ppm at present. The average levels of carbon dioxide in the atmosphere have risen at a rate of 1.4 ppm per year over the period 1960–2005, but this rate is increasing. For example, the increase in CO_2 levels in the decade 1995–2005 was 19 ppm. When the global warming effects of the other greenhouse gases (primarily methane and nitrous oxide) are also taken into account, the level is around 430 ppm of carbon dioxide equivalents. The levels of greenhouse gases are presently higher than they have been for any time in the past 650,000 years. These contributions to the increase in atmospheric CO_2 concentration since the industrial revolution come mainly from the combustion of fossil fuels, gas flaring and emissions associated with cement production. Other sources include deforestation, land use change and biomass burning (contributing about 20 %) (IPCC 2007).

As global temperatures rise, there will be increasing numbers of severe storm events, resulting in greater economic burden associated with the resultant infrastructure repair costs (Stern 2006). The Stern review originally concluded that in order to avoid the most serious consequences associated with climate change, it would be necessary to keep the atmospheric concentration of greenhouse gases

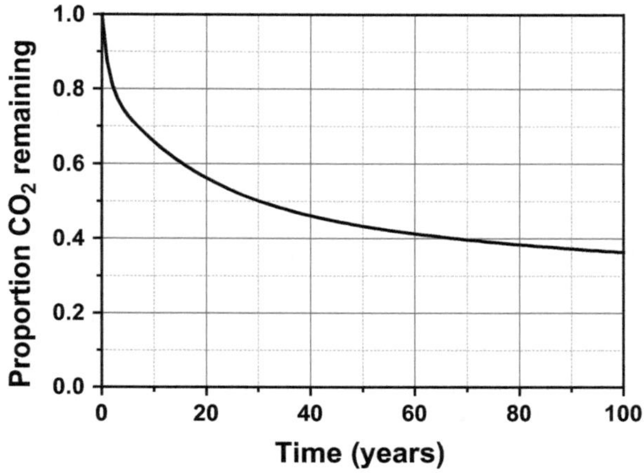

Fig. 10.1 The Bern Carbon Cycle showing the gradual loss of a pulse of fossil carbon dioxide from the atmosphere

below 550 ppm of CO_2 equivalents (CO_2e), although more recently this figure was revised downwards.

There are various natural processes by which atmospheric carbon dioxide is removed from the atmosphere. These are:

- Photosynthetic production of biomass (terrestrial and aquatic);
- Weathering of silicate rocks;
- Dissolution in the oceans.

If all human additions of carbon dioxide to the atmosphere ceased immediately, the atmospheric concentration would gradually return to the pre-industrial levels. This is simulated by the Bern Carbon Cycle Model (shown in Fig. 10.1) (IPCC 2007). About 50 % of the increase above the background level of 280 ppm would be removed in 30 years. This is assuming that anthropogenic interference in the climate does not lead to irreversible effects, such as melting of methane clathrates, or oxidation of peat. It is very important to note that this model describes the removal of carbon dioxide derived from fossil carbon sources. It is not connected with carbon dioxide where the carbon is derived from a biogenic origin. However, where the biogenic origin is from permanent land use change (e.g., destruction of forests) then this has the same effect of raising the atmospheric carbon dioxide levels and should therefore be treated as if it were from a fossil origin.

Forests have a very important role to play in helping to reduce the amount of carbon dioxide in the atmosphere. In order to do this, two strategies must be employed: (a) halting deforestation of virgin forests, (b) increasing the area of forests. Both of these require economic incentives to protect existing and encourage the planting of new forests. This necessitates putting an economic value on forests. This can be achieved by valuing the ecosystems services that forests provide and also providing a market for forest products that are produced sustainably from plantation forests.

A further point should be made with respect to the ability of forests to store atmospheric carbon dioxide. A mature forest is in an equilibrium state with the atmosphere, the rate of release of carbon dioxide from a mature forest is equal to the rate at which the carbon dioxide is sequestered. A mature forest is a carbon pool, but it is not a sink. In contrast, plantation forests are seldom left to reach full maturity (which can take over a 100 years for a coniferous plantation) but are usually harvested with a shorter rotation cycle. From the point of view of militating against climate change, it is important that a forest is managed so that carbon dioxide is constantly sequestered. This is achieved through sustainable harvesting practices; no more timber is extracted from the forest than is produced each year through the growth increment. This can be achieved through a variety of silvicultural management systems including compartmental felling with rotation, or continuous cover forestry. It is essential that the management practices do not lead to the release of carbon that is built up in the soil over the years. This is a subject of much research and is an important consideration. With sustainable management, the forest becomes a sink for carbon in perpetuity. What happens to the timber subsequently is important. If the harvested biomass is immediately burnt for its energy content, then this is only beneficial if substituting for fossil fuel use. A far better approach is to use the timber in long-life products, then at the end of life cascade the material down the value chain until it is eventually incinerated with energy recovery, finally returning the embedded carbon back to the atmosphere. One of the best ways of using timber in long-term products is in built environment applications.

This requires recognition of the value of such a practice. The value placed upon the storage of atmospheric carbon has to be represented in the market. Although the IPCC recognises the importance of the built environment, its mitigation strategies listed in the fourth assessment report (IPCC 2007) are almost exclusively concerned with energy consumption. Mention is made of the issue of trade-offs between embodied and operating energy when building design is considered. The use of wood as an example of a low embodied energy material is mentioned, but there is no consideration given to the potential for timber and other plant derived products to act as carbon stores in the built environment. The IPCC fourth assessment report states that although the forestry sector has much to contribute in terms of mitigation, there is a lack of political will to implement the necessary strategies. It is noted that increasing carbon stocks in wood products can contribute towards a mitigation strategy, but this idea was not taken up in the chapter dealing with the built environment.

In its fourth report, the IPCC considered various mitigation strategies to reduce the levels of carbon dioxide in the atmosphere. Noting that current mitigation strategies and reductions in energy intensity have been inadequate and that carbon dioxide levels are actually rising faster now than at the end of the last century. The ultimate aim of the United Nations Framework Convention on Climate Change (UNFCCC) is to stabilise the build up of greenhouse gases in the atmosphere before this leads to dangerous interference (i.e. a change in average global temperature greater than +2 °C) with the global climate system. It is predicted that, unless there is a substantial change in policies, the energy mix that will be used over the next 20 years or so will be largely unchanged from the present (i.e. about 80 % based upon fossil fuels). Consequently, the energy-related emissions of CO_2 by 2030 will be 40–110 % higher than at present. The combined effects of population growth, economic development, consumption behaviour and technological activities have thus far overwhelmed any reductions in anthropogenic greenhouse emissions that have been achieved through efficiency gains. It is generally agreed that the level of all greenhouse gases in the atmosphere should not exceed 550 parts per million (ppm) CO_2e (which means a CO_2 level no higher than 450 ppm (IPCC 2007) in order that the average global temperature increase does not exceed 2 °C, although more recent evidence indicates that these levels may actually need to be lower. It is recognised that countries that have not contributed in any significant way to historic emissions of greenhouse gases have a right to follow the development path enjoyed by the more prosperous nations of the planet and that this will almost certainly lead to increased emissions of greenhouse gases by those countries. This then requires even greater efforts on the part of developed countries to reduce their emissions of GHGs.

Land use mitigation strategies could contribute as much as 15–40 % towards cumulative abatement in the twenty-first century (IPCC 2007). Both agricultural and forestry mitigation options are considered to be cost effective abatement strategies. The question asked in this chapter is how do we develop effective standards in order to encourage the use of timber products in the built environment? We begin by considering the current framework for determining environmental impact.

10.3 Type III Environmental Declarations and Standards

The procedure for the development of programmes to produce Type III environmental declarations (EPD) is enshrined within ISO 14025 'Environmental labels and declarations – Type III environmental declarations – Principles and procedures'. The aim of such declarations is to allow for comparisons between the environmental performance of products that fulfil the same function. Such comparisons are based upon independently verified data using life cycle assessment methodology. This is part of a concerted move by national governments and other agencies to allow for informed decisions on the use of products and materials to be made which are based upon quantifiable data.

There is an increasing awareness of environmental issues amongst the business community and the general public and this is leading to a desire to make environmentally-responsible decisions regarding purchases of goods and services. The providers of such goods and services are well aware of this trend and environmental claims can form an important part of their marketing strategy. Regrettably, such claims are often not justified and there has accordingly been a need to develop methodologies that allow for informed choices to be made when it comes to purchasing decisions.

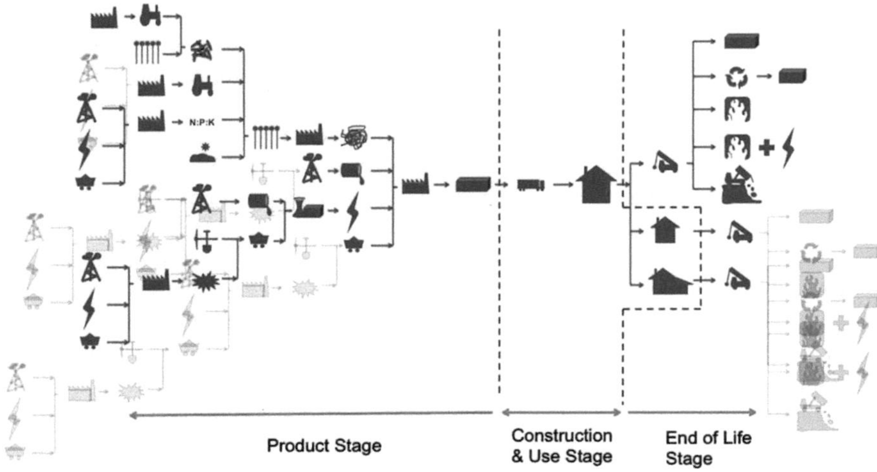

Fig. 10.2 An LCA as a complex analytical process. The quality of an LCA is highly dependent upon the quality of information that is used to construct it. For the product stage of the life cycle it is possible to obtain reasonably accurate data which can be audited, but as one moves further along the life cycle, it is increasingly necessary to make assumptions, which can have very significant impacts on the LCA. With increasing use of chain of custody procedures throughout the whole life cycle, it will become possible to obtain more accurate LCA data. The shading in the diagram represents the relative impacts of the different processes. Behind each process there lies a chain of other processes and decisions have to be made where to draw the system boundary and where to apply cut-offs (copyright Renuables)

Life cycle assessment (LCA) is a tool that has been developed in order to analyse and quantify the environmental burdens associated with the production, use and disposal of a material or product and is arguably the best way of quantifying this information (Hill 2011). The methodologies of LCA are based upon thermodynamic principles. The system that is being studied is defined and a system boundary is drawn around it. Mass and energy flows across the boundary are then quantified. The environmental impacts associated with these flows are determined. Although the methodology is, in principle, the best approach, the details of how such a series of calculations can accurately reflect the environmental burden often become exceedingly complex (Fig. 10.2).

Crucial factors affecting the outcome of an LCA are the choice of functional unit, system boundary, various assumptions made with respect to the product life cycle, data quality, and the source of generic data. This makes it exceptionally difficult, or very often impossible to compare the environmental performance of products that perform a similar function. For example, it might be decided that a functional unit is a window of certain dimensions for a building project. It is very unlikely that different manufacturers will have chosen the same functional unit, system boundary, life cycle scenarios, etc.; often making informed choices extremely problematical. Other problems arise when manufacturers wish to present their product in the best possible way in terms of environmental performance. There are many ways of doing this, through (for example) judicious choice of the system boundary, or making favourable assumptions regarding the product performance during the lifecycle and especially with respect to end of life scenarios.

In order to develop a framework that allows for comparability of environmental performance between products, ISO 14025 was introduced. This describes the procedures required in order to produce Type III environmental declarations (EPD). This is based on the principle of developing product category rules (PCR) which specify how the information from an LCA is to be used to produce the EPD. A PCR will typically specify what the functional unit is to be for the product. Within the framework of ISO 14025, only the production phase (cradle to gate) of the lifecycle has to be included in the EPD, forming what is known as an information module. It is also possible to include other lifecycle stages, such as the in-service stage and the end of life stage, but this is not compulsory. ISO 14025 also gives guidance on the process of managing an EPD programme. This requires programme operators to set up a scheme for the publication of a PCR under the guidance of general programme instructions.

There has to be transparency as to how the programme works and there must be a mechanism for the verification of a PCR as well as the means to allow for consultation with interested parties. The programme operator provides a repository for the store of the general programme instructions, the PCR and EPD, although an EPD is owned by the manufacturer(s) of the product. Clearly, there is the distinct possibility that different programme operators will produce a different PCR for the same product category and ISO 14025 encourages programme operators to harmonise their product category rules. Some countries have taken a lead in developing national EPD programmes, which although a positive move in terms of providing an incentive for the improvement of the environmental profile of goods and services, was viewed as being a potential barrier to trade within Europe. In March 2011, the Construction Products Regulation (305/2011) was introduced, replacing the Construction Products Directive (89/106/EEC). The Construction Products Regulation states that where a European standard exists then this has to be used. In addition it states that 'For the assessment of the sustainable use of resources and of the impact of construction works on environment Environmental Product Declarations should be used when available.'

There have been standards issued that apply to the construction sector in order to ensure greater comparability of the environmental performance of products. ISO 21930 gave some guidance on both PCR and EPD development, but this was recently replaced in Europe by EN 15804, which is a core PCR for building products and it is therefore considerably more detailed and prescriptive. Further guidance is given in EN 15942, which gives information regarding the format of an EPD for business to business (b2b) communication in the construction products sector. The primary purpose of an EPD according to ISO 14025 is for b2b communication, but an EPD can be used for business to consumer (b2c) communication. In the latter case, there are further requirements upon the process, which apply especially to the verification procedures. In any case, ISO 14025 encourages those involved in the production of an EPD to take account of the level of awareness of the target audience. Standards are increasingly removing the flexibility that was once available when determining the environmental performance of products and services.

The issue of carbon storage in products for use in the built environment is not dealt with directly in EN 15804, although it does not rule out the use of other methodologies for calculating the benefits of the storage of sequestered carbon dioxide. In Section 6.4.3.2 of EN 15804 it states that: ‘Materials flows carrying specific inherent properties, e.g. energy content, elementary composition (e.g. biogenic carbon content), shall always be allocated reflecting the physical flows, irrespective of the allocation chosen for the processes’. At the time of writing this chapter there is a pre-standard pr-EN16485 ‘Round and sawn timber – Environmental Product Declarations – Product category rules for wood and wood-based products for use in construction’ which is out for consultation, which does include sequestered carbon in timber products. It remains to be seen if this remains in the final version.

10.4 Principles of Life Cycle Assessment

Since LCA methodology underlies the production of an EPD, it is instructive to consider the process of producing an LCA in more detail. The first part of an LCA is to determine the goal and scope of the study; that is the reason that the LCA is being performed in the first place. Based upon this, the next step is usually the consideration of a functional unit. In some ways this can be a relatively simple and straightforward process, for example the functional unit is a door, but this does not necessarily allow for comparability between products. For this reason, it is important to be very specific about the functional unit, specifying dimensions and other data to ensure that the LCA is realistic, e.g. are the door fittings included or not? It is also necessary to specify a system boundary, in other words what is included in the analysis and what is not, but just as important which parts of the process for which data is collected, where generic data is used and where data is based upon estimations, or assumptions. Some of the assumptions made may have a significant impact on the LCA and in order to test this, a sensitivity analysis is performed. An important aspect of any LCA is transparency, something that can be very difficult to achieve, especially if the LCA is complex and/or uses highly sensitive production data that (for understandable reasons) the manufacturer may not wish to disclose. The issue of data sensitivity can be dealt with by the LCA review process which is an important aspect in proving the credibility of the assessment.

The methodology for performing an LCA is now enshrined in two standards (ISO 14040, 14044) and the procedures follow a series of well defined steps. The first of these is ‘Goal and Scope Definition’. The goal is the reason for carrying out the LCA, the intended audience and the application of the study. The scope definition is concerned with the determination of the functional unit to be studied, the delineation of the system boundary, allocation procedures, assumptions, limitations and all the other technical aspects concerned with deciding what is to be included in the LCA. Once the goal and scope are defined, the next stage is an 'Inventory Analysis' which is concerned with various aspects of data collection, analysis and validation as well as refining the system boundary, determination of mass flow inputs and outputs and allocations. This stage involves an iterative process with the 'Goal and Scope Definition' phase. Having gone through this bookkeeping process, the LCA then moves on to the 'Impact Assessment phase'. This involves the mandatory elements of the selection of the impact categories, classification and characterization. There may also be other stages where the various impacts are normalized, weighted and ranked. Some sort of data quality analysis will also be included. The last process is 'Interpretation', where the significant issues are identified and the data is evaluated for completeness and for its sensitivity to variables. This stage will often involve peer review. Finally, conclusions and recommendations are made.

10.4.1 Goal and Scope Definition

The results obtained from an LCA study are wholly dependent upon what the goal and scope of the study is. It is extremely important that this is defined so that the LCA is not quoted out of context. There are many reasons for performing an LCA; it may be desired to determine where the greatest environmental impacts (hotspots) of a process are, a company may wish to provide data to customers, or it may be used as a marketing tool. It is at this stage that the system boundary of the study is decided. This can be quite a difficult thing to do in practice, especially for complex products or services and it is quite usual to define what is termed a foreground system and a background system. The foreground system is where the LCA practitioner will gather real data from the factory, or building site or anything else needed for the processes that are closest to the subject of the study, whereas the background system might represent grid electricity, or road transport, or some other generic input that is taken from existing data sets. In most cases an LCA will represent a 'cradle-to-grave' study, from inception to disposal, but in other situations only a small part of the life cycle may be studied and it is obviously very important to state this. It is also very important to state what has been included and what has been left out of an LCA, especially if the merits of differing products are being compared. An example might be ignoring the environmental impact of the disposal of a favoured product, or an unrealistic assumption for a product lifetime to give a better LCA result. If the LCA is to have any virtue then the assumptions made must be stated, it must be transparent or it has no value.

One of the most important components of an LCA is the functional unit. This represents a quantitative measure of a product or service that is the subject of the study. An example of a functional unit may be 20 m^2 of a wall for a suburban domestic house and the object of the LCA study may be to compare the relative merits of timber, brick, or concrete as building materials. The functional unit is a fair comparison, since this is what will be used in service. If this was not the determinant, then unscrupulous vested interests might choose to use volume or mass or any other measure that gave their product a competitive edge.

10.4.2 Inventory Analysis

The life cycle inventory (LCI) analysis involves the collation of data about the inputs and outputs associated with the various substances and energy flows into and out of the system of interest. By measuring the flows of matter into and out of the system, it is possible to do a mass balance calculation to show if anything is missing. Matter flowing into the system should equal matter flowing out.

The whole system is broken down into a series of subsystems and the inputs and outputs associated with each of these are determined. Problems can arise at this stage because the data used may be confidential and hence unverifiable, or it may be open source data that is a poor representation of the real process. It is important that data sources are revealed in the final report. This process becomes much more complicated if the subsystems of interest fulfill more than one function. This is known as the allocation problem. Examples of multifunctional systems are:

- The material of interest is but one product from a complex manufacturing process. An example of this is a product from a petrochemical plant where all of the processes are interlinked;
- The waste generated by the process is one of many inputs into a waste management system which produces emissions and maybe energy as a by-product;
- The material may be one part of a cascading recycling system which requires material and energy inputs and has emissions and wastes at each stage of the downcycle, but the material of interest may be obtained from one part of the process only.

How can allocations be made in the case of such multifunctional processes?

- The best strategy is to avoid allocations wherever possible. There are two ways to do this, either by breaking the subsystems down further so that the associated environmental burden becomes apparent, or by expanding the systems under study so that the same functions are incorporated by all of the systems;
- If allocations cannot be avoided, then it is best to make an allocation on the basis of how a quantitative change affects the environmental burden in a meaningful way. This might mean a mass allocation an economic allocation or some other appropriate physical parameter. Economic allocations tend to be favoured because the reason why the material or product is required is driven by economic factors. However, with economic allocations problems do arise in accounting the impacts or benefits physically associated with the product, such as beneficial carbon storage in a wood product not in its associated offcuts.

10.4.3 Impact Assessment

The Life Cycle Impact Assessment (LCIA) phase involves multiplying the various environmental burdens (outputs) by factors representing the environmental damage or extent of resource depletion that these different substances exhibit. The way that this is done depends upon the goal and scope of the LCA, and reflects the environmental issues that the LCA is designed to address. The process also involves the clustering of the data into a relatively small number of environmental impact factors. Other components of this part of the analysis may include sorting the impacts into a hierarchy of importance, or the weighting of impacts; which may involve some kind of qualitative assessment. The problem with these sorts of activities is that the aggregation of complex phenomena into a few impact factors, although desirable from the point of view of making the LCA understandable, may end up producing misleading (or even meaningless) data. There have been hundreds of different impact categories presented in LCAs, with some being more commonplace [e.g. Leiden University Institute of Environmental Sciences (CML) baselines, or the International Reference Life Cycle Data System (ILCD)]. Notwithstanding the problems that may arise from aggregating and simplifying disparate data into relatively few impact factors, it is usual to report environmental impact in terms of the following factors:

- Abiotic resource depletion: which includes depletion of fossil fuels, minerals and metals and is reported either in terms of kg of antimony equivalent or kg of oil equivalent;
- Land use impact: this is the area of land used multiplied by the occupation time and is expressed as $m^2\ yr^{-1}$;
- Global warming potential: which is the warming potential of the various emitted gases expressed as kg CO_2 equivalents (CO_2e) over a period of time (20, 50, 100, 500 years) and indicated as GWP100, for example. This time frame is very important because different gases have various residence times in the atmosphere. For example methane, which has a greater global warming potential than CO_2, has an average residence time of 12 years before it is oxidized to CO_2 and water, with the result that it has a GWP20 of 72× CO_2e, but a GWP100 of 25× CO_2e;
- Stratospheric ozone depletion potential (ODP): this is expressed in terms of kg of CFC-11 equivalents;
- Human toxicity potential (HTP): this is expressed in terms of the toxicity with respect to the reference substance 1,4-dichlorobenzene (kg 1,4-DB eq.);
- Eco-toxicity potential (ETP): is also given as kg 1,4-DB eq., and is a factor that includes the toxicity potential for organisms in the air, soil and in water;
- Acidification potential (AP): is the contribution of gases such as SO_2 and NOx to acid deposition and is expressed in SO_2 equivalents;
- Photochemical oxidant creation potential (POCP): is related to the emissions which are capable of causing photochemical smog and is usually expressed as kg ethylene eq.;

- Eutrophication potential (EP): is related to the ability of nutrients such as nitrate to cause over-fertilization of soil or water and is expressed in terms of phosphate equivalents.

10.4.4 Interpretation

This phase of the LCA involves analyses of the results obtained from the LCI and LCIA, gives conclusions and explains what the limitations of the study are. Prior to this, a sensitivity analysis is carried out in order to examine the influence of data quality, variability and availability upon the LCA to indicate the level of reliability of the study. Continual feedback is strongly advocated in ISO 14040 and the sensitivity analysis is one of the major stages where the LCA practitioner may question the original data collection, or the appropriateness of secondary data used to ensure the most accurate study. If the study is to be used externally it is also necessary to subject the analysis to critical review by a suitably qualified third party as a check on the veracity of the work. Finally, the findings and conclusions are reported with reference to the intended use of the study. The LCA in total should be a complete, transparent and unbiased account of the study. LCAs not meeting these essential criteria should be treated with extreme caution at the very best.

10.5 Principles of Atmospheric Carbon Storage

Trees capture atmospheric carbon dioxide via photosynthesis and a proportion of this sequestered carbon is stored in the above-ground woody biomass. Wood is composed of three main biopolymers (cellulose, hemicellulose and lignin) and to a first approximation it can be assigned a stoichiometric ratio of CH_2O. This means that atmospheric carbon comprises a minimum of 40 % of the dry wood mass (increasing somewhat with increasing lignin content). Each tonne of dry wood therefore equates to the removal of approximately 1.5 tonnes of atmospheric carbon dioxide. The advantage of this ability to store atmospheric carbon depends upon the length of time before the material is oxidised and the carbon released back to the atmosphere. In all situations where carbon flows are considered it is essential that a distinction is made between biogenic and fossil carbon sources. However, even with biogenic carbon it is also important to differentiate between carbon that is held in long-term storage (such as old-growth forest) and that derived from newer managed, or plantation forests.

Consider the different scenarios illustrated in Fig. 10.3. In Scenario (a), old growth forest is burnt and the land cleared for alternative use. The result in terms of GWP100 is a release of atmospheric carbon dioxide. This is the carbon content was previously held in long term (historical) storage. Thus although technically this is biogenic carbon, it represents carbon that would have been in storage prior to the

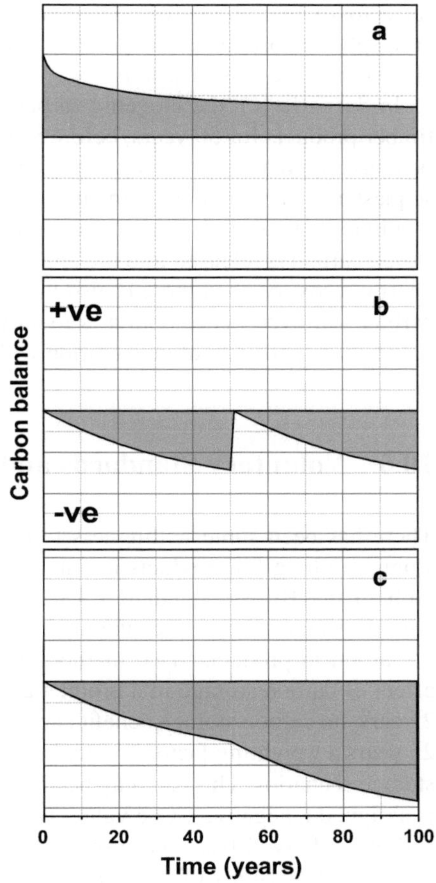

Fig. 10.3 Effect on carbon balance of burning old growth forest (**a**) burning plantation forest with a 50 year rotation (**b**) and using timber in long life products (**c**)

industrial revolution was part of the natural biogenic cycle and can be viewed as being the same as fossil carbon.

In Scenario (b), a new forest plantation is established and the trees are allowed to grow for 50 years before harvesting and restocking. Carbon is removed from the atmosphere as the atmospheric carbon dioxide is photosynthetically bound in the biomass. The overall result in terms of GWP100 is a benefit because atmospheric carbon dioxide has been sequestered. If the forest biomass is burnt with energy recovery after 50 years, then the above-ground biomass is oxidised and the accumulated atmospheric carbon is lost. The overall result is nonetheless, still a benefit in terms of carbon sequestration. This is because there has been removal of atmospheric carbon dioxide during the 100 year period of consideration and when the above ground biomass is burnt; this results in the return of atmospheric carbon dioxide. This only applies because new forest was created. However, the burning of woody biomass cannot seriously be considered an effective mitigation strategy. Far better, is one in which the calorific value of the biomass is utilised and substituted for a fossil fuel alternative. The benefit then arises not only from the storage of atmospheric carbon in the growing biomass, but additionally from the avoided emission of the fossil carbon.

In Scenario (c), the biogenic carbon stored in the plantation forest is stored in timber products for 50 years, before it is used to generate energy. In this way three benefits are realised. During the growth phase of the forest carbon dioxide is sequestered due to the incremental growth of the trees. After harvest, the carbon continues to be stored in the timber products. It is only at the end of the life that this stored carbon is released into the atmosphere. Once again, if the wood is burnt with energy recovery, then there is also the benefit of the avoided emission of the fossil carbon. An even better option is to cascade the wood material down the product value chain before final incineration with energy recovery.

10.6 Voluntary Standards and Carbon Sequestration

There has been some willingness to deal with the evaluation of biogenic carbon storage in long-life products in national standards. In the UK this issue was dealt with in the British Standard PAS 2050. This considers a 100 year assessment period. Annexe C describes the methodology to be used for calculating the storage of carbon in products. Two methods for calculating the weighted average of the effect of carbon storage in a product are given. For a product with a life less than 2 years, no carbon storage benefit can be assigned, but for products with a life of 2–25 years a weighting factor is calculated, with a different weighting factor for other storage scenarios. This can only be applied to the storage of biogenic carbon, which is assigned a negative CO_2e value. However this cannot be applied if the biogenic carbon is derived from old growth, or native forests, where land use change has occurred. Emissions of biogenic carbon are not considered, since the origin of biogenic carbon is atmospheric carbon dioxide. Weighting factors are also applied for delayed release of GHGs.

Methodologies for accounting for the carbon stored in products are given in the International Reference Life Cycle Data (ILCD) Handbook, published by the European Commission Joint Research Centre (Institute for Environment and Sustainability). This also considers a 100 year assessment period. For carbon storage in products, the relevant sections are 7.4.3.6.4 and 7.4.3.7.3. It is recommended that fossil and biogenic carbon releases (as CO_2 and CH_4) should be differentiated. Furthermore, all carbon emissions associated with land use changes and from biomass associated with virgin forests should be treated as fossil carbon. Emissions associated with plantation forests are to be inventoried as biogenic carbon. Uptake of atmospheric carbon dioxide is inventoried as 'resources from air'. A methodology is given for accounting for the removal and storage of atmospheric carbon dioxide. One of the issues discussed is that of carbon storage for a long period of time (e.g. 80 years) and how this then relates to the commonly used GWP100 parameter. GWP100 is a value given to the result of the emission of a pulse of a global warming gas in terms of its effect upon the environment for 100 years. Thus if there is an emission of fossil-derived carbon dioxide into the atmosphere, its radiative forcing effect over a period of 100 years will gradually decrease as it is taken up by various natural sinks (the Bern cycle referred to earlier). For this reason the parameter GWP100 is used (the global warming potential over a 100 year period).

However, in the case of carbon storage in a long life material for 80 years, it would be incorrect to show the emission at end of life in terms of a GWP100 value, since the total accounting time being considered is now 180 years. The ILCD methodology deals with this in the following way. The uptake of atmospheric carbon dioxide is inventoried as 'Carbon Dioxide – Resources from Air' and the emissions as 'Carbon Dioxide (biogenic) – Emissions to Air'. These two flows then cancel each other out. Meanwhile, the issue of the storage in the product is calculated by declaring a correction flow for delayed emission of the carbon dioxide and giving it a value of $0.01 \times$ the CO_2 equivalent mass stored per year. The same method is used to calculate the storage of fossil carbon in a long life product, except that there is no consideration given to the category 'Carbon Dioxide – Resources from Air'. Thus, there is a net effect of the release of the fossil derived CO_2 at the end of life, but the compensatory effect of the delayed emission of the fossil carbon is taken account of.

10.7 Forestry and Chain of Custody Certification

With proper management practices and with the use of forest products in long term applications the forestry sector can make a significant contribution in sequestering atmospheric carbon dioxide. An essential component of such a strategy is the use of robust certification and chain of custody schemes. The forestry sector has long been regarded as being a contributor to climate change because of considerable adverse publicity regarding deforestation. At the present time, it is estimated that deforestation contributes more than 17 % of total anthropogenic GHG emissions (100 year GWP in CO_2e). In order to address negative public perceptions regarding the use of timber products a range of certification schemes have been introduced. Up until the present time, these have been voluntary in nature and have been managed by various organisations. The most commonly encountered are the Programme for the Endorsement of Forest Certification (PEFC) scheme and the Forest Stewardship Council (FSC) scheme. PEFC is an umbrella international organisation that endorses national or sub-national schemes. In the United States of America and Canada the Sustainable Forestry Initiative (SFI) programme is used and within Canada there is also the Canadian Standards Association (CSA) Sustainable Forestry Management Standard, both of which are recognised by the PEFC. PEFC adopts three approaches for chain of custody certification:

- Clients can physically separate and segregate certified wood from different sources during all stages of the value chain, or
- A complete batch of products can be certified if the amount of certified material exceeds a defined threshold, or
- A specific amount of the batch can be labelled as certified which equals the percentage of material obtained from a certified source.

FSC is an international organisation consisting of over 800 representatives worldwide. The organisation is diverse and has representatives from the industry, indigenous groups, community forestry groups and forest certification organisations. FSC forests meet approved standards for forest management, which is backed up by chain of custody certification.

In July 2002, the European Parliament produced the Sixth Community Action Programme, which indentified 'as a priority action the examination of the possibility of taking active measures to prevent and combat trade in illegally harvested wood. . .'. Following on from that, in May 2003, there was a report produced by the European Commission entitled the 'Forest Law Enforcement, Governance and Trade (FLEGT): Proposal for an EU Action Plan'. As a result, the European Union attempted to negotiate Voluntary Partnership Agreements (FLEGT VPAs) with timber producing countries in order to introduce a licensing scheme and regulate trade. To date, six countries have finalised VPA negotiations with the EU, with four more in negotiation. None of these has so far (2012) fully implemented the licensing system. Subsequently, the EU Timber Regulation came into force in December 2010, making it illegal to place illegally harvested timber and timber products in the EU market as of 3rd March 2013. This requires that due diligence is applied at all stages of the supply chain. This will presumably provide the imperative that has, so far, been lacking in implementing the FLEGT VPAs.

10.8 Carbon Valuation and Forestry

Another aspect of ensuring that forestry services are embedded within our economic system is the issue of carbon valuation. At the present time, such schemes are linked to emissions of carbon dioxide and there is currently no scheme operating that is able to place a valuation upon the long-term storage of biogenic carbon. The Kyoto Protocol bound the signatory Annexe 1 countries to participate in a cap and trade scheme for the six major greenhouse gas (GHG) groups (carbon dioxide, methane, sulphur hexafluoride, nitrous oxide, chlorofluorocarbons, hydrochlorofluorocarbons). Nations have been given emissions quotas within this scheme and those that emit less than their assigned quota are permitted to sell these credits to nations that exceed their quota. It is also possible for the developed countries within the scheme to sponsor projects in developing countries that result in GHG emissions and to earn tradable credits in this way. This is achieved through the use of joint implementation (JI) projects within the clean development mechanism (CDM). There have been a number of GHG trading schemes implemented throughout the world.

The European Union Emissions Trading Scheme (EU ETS) is currently the largest international GHG trading scheme in the world. It was created in conjunction with the Kyoto Protocol and started operating in January 2005. Caps were applied to the amount of carbon dioxide emitted by industrial plants with a power supply in excess of 20 MW thermal. Covering nearly half of the carbon dioxide emissions of the EU, the first phase of this scheme, running from 2005 to 2007, attracted much criticism. This was because an oversupply of credits led to a collapse in the market price. In phase two of the scheme, the European Commission claims to have been much tougher on emissions permits, but the introduction of a carbon offsetting programme allows for the possibility that the reductions in the cap could be met by offsets alone.

Despite the non-participation of the US in the Kyoto Protocol, several US states began to develop carbon trading schemes of their own. The State of New York developed the Regional Greenhouse Gas Initiative (RGGI) in conjunction with nine north eastern US states, with the aim of reducing the carbon emissions of the electricity generating sector by 10 % below 2009 levels by 2018. This programme was launched on 1st January 2009. The Chicago stock exchange rolled out a CO_2 emissions trading scheme in 2003 and in 2007 and it created a mechanism for emission offsets by involving projects that cleanly destroy CFCs. In 2007, four Canadian provinces and seven US states joined together to form the Western Climate Initiative which is a regional carbon trading system.

The New Zealand Emissions Trading Scheme is being gradually phased in, with a transition period operating from July 2010 until December 2012. During this period, a New Zealand Emissions Unit will need to be surrendered for every two tonnes of carbon dioxide equivalent emissions, at a fixed cost of NZ$12.5 per tonne CO_2e.

We are still very far from the situation where there is a global GHG trading market and the patchy coverage that we have at the moment allows for the possibility of carbon leakage. This can take the form of direct leakage, where production and hence the environmental burden is shifted out of a country to one that is not compelled to reduce or fix its carbon emissions. Indirect leakage occurs because the externality costs are being applied to emissions rather that the fossil fuels directly, allowing non-participating countries to purchase these cheaper feed stocks on the open market. One way to even-out the market is for import tariffs to be set which reflect the extra costs borne by industry in participating countries. However, this type of mechanism is likely to fall foul of the General Agreement on Tariffs and Trade.

There has been an introduction of schemes such as the Verified Carbon Standard, which claims to be one of the major agencies used by agencies issuing credits in the voluntary carbon market. The website of the Environmental Finance Directory lists 36 such schemes operating worldwide at present. Such schemes are able to provide verification programmes for carbon offsetting through forestry projects. However, the extension of the carbon credit value chain to the use of renewable materials in the built environment has not yet been included.

10.9 Conclusions and Recommendations

In terms of the use of materials in the built environment and evaluating their environmental impact, we are still in a situation where voluntary standards are being used, where they exist. There has been action to make these standards more rigorous and prescriptive with the introduction of EPDs and within Europe with the introduction of a core PCR for materials used in the built environment and the introduction of the 'Construction Products Regulation'. Although the production of EPDs is presently voluntary, there will rapidly be a necessity to produce EPDs in order to meet the requirements of procurement. The certification of timber products has, until now, been voluntary, but take up has been patchy. Europe has encouraged the wider use of certification on a voluntary basis, but with relatively little success. There is a tendency to use voluntary schemes only when they are perceived to be of benefit, either for marketing purposes, or when there is some coercion, through the enactment of appropriate legislation. If we are to create carbon markets that are able to assign a monetary value to sequestered carbon stored in the built environment, it will become necessary to move towards a system where it is a legal requirement to have proper certification of the carbon footprint of products. The formalisation of procedures related to the chain of custody of forest products provides an opportunity for simultaneously incorporating LCA data. This represents an opportunity for the forest products sector that should be addressed. One of the problems with this sector is the diversity of sources, heterogeneity of material and huge range of products that are produced. This is a much more complex situation than that faced by the concrete, steel and polymer sectors. It is essential that the forest products industry uses adopts chain of custody systems that are integrated with LCA tools. The ability to track products through the value chain when they are used in buildings will be possible with the increasing adoption of BIM. It will be necessary to extend chain of custody through first life and on to subsequent lives, as the material is cascaded down the value chain and at end of life when the sequestered carbon is finally returned to the atmosphere. This will allow for a really effective and accurate tool for informing LCA, policy makers and the public. The forest products industry has considerable experience in chain of custody certification, this expertise should be harnessed in the future to use chain of custody procedures to 'pull through' environmental information.

References

Council Directive 89/106/EEC of 21 December 1988 on the approximation of laws, regulations and administrative provisions of the Member States relating to construction products. Official Journal of the European Communities No L 40 of 11.2.1989, pp 12–26

EN 15804:2012 Sustainability of construction works – environmental product declarations – core rules for the product category of construction products

EN 15942 Sustainability of construction works – environmental product declarations – communication format business-to-business

Environmental Finance Directory (2012) Voluntary carbon. http://www.efdirectory.com. Last accessed 20 Mar 2013

Hill C (2011) An introduction to sustainable resource use. Earthscan, London

IPCC – Intergovernmental Panel on Climate Change (2007) IPCC fourth assessment report: climate change 2007. Cambridge University Press, Cambridge

ISO 14025:2006 Environmental labels and declarations – type III environmental declarations – principles and procedures

ISO 14040:2006 Environmental management – life cycle assessment – principles and framework

ISO 14044:2006 Environmental management – life cycle assessment – requirements and guidelines

ISO 21930:2007 Sustainability in building construction – environmental declaration of building products

Regulation (EU) No 305/2011 of the European Parliament and of the Council of 9 March 2011 laying down harmonised conditions for the marketing of construction products and repealing Council Directive 89/106/EEC. Official Journal of the European Union No L 88 of 4.4.2011, pp 5–43

Stern N (2006) Stern review: the economics of climate change. http://www.hm-treasury.gov.uk/sternreview_index.htm. Last accessed 20 Mar 2013

Chapter 11
Exploring Market Strategies Based on Voluntary Environmental Certification in a Post-Soviet Transition Economy

Ludmila Palekhova and Gennadiy Pivnyak

11.1 Introduction

Unlike the relatively static form of regulation that is associated with government controls, voluntary certification, including environmental certification, is based on market mechanisms promoting the development of free and fair competition. Voluntary environmental certification and the use of eco-labels demonstrate the degree of a participating producers' responsibility regarding the environmental safety of their products and production processes. This invites an improvement in consumer confidence and potentially leads to increased domestic and foreign sales.

It should be noted that voluntary application of international standards, such as ISOs and sector specific voluntary sustainability standards, is a new phenomenon for post-Soviet states which have endured the complex realities linked to a transition economy. Against the background of total economic recession, these states maintained their long held principles of 'command-and-control' economic administration with strict control upon all forms of production activities.

Russia, Ukraine, Belarus, Kazakhstan, and other newly independent states adopted national certification systems for products and production processes; however, they all inherited the main disadvantages of the Soviet system of technical regulation—compulsory nature, complexity and length of certification procedures, unwieldiness, and chaotic character of outdated standards (Nikiforov and Bakiyev 2005).

L. Palekhova (✉)
Department of Marketing, Institute of Economics, National Mining University (NMU), Karl Marx Av., 19, 49005 Dnepropetrovsk, Ukraine
e-mail: pall@hotmail.ru

G. Pivnyak
Rector of the National Mining University (NMU), Karl Marx Av., 19, 49005 Dnepropetrovsk, Ukraine
e-mail: rector@nmu.org.ua

C. Schmitz-Hoffmann et al. (eds.), *Voluntary Standard Systems*, Natural Resource Management in Transition 1, DOI 10.1007/978-3-642-35716-9_11,

In recent years, under the pressure to adapt to globalisation of trade and financial relations, the terms of doing business within the post-Soviet space have undergone profound changes. An important factor here is that large industrial states, such as Ukraine and Russia, joined the World Trade Organization (WTO). Additionally, in 2011, eight newly independent states signed the Free Trade Area Treaty in which trade relationships within the Commonwealth of Independent States (CIS) is regulated based on WTO standards. Along with this, post-Soviet countries are also implementing their own vector of integration development. For example, Ukraine strengthens its European orientation with a view to realisation of the EU Association Agenda, while developing bilateral cooperation and forming free trade areas with international partners throughout the globe.

To comply with international agreements and to promote both global and domestic market relations, some post-Soviet countries, e.g. Ukraine, implemented reforms to reduce the abundance of technical regulations in trade through harmonisation of certification procedures with those procedures which were established in the EU, in accordance with provisions of EU Directives. The reforms also should help to implement the recent field practices of voluntary certification based on the voluntary sustainability standard systems. As a result, in the majority of post-Soviet countries the list of products subject to compulsory certification has been significantly reduced: in Russia by 25 %, in Belarus by 40 %, and in Ukraine by 65 % (UkrSREC 2013).

However, it should be noted that the majority of manufacturing enterprises cannot easily adapt to new business conditions and therefore suffer severely from the 'crisis of liberalisation'. Formerly, the state was charged with performing a broad spectrum of strategic tasks including those related to the environment, i.e. setting compulsory rules and requirements for every kind of production and product. Today business executives facing free choice have to make independent decisions regarding transition to meeting high environmental standards. In particular, the problem they face is the selection of market strategy based on voluntary environmental certification and associated eco-labelling.

Therefore, the main aim of this chapter is to explore the potential of market strategies for the development of companies that are based on voluntary environmental certification, under the conditions of transition economy. To achieve this aim, the following research objectives were formulated:

- to study the key drivers for using the voluntary environmental certification by manufacturing enterprises in the post-Soviet countries;
- to identify the main benefits of voluntary systems for environmental certification and their contribution towards an improved positioning on the market;
- to suggest generic marketing strategies based on the voluntary environmental certification for producers in countries with economies in transition.

11.2 Analysis of Key Drivers for Using Voluntary Environmental Certification in Countries with Transition Economies

Economic theory and standard global practice of doing business have already recognised that voluntary environmental certification is a valuable and flexible tool in facilitating company development and stimulating their respective market activities (Thompson et al. 2009). Experience gathered from numerous studies have shown that the key motivations for producers applying different systems of voluntary environmental certification are: achieving market targets, including the development of competitive advantage; image improvement with particular focus on 'green' credentials in target markets; demonstration of high eco-efficiency of processing and packaging, etc. Considering the environmental component as a tool for development rather than unavoidable costs, producers have an opportunity to access the profitable international markets of ecologically valuable products (González-Benito and González-Benito 2005).

At the same time, companies endeavour to ensure long-term economic benefits from the environmental certification programmes owing to more efficient production systems. In this aspect, economically quantifiable benefits include increasing the capacity, energy and resources saving, reduction of waste treatment costs, raw materials savings, insurance cost reductions, etc. (Matuszak-Flejszman 2009). As a result, a producer can develop a more efficient business model than that of their competitors, and subsequently capture and maintain higher added value than competitors are capable of.

In the post-Soviet space, the voluntary environmental certification is based mainly on the ISO 14000 family of international standards—this applies to Russia, Ukraine, Kazakhstan, and other newly independent states. However, this instrument has not become a significant trend in any of them (Cherp and Vynychenko 2012). Particularly in Ukraine, as of 01.01.2012, only 92 companies were certified for compliance with ISO 14001; this certification level is at the 2006 level (UkrSREC 2013). According to information provided by the Ukrainian non-governmental organisation Living Planet, up till now, only 60 Ukrainian manufacturers have eco-certified their products, which includes approximately 230 brands.

For the purpose of this research we have studied 11 Ukrainian companies to identify factors that determine the reasons for using voluntary environmental certification under the conditions of a transition economy. These industries included: coal production, tires, aggregate, batteries, brick, cement, compressors, pipes, a metallurgical plant, juice factory, and a vineyard. Two of them conducted eco-certification of their products (piston compressors and clinker bricks); two companies have ISO 14001 certification: the cement plant of the HeidelbergCement Group (Dneprodzerzhinsk), and the battery plant 'Vesta-Dnepr' (Dnepropetrovsk). The juice factory of the Sandora group (Nikolaev region) is certified for compliance with ISO 14001 and Forest Stewardship Council (FSC) standards.

Anonymous questionnaires and in-depth interviews with top-managers were the methods of information gathering employed for this study. The questionnaire used in surveying consisted of ten open-ended questions to clarify motives for application of voluntary environmental certification of products and management systems. Interviews were conducted with the help of the laddering method, which refers to an in-depth, one-on-one interviewing technique used to develop an understanding of how consumers translate the attributes of products into meaningful associations with respect to self, following Means-End Theory (Gutman 1982). We used a sequence of partially standardised questions to clarify hidden causes which slow-down the development of voluntary environmental certification. The questions were formulated in such a way to avoid stereotyped statements, and not to make respondents answer in one particular way or another.

As a result of the survey, we determined the following motives in favour of voluntary environmental certification (listed in-order of increasing endorsement by responding subjects):

- *The best environmental characteristics of products and management entail a more favourable condition for business (two respondents).* It is easier for environmentally friendly companies to obtain licensing for new construction activities; they have more favourable terms of insurance, crediting, etc.
- *Environmental certification of products and management systems opens possibilities for access to mechanisms of government support (three respondents).* Companies having environmental certification may participate in government programmes of 'green' procurement; they can have tax benefits and state support in the implementation of environmental innovation projects, etc.
- *Good environmental performance of products and processes reduce the environmental costs (three respondents).* Some environmental costs are covered by the profits of the company, i.e. payments for excessive emissions, environmental penalties, sanctions, etc.; environmental orientation of management can significantly reduce these expenditures, and therefore increase net profits.
- *Environmental certification is considered as an important stage in development of integrated management systems (six respondents).* The consistent implementation of management standards, e.g. ISO 9001, ISO 14001 and OHSAS 18001, reduces system conflicts and increases the efficiency of the entire management system.
- *Environmental management systems and ecological certification of products are used for positioning in the target markets (eight respondents).* As for certain market segments, company must take a position that distinguishes it amongst its competitors; demonstration of environmental policy may be such a position.
- *Environmental certification is an effective marketing tool to remove barriers related to entering specific markets (ten respondents).* Environmental certification demonstrates to potential customers that the manufacturer focuses on aspects of environmental responsibility and compliance with higher environmental standards.

- *Environmental certification is the condition for products distribution primarily in the foreign markets (eleven respondents).* Exporters have to reckon with international environmental standards and higher demand for environmentally certified products in the foreign markets.

As revealed by the survey results, the majority of managers noted that voluntary environmental certification has potential for opening up new ways of business development in the domestic market; that it is essential for establishing partnership relations with suppliers, consumers, and other stakeholders.

In addition, each manager indicated the likely necessity of environmental certification for entering the global market or for fulfilling obligations to investors. This can be confirmed by the analysed case studies: "Vesta-Dnepr" battery plant exports 75 % of its products and has a share of 1.4 % in the global market for accumulator batteries; Dneprodzerzhinsk cement plant was certified after it had been taken over by its investor HeidelbergCement Group. A company such as "Kerameia" is the only environmentally certified enterprise in the domestic market of clinker bricks. This company was established with the support of the American investment venture fund "Horizon Capital".

The results of in-depth interviews made it possible to clarify obstacles to the broad implementation of voluntary environmental certification as a factor of success in attaining of the competitive advantage. They are as follows:

- *Lack of wide-scale integration of sustainable development principles into the existing systems of management and marketing.* Implementation of international environmental standards should be supported by the general concept of integrating environmental concerns into the management functions of a company and strategies for its business development.
- *Poor understanding of tendencies for increased environmental awareness and their effect on market relations.* Voluntary environmental certification is the tool for harmonisation of policy regarding market development of a company with consideration for growing environmental awareness, also influencing market relationships.
- *Foreign customers and investors do not trust certificates issued by national certification bodies.* Accreditation of the Ukrainian eco-labelling programme through the international eco-labelling system GENICES gave the right to declare environment-related claims for Ukrainian products in accordance with ISO 14024. In reality though, 'western' consumers are not always aware, nor trusting, of Ukrainian eco-certification and labelling.
- *Voluntary environmental certification requires significant financial investment.* As a rule, companies belonging to 'dirty' sectors of the economy are faced with deep crisis situations; for them the shift of production towards environmentally-friendly products and implementation of environmental management systems is a difficult challenge.
- *Environmental management systems based on general standards of the ISO 14000 family are too uniform and are not always sufficient for companies in certain branches.* In newly independent states the implementation of voluntary

environmental certification is done almost exclusively based on standards of the ISO 14000 family. Whilst this is excellent in certifying a company and their environmental management system, it does not provide significant information regarding the eco-friendliness of products. Customers may be turned-off to the management system situation, when they actually demand product eco-quality and the respective eco-certification of products. Furthermore, product eco-quality certification should follow after the implementation and certification of an environment management system.

Therefore, as confirmed by the results of interviews, extending practices of voluntary environmental certification it is necessary to develop the whole management system of companies in accordance with the principles of sustainable development. These take into account the increasing environmental awareness of customers (including lucrative government authority contracts, i.e. green procurement schemes) and its influence on the market relations, potentially filtering through to influence the business at different levels. Also, there are no doubts concerning the benefits of voluntary certification of environmental management in companies, but for some businesses certification based only on the ISO 14000 standards family may be not sufficient. In particular, the manager of the coal company "Pavlogradugol" made a clear statement that formal standards ISO 14001 are set for a generic enterprise without taking into account the specifics of the mining industry. Furthermore, certification of a company itself (i.e. its management schemes) will not provide sufficient evidence about the environmental quality of its products (as outlined in previous bullet-point).

It is tempting to management to consider selecting from the various international industry-specific voluntary sustainability standards to certify their products (e.g. FSC for wood and paper products), in addition to the general standards like ISO 14001, ISO 14024, etc. But in any case, managers of enterprises should only select from such available standards and systems of environmental certification those which would help to form the most advantageous strategy for market behaviour.

11.3 Market Strategies Based on Voluntary Environmental Certification

As argued by Porter (1998), any business needs an effective market strategy for securing an advantageous market positioning and ensuring a stable competitive advantage. In this context, the results of questionnaires and interviews confirmed that the decision regarding voluntary environmental certification is developed as a response to the strategic targets of an enterprise, and may be considered as a tool to reach them. Moreover, some enterprises may be competitive and attractive at national and international levels only if their market strategies take into consideration the international environmental standards developed for sustainable development of particular sectors of the economy.

For example, for different businesses, strategies based upon compliance with one of the following standards may be essential for their effective market positioning: Forest Stewardship Council (FSC) which ensures sustainability of forests and wood-processing industries; Marine Stewardship Council (MSC) forming rational fishing quotas and traceability within supply chains of seafood; Sustainable Agriculture Network (SAN) standard applied to farms and farmers' cooperatives; or Responsible Mining Assurance (IRMA) standard for mining industry, etc.

Using such voluntary sustainability standards, companies may supplement their market positions, primarily through the completely new principles of doing business: increased consideration for the sustainability of certain raw materials at their point of origin, production processes and traceability of supply chains, as well as demonstration of corporate environmental responsibility. It is possible to distinguish two types of competitive strategies based on the extensive use of voluntary environmental certification in order to gain a competitive advantage: differentiation and focus strategies.

As is widely accepted in strategic management theory (e.g. Thompson et al. 2013; Hill and Jones 2012; Wheelen and Hunger 2012; David 2011; Porter 1998; Hitt et al. 2007), the differentiation strategy is the ability of a company to provide a unique offer with a special value to the customers in terms of its quality or special features. Using a differentiation strategy allows a company to promote its brands and charge a premium price for its products (Hill and Jones 2012). Market strategies based on the use of voluntary environmental certification place a special emphasis on the environmental sustainability of a product or a company within the specific sector of the economy. The target consumers may consider this approach as something unique, and attribute added value to it.

Depending upon the type of voluntary standards and certification being adhered to, a company can implement various differentiation strategies; divided into two categories—product differentiation and image differentiation.

In general, the strategy of environmental differentiation of the product and services portfolio has proven its worth for sustaining competitive advantage and is a significant success factor. The potential buyer selects the product from among a variety of other similar products, being ready to pay a premium price for its environmental uniqueness which is confirmed by an appropriate certificate and visual packaging eco-label. However, it is difficult to apply product differentiation strategy if environmental features of products are not well-known or are a subject of indifference for a target consumer group.

The aim of environmental image differentiation is to create a corporate image of a company as being environmentally sustainable, thereby distinguishing it from competitors. This strategy can also be called an environmental branding strategy. From the perspective of this strategy, environmental certification and eco-labelling are essential brand elements which can act as a handicap for non-certified competitors, as well as generate additional (i.e. monopoly) profits.

The focus (or concentration) strategy is based upon selection of a narrow competitive scope within an industry (Porter 1998). This strategy is aimed toward serving the needs of a limited customer group or segment, where a company may choose to concentrate on a particular market niche, which can be defined geographically, by type of customer, or by a segment of the product line (Hill and Jones 2012, p. 125). With regards to using voluntary sustainability standards, a company may choose a particular market segment or a niche, in which there is a demand for eco-certified products; then, it focuses activities on serving the needs of this market segment. Integrating voluntary environmental certification into the focus strategies may provide a competitive advantage in the target segments due to unique features attained by goods or services through such certification, although it does not create a competitive advantage overall.

It should be noted that in CIS countries the experience with strategic management based on environmental certification is very limited. At the same time there are examples of successful implementation of competitive strategies using, in particular, FSC standards. In the following paragraphs we will analyse two examples of companies which have made recent gains in attaining significant market share of their respective sectors of the economy, characterised by a high level of management and are certified with ISO 9001 and ISO 14001 standards. Nevertheless, to fully achieve their marketing objectives they had to fully consider the new format for environmental trends in the society.

The first company, OJSC[1] Mondi Syktyvkar Timber Processing Complex is one of the leaders in pulp and paper production and the biggest paper producer in Russia. It is specialised in the production of office paper, offset paper, newsprint paper, and white top liner. The company strives for leadership in the domestic market and towards expansion into international trade. To implement these strategic targets, the company adopted product differentiation strategy: production of special products which are designed, produced, and adhere to environmentally responsible processes.

Taking into account the current tendencies of increasing environmental awareness, the Mondi company based its strategy on conformance to such FSC standards as FSC-STD-40-004 V 2-0 "Chain of Custody standard for companies supplying and manufacturing FSC-certified products" and FSC-STD-40-005 V 2-1 "Company evaluation of FSC Controlled Wood". As a result, office paper "Snegurochka"[2] by Mondi is the first Russian brand of office paper to be FSC certified. Today the product takes the market lead among office paper produced in Russia; it is a four-time winner of the 'Brand of the Year' Award. Moreover, it is successfully exported to foreign markets (Mondi Syktyvkar 2013).

---

[1] An Open joint-stock company, abbreviated to OJSC, is a type of company in many post-Soviet states, in particular in Russia and Ukraine. Its distinguishing feature is the right of stockholders to trade in stocks without the permission of other stockholders.

[2] The Snow Maiden, a character in Russian fairy tales.

For more than 18 years, the Ukrainian company "Sandora" has maintained leadership of the domestic juice market. The company has three production plants with the total estimated capacity of approximately 1.5 billion packs of juice and juice products per year, and total daily productivity of more than four million packs. Sandora exports about 20 % of its products, representing 60 % of Ukrainian juice exports (Sandora 2013). However, year by year, it has become increasingly difficult to lead this market as it approaches saturation, and strong competitors offer products similar in assortment, quality, and price.

To reach a significant competitive edge, in 2012 Sandora moved to the image differentiation strategy. It developed an image of a manufacturer producing beverages with the use of natural raw ingredients, and packing them into environmentally sustainable packages. The company is planning to switch to using only FSC-certified Tetra-packs which show the environmentally sustainable origin of the forest resources used to produce the packing.

Practices by companies such as Mondi Syktyvkar and Sandora, including a number of others, prove the success of market strategies based on use of voluntary environmental certification in the context of transitional economies. During the study it became clear that some industry leaders take on the role of innovators. Therefore, they can obtain the positive effect of the strategies based on voluntary environmental certification before their competitors; and of course this can lead to an increase in their profits. Such leaders can build a new type of network throughout the chains of business relations within an industry and beyond. In time their partners, competitors, and other market players will adopt a successor strategy in an effort to preserve their market share or to step it up. It is particularly important that leaders involve small and medium-sized businesses in the process.

11.4 Conclusions and Recommendations

In terms of an economy in transition, certification based on international voluntary sustainability standards may become an effective marketing tool to build trust and to gain credence among the desired target consumers, and, therefore, gain and maintain a competitive advantage. However, voluntary environmental certification has not become a significant trend in any of the post-Soviet countries.

The study showed that the majority of Ukrainian managers generally understand the significance of voluntary environmental certification as an instrument for companies' development; but the underlying motivation for using them comes from the market strategic objectives of their company. Today such motivation is mainly related to problems of entering international markets or pre-existing obligations to investors. In addition, with few exceptions in post-Soviet countries, voluntary environmental certification is implemented only in a format of the general standards of ISO 14000 family. Nonetheless, some strategies based on conformance to requirements of sector-specific standards such as FSC, MSC, SAN, etc., can be useful in certain businesses.

Generally, there are the two types of competitive strategies based on voluntary environmental certification—differentiation and focus. In this context environmental certification may support strategies of product differentiation and image differentiation. A focus strategy based on environmental certification is aimed at meeting the narrow scope of demands in environmentally-sustainable products and production.

However, even limited experience in application of such strategies in post-Soviet countries could demonstrate their high efficiency in the context of a transitional economy. Of course, strengthening principles of market economy as well as narrowing of the scope of mandatory certification should increase interest in strategies based on voluntary certification; it particularly concerns certification based on sector-specific voluntary sustainability standards. However, it should be noted that it is of primary importance for transition countries that the process should be headed by industry leaders. This would improve the position of domestic producers in home and international markets, and will provide rapid propagation of new approaches to management within their industries and within other market segments. In this respect governmental authorities and municipal authorities should support a movement in support of voluntary environmental certification; particularly, they should contribute to the dissemination of knowledge on best practices to different branches of activity.

References

Cherp A, Vynychenko V (2012) International standards on environmental management systems (in Russian). http://www.promwood.com/raznoe/ekonomyka_y_rynky/standarti/evropejskye_standarty_kachestva/1499.html. Last accessed 18 Sept 2013

David FR (2011) Strategic management: concepts and cases, 13th edn. Pearson Education/Prentice Hall, Upper Saddle River

González-Benito J, González-Benito O (2005) An analysis of the relationship between environmental motivations and ISO14001 certification. Br J Manage 16(2):133–148

Gutman J (1982) A means-end chain model based on consumer categorization processes. J Mark 46(2):60–72

Hill CW, Jones GR (2012) Essentials of strategic management, 3rd edn. South-Western/Cengage Learning, Mason

Hitt MA, Ireland RD, Hoskisson RE (2007) Strategic management: competitiveness and globalization (concepts and cases), 7th edn. Thomson/South-Western, Mason

Matuszak-Flejszman A (2009) Benefits of environmental management system in polish companies compliant with ISO 14001. Pol J Environ Stud 18(3):411–419

Mondi Syktyvkar – Mondi Syktyvkar OJSC (2013) About Mondi Syktyvkar. http://www.mondigroup.com/en/desktopdefault.aspx/tabid-351. Last accessed 18 Sept 2013

Nikiforov A, Bakiyev T (2005) Metrology, standardization and certification, 3rd edn. Publishing House for Higher Education (Vysshaya Shkola), Moscow (in Russian)

Porter ME (1998) Competitive advantage: creating and sustaining superior performance: with a new introduction. Free Press, New York

Sandora (2013) About the company: business card (in Russian). http://www.sandora.ua/index.php?id=23. Last accessed 18 Sept 2013

Thompson DW, Anderson RC, Hansen EN, Kahle LR (2009) Green segmentation and environmental certification: insights from forest products. Bus Strategy Environ 19(5):319–334

Thompson AA Jr, Peteraf MA, Gamble JN, Strickland AJ III (2013) Crafting and executing strategy: the quest for competitive advantage (concepts and cases), 19th edn. McGraw-Hill Higher Education, New York

UkrSREC – State Enterprise "Ukrainian Scientific-Research and Educational Center for Standardization, Certification and Quality" (2013) Monitoring data of the certified management systems (in Ukrainian). http://www.ukrndnc.org.ua/index.php?option=com_content&task=category&sectionid=8&id=46&Itemid=54. Last accessed 19 Sept 2013

Wheelen TL, Hunger JD (2012) Strategic management and business policy: toward global sustainability, 13th edn. Pearson Education/Prentice Hall, Upper Saddle River

Part IV
Implementation and Impact of VSS

Part IV features arguably the most diverse range of global examples within this publication of VSS in operation. The following chapters comprise of voluntary standards as applied to the production of alternate fuel sources, building material, food and other commodities. A few notable topics are featured as follows in this introduction.

Wood, and its processed derivatives, features frequently in regard to the management of timber production and subsequent export to EU and US markets (Chaps. 12 and 15). Here the premium on sustainably produced wood is realised by the increased demand by environmentally (and to a lesser promoted extent, socially) aware consumers. With EU regulatory standards inducing pressure on exports from producing countries, the level of uptake of VSS and obstacles to such systems are discussed.

China, often regarded as a major polluter as the nation raced to embrace industrialisation and explosive growth, is presented as an unlikely partner in embracing voluntary standards (Chaps. 16 and 17). Known as an exporting superpower, VSS implementation serves as a potential remedy to the historical record of environmentally damaging production, with national objectives shifting towards sustainability in an attempt to overcome international green barriers in trade and repair some of the damage done to the 'Made in China' brand. The Chinese paper making industry showcases the switch over from distributed outdated production systems to larger centralised producers using contemporary technology and production methods. However, lack of familiarity and proper guidance with VSS as well as conflicting decisions regarding choice of certification system are highlighted as some of the remaining roadblocks to progress.

Voluntary standards, as applied to aquaculture and fisheries, are explored with some surprising results in Chap. 22. We discover how innovative producers can actually initiate significant change in standard industry practices to yield sustainable results, motivated both by adding value to their product and in stabilising producer income. Such initiative is shown to serve as the template from which standards can be developed and to which other producers may aspire.

Of course no discussion on sustainable production would be complete without visiting VSS as it is applied to cocoa and coffee production, two globally popular commodities with which western consumers are obsessed (Chaps. 19–21). The prevalence of small-holder producers within these production chains gives added difficulty and also enhanced benefits with the use of VSS. However competing land-use issues, farmer training, familiarisation with sustainable working practices, exclusion and barriers to certification, land ownership issues, and lack of administrative experience all form barriers to VSS penetration into such areas. Some relief is shown in the form of collaboration between certification bodies, where areas of synergy in their standards are acknowledged and accommodated for to compensate for the confusion of standard multiplicity which confused potential participants.

Chapter 12
Evaluation of the Interrelation Between Voluntary Standard Initiatives and Regulatory Approaches Relevant to Forest Management

Berthold Hansmann, Stefan Essel, and Sophie Klose

12.1 Introduction

The biodiversity of tropical forests represent the richest form of terrestrial ecosystem worldwide, being home to 50 % of all species, with the majority located in South America, the Congo Basin and the Asia-Pacific. Against this background, the development of new forest relevant standard initiatives becomes increasingly important for the maintenance of tropical forests and their biodiversity, especially for the interface between forest area and agricultural land.

This chapter will discuss the significance of forest related standard initiatives and regulatory approaches to contribute to the maintenance of tropical forests and the concrete challenges. The main focus will be on the evaluation of synergies between voluntary and regulatory approaches to overcome challenges in order to maximise the benefits from such interrelations.

The implementation of voluntary standard systems (VSS) in the tropics increased during the last 10 years especially in the Congo Basin (more than 5 million hectares FSC certified concession forests) and South America. VSS are becoming increasingly relevant for the interface between forests and agricultural

---

This chapter is based on the GIZ Study: Scaling up impacts of forest relevant initiatives to the maintenance of tropical forests and their values, by Alexander Hinrichs, Heiko Liedeker, Liviu Amariei.

B. Hansmann (✉) • S. Essel
Programme for Social and Environmental Standards, Deutsche Gesellschaft für Internationale Zusammenarbeit (GIZ) GmbH, P.O. Box 5180, 65726 Eschborn, Germany
e-mail: berthold.hansmann@giz.de; Stefan.Essel@giz.de

S. Klose
Programme for Social and Environmental Standards, Deutsche Gesellschaft für Internationale Zusammenarbeit (GIZ) GmbH, P.O. Box 5180, 65726 Eschborn, Germany

Department of Environmental Planning, Brandenburg University of Technology (BTU) Cottbus, P.O. Box 101344, 03013 Cottbus, Germany
e-mail: sophie.klose@tu-Cottbus.de

C. Schmitz-Hoffmann et al. (eds.), *Voluntary Standard Systems*, Natural Resource Management in Transition 1, DOI 10.1007/978-3-642-35716-9_12,

land and for the maintenance of tropical forests and their biodiversity. Voluntary sustainability standard initiatives for agricultural commodities like soy, palm oil or sugar cane have significant contribution to the maintenance of tropical forests. These initiatives are therefore categorised as 'forest related' in this chapter.

The chapter starts by exploring the concept of voluntary and regulatory approaches and their relevance for the maintenance of tropical forests and presents important stakeholder groups which could be involved in scaling up the share of certified wood products. In Sect. 12.2.1, an overview of selected forest related voluntary standard initiatives and a comparison between different criteria for the respective initiative will be provided.

Achieving sustainable agriculture and forest management[1] remains a huge challenge in certain regions. A major issue is the clearing of land, in particular for agricultural use, since it competes with forest land use. This increases pressure on tropical forests, in particular in South America, Africa and Asia-Pacific region. Due to the absence of well-defined or enforced property and user rights in many tropical countries, the destruction of natural resources in return for short-term economic gains is continuing. Another key driver of degradation in the tropics is illegal logging. According to the World Bank (The World Bank 2006), it is responsible for a loss of public assets in developing countries in excess of US $10–15 billion annually in illicit earnings, including lost taxes and royalties. According to the 'Forests of the Congo Basin – State of the Forest 2010 report' (de Wasseige et al. 2010), the net annual deforestation rate was 0.09 %, based only on detected change in forest cover.[2]

Section 12.2.2 highlights the role of intergovernmental processes which aim to stop further deforestation and illegal logging while strengthening forest governance structures. Among these processes are the Forest Law Enforcement, Governance and Trade (FLEGT) Action Plan and Reduced Emissions from Deforestation and Degradation (REDD+). Additionally, the new EU Timber Regulation and the U.S. Lacey Act prohibit the import of illegal timber and associated timber products. On the demand side, public procurement policies for wood and wood-based products oblige bidders to demonstrate that timber products come from sustainably managed forests by recognising certified timber (e.g. FSC, PEFC).

Section 12.3 points at the existing synergies between forest related standard initiatives and regulatory approaches and discusses the most prominent common

[1] The International Organization for Standardization (ISO) recommends that in the context of conformity assessment the term 'sustainability' should not be used, and no claims should be made with regards to achieving sustainability. This is the reason why some forest related standard initiatives have opted for using terms like 'responsible forest management' rather than 'sustainable forest management'. However, in this context, the term 'sustainable forest management' is used in accordance with the German Strategy on Forests and Sustainable Development (BMZ 2002).

[2] It is important to note that FAO considers forest degradation one of the topics where no agreed definitions or assessment methodologies exist. Therefore, the FAO Global Forest Resources Assessment 2010 does not present figures on forest degradation. This is one of the topics that are subject to focused studies, with more results expected during 2011 (FAO 2010).

features and differences. The chapter ends with conclusions in Sect. 12.4 summarising important aspects which need to be addressed to mobilise the full potential of voluntary standard initiatives relevant to forest management as well as regulatory initiatives on tropical forests before finally providing some recommendations.

12.2 Voluntary and Regulatory Approaches and Their Relevance for the Maintenance of Tropical Forests

12.2.1 Forest Related Voluntary Standard Initiatives

Over the last decade, market-based standard initiatives have demonstrated that they are an important driver to improve performance in certified agriculture and forest units and contribute significantly to reducing pressure on tropical forests. However, the expansion of sustainable management practices depends on favourable governance structures. In tropical countries forests are mostly owned by the state with widespread poor governance, limited management capacities and prevailing corruption.

Important stakeholder groups for scaling up the share of certified products are:

- public and private procurers
- sustainable sourcing demands from timber dependent industries
- the retail sector
- environmental and social organisations
- the consumer in their daily buying decisions.

Voluntary standard initiatives developed clearly defined tools to work out high performance in all relevant parts of standard setting. The High Conservation Value (HCV) or Free, Prior and Informed Consent (FPIC) concepts are generally applied, but also multi-stakeholder based decision-making is central. The concept of HCV was first introduced in forest conservation in 1999 by FSC. Its system is based on the identification of values that are important from an environmental or social perspective. Once these values have been defined and identified in the respective areas, a management plan is set to maintain identified core values. Some countries use the HCV concept (e.g. China) in land use planning (e.g. forest planning) and as condition in environmental finance.

The Programme for the Endorsement of Forest Certification (PEFC) and the Forest Stewardship Council (FSC), that developed specific guidance on implementing FPIC, are the most widely used forestry standards systems worldwide. FPIC plays a significant role in minimising conflicts with local and indigenous populations and is therefore highly relevant to success in implementing land-use based processes such as certification of management units as well as FLEGT and REDD on national scale.

The Roundtable on Sustainable Palm Oil (RSPO) and the Roundtable on Responsible Soy (RTRS) and other agricultural standards such as RA–SAN (Rainforest Alliance–Sustainable Agriculture Network), UTZ Certified and the RSB (Roundtable on Sustainable Biofuels) also play important roles in maintaining tropical forests, on the interface between agriculture and forestry. Many agricultural areas managed under these standard initiatives are located close to tropical forests, or even contain remnants of tropical forests. In the absence of functional international and national mechanisms that balance the value of tropical forest ecosystems against the value of agricultural land, tropical forests will continue to be converted.

The following Table 12.1 shows the comparison and analysis of the major forest related voluntary standard initiatives. Several criteria that are considered most relevant from the perspective of this chapter are reflected and demonstrate similarities and differences. A standard initiative is considered to be 'highly relevant' in cases where objective evidence exists, showing that the initiative is having an impact on tropical forest management and conservation. An initiative is considered to be 'relevant' if it is present in tropical forest areas, even if the extent of this presence is limited, provided it has good perspectives to develop in these areas.

The following abbreviations were used in Table 12.1: *CCBA*—Climate, Community and Biodiversity Alliance; *REDD+*—Reducing Emissions from Deforestation and Degradation and contribution to conservation, sustainable management of forests and enhancement of forest carbon stocks; *SES*—Social and Environmental Standards; *ITTO*—International Timber Trade Organization, *IUCN*—International Union for Conservation of Nature; *EU RED*—EU Renewable Energy Directive.

12.2.2 Regulatory Approaches Relevant to Tropical Forests and the Public Sector

There are clear expectations and initial indications that the regulatory approaches described in this chapter will have a significant impact on the maintenance of tropical forests, due to their broad scope (national level), high national and international visibility, legally binding enforcement and the particular focus on transparency. However governments, especially in tropical forest rich countries, will be challenged to enforce the laws and norms applicable in the regulatory approaches.

REDD+

Reducing emissions from deforestation and degradation with ecological and socio-economic safeguards (REDD+) is a mechanism targeted at mitigating climate change by reducing emissions from deforestation and forest degradation. REDD attributes a quantifiable value to forest ecosystems based on the ecosystem services forests provide—the quantifiable storage of carbon and the ability to sequester

Table 12.1 Overview of selected forest related voluntary standard initiatives

Standard initiative/criterion	FSC	PEFC	CCBA and REDD SES	Voluntary legality standards
Overall relevance for tropical forest conservation and management	Highly relevant, increase in certified tropical forests: concessions in natural forests and plantations, impact on tropical forests biodiversity	Relevant, but not well represented in tropical areas so far; certified forests in Brazil, Chile and Malaysia	Potentially highly relevant, involvement in REDD+ implementation, potential for integrating different approaches in land-use planning	Relevant, but scope limited to legality aspects
Scope and applicability	Global applicable principle and criteria	Global certification, of approved national schemes	Global, operates at project level, in many countries that are relevant for the maintenance of tropical forests	Global, limited by business scope of certification body
Participation in inter-governmental regulatory processes	Engaged in EU Timber Regulation and in observing other inter-governmental processes; attempt to position itself in the REDD and Carbon debate	Engaged in EU Timber Regulation and in FLEGT as well as in public procurement regulations	Engaged in standard development relating to REDD+; used at project level in connection with REDD+ (i.e. for carbon projects) and testing of REDD SES on national scale	Some schemes engaged in FLEGT implementation process
Use of High Conservation Value (HCV) concept	Mandatory use of HCV concept in certified Forest Management Units	Similar concept, but reference to ITTO/IUCN guidelines for the conservation and sustainable use of biodiversity	Comprehensive use of HCV concept mandated	No use of HCV concept
Use of Free Prior and Informed Consent (FPIC) concept	Mandatory use of FPIC concept; is further developed during revision process of the FSC principles and criteria	Mandatory use of FPIC concept shall be addressed by endorsed national initiatives	Mandatory use of FPIC concept	Use of the FPIC concept only mandated as legally required
Dialogue with governments	Dialogue with some governments mainly at project level, on issues of common interests	Dialogue with many national governments	Emerging scheme, involvement of various government institutions in REDD SES standard process	Limited dialogue with governments on definition and scope of legality

(continued)

Table 12.1 (continued)

Standard initiative/criterion	RSPO	RTRS	RSB	UTZ	RA–SAN
Overall relevance for tropical forest conservation and management	Relevant in relation to conversion and high conservation value areas	Esp. relevant in relation to conversion and high conservation value areas	Esp. relevant in relation to conversion and high conservation value areas	Relevant, but relating only indirectly to forests and limited to some crops (i.e. coffee, cocoa, tea)	Relevant, due to good representation in tropical countries and high number of products related directly or indirectly to forests
Scope and applicability	Limited by its scope to countries and regions where oil palm is cultivated	Limited by its scope to countries and regions where soy bean is cultivated	Global scope as it refers to biomass and biofuels independent of agricultural crops	Scope limited to countries where coffee, cocoa, and tea are cultivated as important crops	Wide scope as it covers more than 100 crops and has activities in 25 tropical countries
Participation in inter-governmental regulatory processes	Engaged in discussions on EU RED	Engaged in discussions on EU RED	Engaged in discussions on EU RED	Observing inter-governmental processes	In dialogue with different inter-governmental processes
Use of High Conservation Value (HCV) concept	Comprehensive use of HCV concept mandated	Comprehensive use of HCV concept mandated	Comprehensive use of HCV concept mandated	No use of HCV concept	Comprehensive use of HCV concept mandated
Use of Free Prior and Informed Consent (FPIC) concept	Mandatory use of FPIC concept	Mandatory use of FPIC concept regarding land use	Mandatory use of FPIC concept	Use of the FPIC concept not explicitly required	Mandatory use of FPIC concept
Dialogue with governments	Recognition by different governments in tropical countries where oil palm is cultivated	Emerging scheme, dialogue available but criteria difficult to evaluate	Emerging scheme, dialogue available, support by governmental institutions, but involvement of governments limited	Limited, governmental support in Netherlands and Ireland	Dialogue with some governments mainly at project level, on issues of common interests

carbon (Pistorius et al. 2010). In late 2008, REDD was amended into REDD+ to include activities aiming at enhancing carbon stocks by sustainable forest management and conservation. In principle, there are four forest related areas that could help reduce carbon emissions and enhance carbon storage: avoided deforestation, afforestation, reforestation of degraded lands and sustainable forest management.

EU: Renewable Energy Directive

The EU Renewable Energy Directive (EU-RED) sets ambitious targets for the EU to reach a 20 % share of energy from renewable sources by 2020 and a 10 % share of renewable energy in the transport sector. The Directive is relevant to the forestry sector where it outlines in its Article 17 (sustainability criteria) clear requirements for the production of biofuels and bioliquids. The raw materials for biofuel production are not to come from high biodiversity areas such as: primary forest, other wooded land, or areas designated, either by law or by the relevant competent authority for nature protection purposes or international agreements. It also prohibits the raw materials to come from high carbon stock lands such as wetlands or continuously forested areas.

The directive aims to reduce conversion of forests to agricultural land and is relevant to bioenergy imports. Nevertheless, it is criticised that the directive is supporting biofuels and bioliquids on the market that have been produced by converting natural forests to agricultural land before 2008, the so-called cut-off date for land conversion. Furthermore, the directive does not include provisions on indirect land use change effects. Moreover, secondary forest with high biodiversity value is not protected. These insufficiencies lead to potentially adverse impacts on tropical forests, such as increased pressure on areas that according to the FAO definition of forests are considered forested lands, but fail to be protected under the EU-RED.

EU Forest Law Enforcement Governance and Trade Action Plan

The EU Forest Law Enforcement, Governance and Trade (FLEGT) is a broad action plan on the supply and demand side of the timber trade, aiming at reducing illegal logging and its associated trade by improving governance and law enforcement in tropical timber producing countries. The FLEGT Action Plan and the EU FLEGT Regulation (2005) can be regarded as a fundamental move towards improved forest sector governance through market incentives, and an important intermediate step on the road to sustainable forest management (BMZ 2007; EC 2007). A central element of the action plan are bilateral *Voluntary Partnership Agreements* (VPAs) between the EU and tropical timber exporting Partner Countries on which basis only licensed legally produced timber (so called FLEGT license) is given a green lane under the EU Timber Regulation. FLEGT VPAs build on a commitment to increase sector transparency, inter-agency coordination and stakeholder involvement, for which the Partner Country needs to develop and implement a national Legality Assurance System (LAS). Under the licensing scheme, certain timber products exported from a partner country and entering the EU at any customs point designated for release into free circulation should be covered by a license issued by the partner country. This license states that the timber products have been produced from domestic timber that was legally harvested or from timber that was legally imported into a partner country in accordance with national laws as set out in the respective Partnership Agreement. Compliance with those rules should be subject to third party monitoring along the trade chain (timber tracking system). So far six VPAs are signed, and the EU is in negotiations with six more countries.

In Ghana, the VPA process has already created a new set of relationships and working practices between the government, its different agencies and key stakeholders (ProForest 2010). Even before the FLEGT licensing scheme commences, the VPA has already improved communication structures, which can be used for wider stakeholder concerns, e.g. with regard to land-use planning, legal reform, and coordination with other forestry related concepts like REDD+. The process of VPA however is criticised for neglecting social safeguards (e.g. impacts on the informal markets) and underestimating the challenge for Partner Countries to build sufficient capacities for a functioning LAS system.

In the FLEGT VPA context, data reconciliation between the forests, import locations, processing sites and export locations is targeted to ensure that volumes traded do not exceed volumes harvested or legally imported, but do not include information from voluntary standard initiatives. On the contrary, whether timber entering the FLEGT Timber LAS originates from a certified or non-certified source is not recorded, and this information will be lost when a FLEGT license is issued and the timber or timber product is exported, in particular when the certified timber has not been labelled. Overall the FLEGT approach has the potential to advance further to promote broad implementation of sustainable forest management practices.

EU Timber Regulation

On October 20th, 2010, the European Parliament and the EU Council published the Regulation No 995/2010 (EC 2010a), laying down the obligations of operators who place timber and timber products on the EU market, now commonly called the *EU Timber Regulation* (EU TR), which applies from March 3rd, 2013 onwards. The EU FLEGT Regulation and the EU TR mutually reinforce each other (EC 2010b). The Timber Regulation prohibits the placing of illegally harvested timber or timber products on the EU market, and requires operators/traders to use a due diligence system in line with minimum requirements established under the legislation. Due diligence is not simply a moral duty of care but a legal requirement for proactive behaviour, which requires traders and operators to comply with a ‘traceability obligation’ in regard to their traded products.

EU TR and the US Lacey Act (see below) will have a significant impact on the way the international timber trade works. Operators in these markets now have to apply due care (Lacey) or due diligence (EU TR) systems to assure market partners and EU/US Government agencies that the timber they trade in or use comes from legal sources.

It can be expected that both regulations will significantly reduce illegal logging and promote enforcement procedures verifying legal origin and compliance and the use of regulatory approaches such as the FLEGT VPAs in tropical countries. At the same time, the application of voluntary standard systems like forest certification and verification of timber legality can be stipulated. The recent interest in VPA negotiations by countries like Vietnam, Thailand, Papua New Guinea, Laos and many more countries serves as a clear indication in this regard.

The US Lacey Act

The *Lacey Act* amendment of 2008 provides the US government the power to fine individuals and companies who traffic in illegally harvested wood products on the U.S. market. Wood products on the U.S. market must now be accompanied by an import declaration, stating scientific name(s), value, quantity and country of origin. Timber legality is based on all laws and regulations in the country of harvest and 'due care' when exploring the origin of a product must be applied all along the supply chain.

In its nature the Lacey Act creates uncertainty for all companies trading on the U.S. market, since the Lacey Act is fact based and not document based. Companies buying certified or legality verified timber therefore have no guarantee to avoid prosecution.

Public Procurement Policies on Wood and Wood Based Products

Public procurement policies have been implemented in 14 countries on wood and wood-based products that require legality and/or sustainability of these products to be established at their origin. Currently, seven public timber procurement policies in the EU recognise forest certification schemes as instruments to ensure that the timber products come from certified sustainably managed forests. Some countries like the UK and the Netherlands have developed their own criteria and indicators for legality and sustainability. They established special assessment institutions to evaluate the compliance with voluntary standards, while others directly refer to voluntary standard initiatives like FSC and PEFC as proof of compliance. In order to harmonise these policies that have been developed and implemented at different government levels, ITTO recommendations on public procurement suggest that central and local governments work together to ensure that the specific requirements of these policies are similar and consistent between the different levels of government (Nielson 2010).

The impact of public procurement policies for wood and wood-based products on the international timber trade is already documented in some studies (e.g. ITTO 2010), even though systematic assessments of these policies are also lacking, with reliable and comparable data not yet available. Their impact on tropical forests depends, among others, on the trade volume affected by the policy and the level of requirements, which varies significantly among countries.

Precise data on the impacts of regulatory approaches like REDD+, FLEGT, EU-TR and public procurement policies for wood and wood-based products for the maintenance of tropical forests is lacking, due to the following aspects:

- Most of the approaches are new or still under development;
- Monitoring of impacts is so far poorly developed within the approaches; and
- Relevant scientific (long term) studies are not yet available.

In any case, the stakeholder-oriented development of regulatory approaches has generated significant expectations amongst governments, NGOs and private sector representatives. The anticipations are in relation to the possible impact of these approaches on reducing illegal activities, improving governance and proven maintenance of tropical forests.

12.3 Synergies Between Forest Related Standard Initiatives and Regulatory Approaches

Forest related voluntary standard initiatives and regulatory trade related approaches share a number of commonalities in achieving the common goal of keeping tropical forests standing. The most prominent common features and differences are the following:

Both approaches are market-based and have impacts on production, processing and trading conditions in countries along supply chains. For liability control, performance needs to be assessed, monitored and conformity enforced. Within both approaches, a wide range of stakeholders are involved in decision-making processes.

With regard to legal enforcement the approaches differ significantly, due to their different constituency. Regulatory approaches are national approaches, aiming for countrywide implementation, whereas voluntary initiatives work on enterprise level. Since regulatory approaches are legally binding and may be sanctioned by authorities, if certificate holders are not sufficiently complying with defined standard requirements of voluntary standard initiatives the certificate can be withdrawn.

Another important issue is the definition of forest conversion. While voluntary standards such as the PEFC and FSC completely prohibit forest conversion to other land uses, regulatory approaches allow forest conversions as long as in compliance with national law requirements. The same can be held for traceability. Voluntary standard initiatives can fully document certified products from origin to the final consumer, whereas scope and requirements of traceability are limited to FLEGT countries and the EU/US market. Legality definitions in national VPA processes require a strong stakeholder involvement during the development phase and result in comprehensive lists of regulations and verifiers (legality grids). Countries like Ghana have used the process to systematically analyse and review their legal framework and reinforce their domestic concerns. Moreover, concerns have been raised that legal verification alone could supplant efforts to move towards sustainable forest management and would be at the end counterproductive ('race to the bottom').

Documenting and enforcing legal forest management and legal timber supply chains should be a comprehensive step towards national implementation of sustainable forest management practices. However voluntary standard systems are not a surrogate for national governance, but can support its compliance on enterprise level and across national jurisdictions.

Voluntary standard initiatives can be recognised by partner countries as being compliant with legality matrix verifiers in African VPAs based on a formal evaluation. Partner Countries, and not the EU, need to conduct rigorous evaluations of such systems to ensure that their standards fulfil the national legality definition, and incorporate sufficient control and transparency: "*Where a legality assurance system includes market-based elements, the Partner Country government, not the EU, will be responsible for approving those elements and ensuring that they remain effective*" (FLEGT briefing note No. 8, P. 2).

Implementing national timber tracking system of a FLEGT LAS must be understood as a very challenging task, since partner countries will be confronted with serious issues of enforcement and control. Experience from timber tracking and chain of custody certification developed under voluntary standard systems can support designing and implementing practical LAS, increasing mutual benefits and credibility for national LAS and voluntary certification in VPA countries with limited enforcement capacities.

Both, the EU Timber Regulation and the U.S. Lacey Act require due diligence/due care and contain a prohibition on placing illegal timber or timber products on the respective markets by making such behaviour a criminal offence. In the regulations, voluntary standard schemes are regarded as one risk assessment and mitigation instrument. The EU argues that certification is voluntary and not regulated by governments and can thus not be automatically accepted (EFI 2010). Till now "*Industry has not felt a severe enough threat from regulation to accept the cost of becoming certified*" (CCIF 2002). However, with the Lacey Act and the EU Timber Regulation in place the conditions have changed and the need for operators to implement better management and control systems has increased.

Besides the already recognised synergies (in U.S. EU TR, EU-RED, FLEGT VPA, public procurement), synergies in implementation of LAS, due diligence/due care, timber tracking, mapping and monitoring implementation of procurement policies as well as on awareness raising and measuring sustainability levels can be maximised, e.g. monitoring systems are needed for implementing national REDD + programmes. Adopting and implementing synergies at a wider scale would facilitate the application of voluntary standards at a larger scale, creating a virtuous mechanism, where regulatory and voluntary mechanisms mutually reinforce each other, aiming at better governance, improved management and conservation of natural resources. Moreover, land-use planning on a large scale context would help to address problems related to the management of the interface between the forestry and agriculture, as agriculture is the main deforestation driver.

12.4 Conclusion and Recommendations

Even though considerable progress has been made over the past few years on illegal logging (Lawson and MacFaul 2010), the overall situation of tropical forests has not improved significantly. Nevertheless, a broad range of public (governmental and intergovernmental) and private (corporate and civil society) initiatives developed aiming at addressing certain issues related to the maintenance of tropical forests, including their inherent social, ecological and economic attributes.

Forest related voluntary standard initiatives, as well as new regulatory initiatives, play an important and, in various respects, mutually beneficial roles for the maintenance of tropical forests. However, the results also indicate that the different voluntary and regulatory instruments are not yet used to their full potential and that implementation is still suffering from a range of deficiencies.

In the following some important aspects are highlighted, which need to be addressed to mobilise the full potential of voluntary standard initiatives relevant to forest management as well as regulatory initiatives on tropical forests.

12.4.1 Data Reconciliation and Systematic Monitoring

Monitoring of market shares of certified sustainable timber and in regard to imports of certified timber volumes and products rely almost entirely on private sector estimates. Facts and figures in voluntary standard initiatives are not flawless as existing schemes focus rather on the number of companies certified, as well as number of certified hectares (in FSC and PEFC focus on Chain of Custody certificates and certified forest area). Similarly, private voluntary systems aimed at agricultural commodities lack systematic data on trade volumes and related information.

Therefore a common understanding on data formats, data structures and necessary monitoring approaches should be established, in order to facilitate consistent data reconciliation on the extent and scope of impacts of private voluntary and public regulatory initiatives and their respective overlaps.

12.4.2 Capacity Building and Awareness Raising at Government, Private Sector and Civil Society Level to Advance Enabling Conditions

Regional organisations and approaches like ASEAN, COMIFAC, regional offices of private voluntary initiatives, etc., play an important role not only in linking global market-based initiatives to national and local approaches but also in addressing trade related interrelations between countries.

Dedicated engagement and investment (e.g. targeted development programs) are critical to enable relevant actors in tropical forests to prepare for participation in both voluntary and regulatory initiatives. Furthermore, mutual understanding and building related capacity for implementation of, for example, VPAs and certification in tropical forest regions is critical. Complementing voluntary and regulatory market-based initiatives with capacity building programmes is a most critical factor in the success of these initiatives.

12.4.3 Horizontal Integration

Forest maintenance is a result of complex local, national, regional and global dynamics across sectors and actors. While different single-issue initiatives are available, there is a lack of coordination of these approaches across sectors. In particular, cross-sector linkages are missing among:

- Different regulatory approaches like FLEGT and REDD+, FLEGT and public procurement, public procurement and sustainable building;
- Standards systems like forest certification and carbon and biodiversity related standards; and
- Between regulatory approaches of different sectors (e.g. agriculture and forestry) and standard systems.

Some regulatory initiatives like on establishing consolidated information on public procurement or cross-sector planning processes like National Forest Programmes aim into this direction. The results are more or less isolated impacts of different initiatives, while often missing the opportunity of achieving 'critical mass' through the identification and efficient translation of potential synergies into concrete actions and impacts on the ground at landscape level.

12.4.4 Development of Effective Impact Assessment Approaches

Due to the lack of appropriate impact measurement systems harmonised across sectors, very little is known about the impacts of both voluntary standard initiatives and public regulatory approaches. Especially the impacts of interactions, synergies and relations between different initiatives and different actors are unidentified. Additionally, in most cases where some form of monitoring has been implemented and applied, social, economic and environmental impacts are not considered equitably.

12.4.5 Enhancing Market Recognition and Market Penetration

In spite of all the progress made through private voluntary and public regulatory initiatives till now, large-scale forest degradation and deforestation have continued. Market-based instruments contributing a transition to a green economy have to significantly increase their outreach; sustainable market demand has to be far larger and more consistent/comprehensive to be able to significantly reduce forest degradation and deforestation in the tropics. In this context it is of particular importance to highlight that only a fraction of tropical timber products currently enters international markets, while a larger part is being used locally, mostly in the absence of requirements relating to legality or sustainability of forest management.

This suggests that forest related private voluntary and public regulatory initiatives in close collaboration need to involve the full range of key market actors including domestic actors, new product initiatives (e.g. environmental services, wood energy products), and emerging markets beyond Europe and the USA.

References

BMZ – German Federal Ministry for Economic Cooperation and Development (2002) Forest sector strategy and sustainable development. Principles and guidelines for German state development with National Forest Programmes. http://www.giz.de/Themen/en/dokumente/en-bmz-23-forest-sector-concept-2002.pdf. Last accessed 28 Mar 2013

BMZ – German Federal Ministry for Economic Cooperation and Development (2007) FLEGT – Combating illegal logging as a contribution towards sustainable development. BMZ Topics No. 180. Bonn

CCIF – Conservation and Community Investment Forum (2002) Analysis of the status of current certification schemes in promoting conservation. Conservation and Community Investment Forum, San Francisco

de Wasseige C, de Marcken P, Bayol N, Hiol Hiol F, Mayaux P, Desclée B, Nasi R, Billand A, Defourny P, Eba'a Atyi R (2010) The forests of the Congo Basin – state of the forest 2010.

Publications Office of the European Union. Luxembourg. http://www.cifor.org/online-library/browse/view-publication/publication/3754.html. Last accessed 28 Mar 2013
EC – European Commission (2007) FLEGT VPA Briefing Notes 1–8, Brussels
EC (2010a) Regulation laying down the obligations of operators who place timber and timber products on the EU market. Regulation No 995/2010. Official Journal of the European Union of 12.11.2010, L295/23-L295/34
EC (2010b) Frequently asked questions on illegal logging and the FLEGT VPA. EC memo/10/331, Brussels, 14 July 2010
EFI – European Forest Institute (2010) The new EU timber legislation. Briefing note of the EU FLEGT Asia office, Kuala Lumpur
FAO (2010) Global Forest Resources Assessment 2010. FAO Forestry Paper 163. FAO, Rome
FSC – Forest Stewardship Council (1999) FSC principles and criteria for forest stewardship, FSC-STD-01-001, Version 4-0
Lawson S, MacFaul L (2010) Illegal logging and related trade. Chatham House, London
Nielson ST (2010) Presentation at the International Timber Trade Federation Day, 6–8 Oct 2010, Geneva
Pistorius T, Schmitt CB, Benick D, Entemann S (2010) Greening REDD+. Institute of Forest and Environmental Policy, University of Freiburg, Freiburg
ProForest (2010) FLEGT licensed timber and EU Member State procurement policies. http://www.proforest.net/publication/bibliog.2011-04-08.7421357592. Last accessed 20 Mar 2013
The World Bank (2006) Strengthening Forest Law Enforcement and Governance: Addressing a Systemic Constraint to Sustainable Development. Report No. 36638. The World Bank, Washington

Chapter 13
Voluntary Standard Systems and Regulatory Processes for Timber Products: Analysis of Green Procurement in Germany

Eike Albrecht, Franziska Rückert, and Michael Schmidt

13.1 Introduction

This chapter aims to discuss voluntary standard systems and regulatory processes for timber products with the example of green procurement policies in Germany. Forests provide numerous ecosystem goods and services which are crucial for humanity. Besides combating climate change and therefore securing basic living conditions for humanity, forests provide an important natural construction material. Therefore, it is necessary to find a way of using the forests while incorporating the needs of present and future generations. Certification of timber is a market tool that shall contribute to this process. Even though certification schemes are in need of schematic improvement (for details see the previous Chap. 12) any approach towards preventing deforestation is urgently required. While Chap. 12 covered the context of international regulatory approaches against illegal logging and voluntary forest relevant standards, this chapter shall describe and evaluate current legal requirements for public procurement of timber in Germany. The question of where restrictions for the use of tropical timber can be detected and under which circumstances its use is legitimate will be answered. Special regard within the topic of national legislation is given towards procurement policies for the public sector. It will be investigated under which conditions the utilisation of tropical wood is permitted within public orders. Therefore, it is to be investigated whether the detected legal requirements from the side of the industrialised countries can be

E. Albrecht (✉)
Department for Civil and Public Law with References to the Law of Europe and the Environment, Brandenburg University of Technology (BTU) Cottbus, P.O. Box 101344, 03013 Cottbus, Germany
e-mail: albrecht@tu-cottbus.de

F. Rückert • M. Schmidt
Department of Environmental Planning, Brandenburg University of Technology (BTU) Cottbus, P.O. Box 101344, 03013 Cottbus, Germany
e-mail: franziska.rueckert@gmx.net; michael.schmidt@tu-cottbus.de

C. Schmitz-Hoffmann et al. (eds.), *Voluntary Standard Systems*, Natural Resource Management in Transition 1, DOI 10.1007/978-3-642-35716-9_13,

considered as a suitable tool for improving the environmental situation in tropical timber producing regions and making forestry more sustainable.

Section 13.2 will describe external and internal forces that pressurise forest certification. This shall introduce the general context of the practice of forest certification in areas of timber supply and legal requirements imposed by timber demanding countries. The exact description and partial analysis of these legal requirements will follow in Sect. 13.3. Provisions for the use of tropical timber within German public orders are covered. Sections 13.3.1–13.3.3 give further insight on the implementation of these national public procurement regulations on a federal state level. Therefore, Berlin, Brandenburg and North Rhine-Westphalia are the chosen reference states. A short comparison to the situation of federal states without timber related procurement rules will be given in Sect. 13.3.4. Section 13.4 discusses inconsistencies between the given legal requirements which may potentially cause conflict. The chapter ends with conclusions and recommendations in Sect. 13.5 summarising the main findings and highlighting areas in need for improvement.

13.2 Driving Forces for Timber Certification

Originally, forest certification had been introduced after concerns about the world's tropical forest situation led to international reactions such as boycotting tropical timber. Large scale production of palm oil and wood as building material or for energy recovery by foreign companies has facilitated massive deforestation in tropical areas of Africa, South America and South East Asia. Deforestation relocated from temperate regions in Europe and North America towards the tropics depending on the respective developmental state of the region. More than half of the former global forest area fell victim to logging while the process is steadily continuing (Blaser et al. 2011). While forests in developed regions of the world have started to recover from former periods of intense deforestation, the problem has now shifted to countries in transition and developing countries. According to the Food and Agricultural Organization of the United Nations (FAO), Africa has lost 10 % of its primary forests in the last 20 years while South America records a loss of 9 % (FAO 2012).

The great areas of native forests are the most effective tool for sequestering carbon dioxide from the air and storing it in the long run. Thus, forests are of utmost importance for meeting the climate goals which are necessary for any future developments. According to the most recent report on the Status of Tropical Forest Management released by the International Tropical Timber Organization (ITTO) in 2011, the total area of tropical forests worldwide only remains 7.6 million km^2 from formerly more than double that amount (Blaser et al. 2011). Besides many important ecological, social and economic functions, climate protection is probably the most recognised service humanity obtains from forests. Counteracting global warming is, therefore, broadly accepted as one of the most significant challenges which current generations have to face.

Many of these ecosystem services and functions are, however, not yet attached to their full importance in terms of financial valuation. This causes exploitation of forest resources without instant negative consequences (Muthoo 2012; Thang 2003). As part of international and national politically agreed plans for reaching climate aims there are several attempts to approach sustainable forestry.

The private sector contributes via environmental labelling schemes for sustainable forest management. The consequent certification is a market tool which aims to combine the valuation and protection of forest resources with its sustainable use. More details about such voluntary forest relevant standard initiatives are given in Sects. 12.1 and 12.2.

From the side of governments, an adequate legal background shall support these private initiatives. The relevant international legislative requirements for timber trade such as REDD (Reducing Emissions from Deforestation and Degradation),[1] the EU Action Plan on Forest Law Enforcement, Governance and Trade (FLEGT)[2] and the derived EU Timber Regulation[3] were covered in Sect. 12.2. These policies shall ensure legal origin of the imported timber and are regarded as being rather successful in bringing together the stakeholders (Beeko and Arts 2010). On the other hand, there are several requirements for public procurement on a national basis in Germany because green procurement policies are a major driving factor for timber certification. The context of forestry certification and procurement rules can be understood as follows: Publicly procured wood accounts for up to 25 % of a country's total wood demand, thus, if major wood consuming countries implement procurement requirements for the use of timber this can be considered as an effective tool for influencing the timber market structure.

Prohibitions on the use of tropical timber were formerly applied in the public sector of European countries such as Norway and Germany (Purbawiyatna and Simula 2008). However, after certification evolved as a tool for sustainable forestry, the legislative situation eased in response. Within the European Union demand for certified timber is steadily increasing, mainly for the reason of stricter procurement policies within the private and the public sector. Legal origin of wood is the minimum standard within the EU import regulations whereas often the member state specific procurement policies additionally demand for certification by voluntary sustainability standards from either the Forest Stewardship Council (FSC) or

[1] Decision 1/CP 16, III C, agreed at Cancun, Mexico, UNFCCC COP 16, 2010; available under: http://unfccc.int/resource/docs/2010/cop16/eng/07a01.pdf, p. 12, last accessed 29 August 2013.

[2] Forest Law Enforcement, Governance and Trade (FLEGT)—Proposal for an EU Action Plan, COM (2003) 251 final; available under: http://www.euflegt.efi.int/files/attachments/euflegt/01flegtactionplanenfinalen.pdf, last accessed 29 August 2013.

[3] Regulation (EU) No. 995/2010 of the European Parliament and of the Council of 20 October 2010 laying down the obligations of operators who place timber and timber products on the market (Text with EEA relevance), OJ EU L 295, 23.

the Programme for the Endorsement of Forest Certification Schemes (PEFC) (UN 2011).

Also increasing awareness and pressure from international NGOs play a contributing role in development of the market structure characterised by increasing demand for environmental labels. The process of advancing corporate social responsibility supports raising the share of certified wood within corporate use. More companies decide on introducing internal strategies for a 100 % usage of certified timber in order to comply with the trend of increasing public awareness. These internal policies and codes secure minimum legal compliance and sustainable development (Purbawiyatna and Simula 2008).

13.3 German Public Procurement Requirements

The underlying document for national and federal state-wide green procurement regulations is the Joint Instruction on the Procurement of Wood Products[4] from 22 December 2010 and the attached Explanatory Notes Regarding the Procurement of Wood Products from 2 December 2010.[5] Both were released by the Federal Ministry of Food, Agriculture and Consumer Protection. Within the document, the federal government commits to the support of sustainably managed forests and will only procure certified timber. The following exact instructions are given for the process of timber procurement:

> Wood products procured by the federal administration must demonstrably come from legal and sustainable forest management. The bidder must furnish proof of this by presenting an FSC or PEFC certificate, a comparable certificate or by producing individual specifications. Comparable certificates or individual specifications are accepted if the bidder can prove that the FSC or PEFC criteria that apply to the respective country of origin have been met.

In the frame of the explanatory notes the procurement regime is annotated. Due to dynamic developments in the area of forest certification there might be changes in certification standards and their execution. Therefore, schemes have to be checked frequently for deficiencies. In case major flaws are revealed within the operation of the accepted certification schemes, the opportunity for improvement will be granted up to a period of 12 months. Otherwise an amendment can be enforced within the regulation and the scheme excluded from the German procurement regime. The main deficiencies that would lead to such action are illegality of the wood, violation against basic principles of the certification scheme itself or lack of transparency.

[4] Joint Instruction on the Procurement of Wood Products (*Gemeinsamer Erlass zur Beschaffung von Holzprodukten*) of 22 December 2010, in force since 17 January 2011, Joint Ministerial Gazette 2010-86 no. 85, p. 1786.

[5] *Begleitende Erklärung zur Beschaffung von Holzprodukten*, annex to the Joint Circular on the Procurement of Wood Products of 22 December 2010; available under: http://www.bmelv.de/SharedDocs/Standardartikel/EN/Agriculture/forestTimberHunting/ProcurementOfWoodProducts-Attachment.html, last accessed 29 August 2013.

248
(Erklärung zur Verwendung von Holzprodukten)

Bieter	Vergabenummer	Datum
Baumaßnahme		
Leistung		

Alle zu verwendenden Holzprodukte müssen nach FSC, PEFC oder gleichwertig zertifiziert sein oder die für das jeweilige Herkunftsland geltenden Kriterien des FSC oder PEFC einzeln erfüllen.

Erklärung zur Verwendung von Holzprodukten

☐ Ich werde Holzprodukte verwenden, die nach FSC und/oder PEFC zertifiziert sind.

☐ Ich werde Holzprodukte verwenden, die nach

______________________________ zertifiziert sind.

Der Nachweis der Gleichwertigkeit - d.h. der Übereinstimmung des Zertifikats mit den für das jeweilige Herkunftsland geltenden Standards von FSC oder PEFC - ist durch eine Prüfung vom Johann Heinrich von Thünen-Institut in Hamburg (vTI) oder dem Bundesamt für Naturschutz in Bonn (BfN) erbracht. Ich werde diesen geprüften Nachweis zu dem von der Vergabestelle verlangten Zeitpunkt vorlegen.

☐ Ich werde Holzprodukte verwenden, die die im jeweiligen Herkunftsland geltenden Kriterien des FSC oder PEFC einzeln erfüllen.

Der Nachweis darüber ist durch eine Prüfung vom Johann Heinrich von Thünen-Institut in Hamburg (vTI) oder dem Bundesamt für Naturschutz in Bonn (BfN) erbracht. Ich werde diesen geprüften Nachweis zu dem von der Vergabestelle verlangten Zeitpunkt vorlegen.

Fig. 13.1 Explanation form for the utilisation of wood products [received from the Public Enterprise on Property and Construction Management Saxony (*Staatsbetrieb Sächsisches Immobilien- und Baumanagement – SIB*), 31 January 2013]

As for the practical application of the instruction, Fig. 13.1 illustrates an abstract of the explanation form for the utilisation of wood products that is to be completed for all construction projects including timber. The form is, however, only available in German language. Besides information on the bidder, the PO number, date and exact building operations, within the process of contract awarding it must be indicated which kind of proof for sustainable origin of the used timber will be adduced. Therefore the following three options are possible:

- the timber has been certified by FSC and/or PEFC;
- the timber has been certified by an equivalent scheme while the proof for equivalence has been carried out by the Johann Heinrich von Thünen Federal Research Institute in Hamburg (vTI) or the Federal Agency for Nature Conservation (*Bundesamt für Naturschutz – BfN*);
- or that the timber has not been certified by FSC or PEFC, however, has been harvested according to the respective sustainability standards of either of the two schemes. Evidence is to be given by vTI or BfN.

In the past there have been several cases where the Joint Instruction on the Procurement of Wood Products was violated and non-certified timber has been used in the public sector (Greenpeace 2002; Robin Wood 2002). These actions reveal weaknesses of the existing legislation. Apparently, the practical implementation is not given sufficient incentive by control and enforcement.

13.3.1 Berlin

On 23 July 2010 the Berlin Tendering and Procurement Act[6] came into effect. In the frame of this law, all procurement agencies are obliged to ensure that negative environmental impacts caused by production, usage and disposal of materials are avoided (Art. 7 para 1 of the Berlin Tendering and Procurement Act). According to Art. 7 para 3 of the same Act, the Senate Administration for Urban Development and Environment (*Senatserwaltung für Stadtentwicklung und Umwelt*) is authorised to enact binding regulations for the procurement of relevant products and services for all public institutions in Berlin (Hermann and Acker 2011). Furthermore a series of circulars on utilisation prohibitions and restrictions for using timber as building material exist.

The circular from 1998[7] by the Senate Administration for Building, Living and Transport (*Senatsverwaltung für Bauen, Wohnen und Verkehr*) contained a ban on tropical timber for use in construction material components and civil engineering. In the explanation of this circular it was stated that the ban is, however, only valid until a reliable system for proof of origin and sustainable forest management has been internationally developed.

After certification schemes such as FSC and PEFC evolved and were internationally established, this tropical timber ban was conditioned by the Senate Administration for Urban Development and Environment in the frame of circular SenStadt VI No. 14/2004[8] from 2004 on "Prohibitions and Restrictions on the Use of Building Materials; here: Components Made of Tropical Wood". The novelty here is that tropical timber for use in construction material components and civil engineering shall be banned, unless it is certified according to FSC or equivalent. The acceptance of equivalent certification schemes is to be applied for at the Senate Administration for Urban Development and Environment.

---

[6] Berlin Tendering and Procurement Act (*Berliner Ausschreibungs- und Vergabegesetz*) of 8 July 2010, Berlin Law Gazette, p. 399, last amended by Art. I of the Act from 5.6.2012, Berlin Law Gazette, p. 159.

[7] Rundschreiben BauWohnV VI Nr. 10/1998 of 30 June 1998. Senatsverwaltung für Bauen, Wohnen und Verkehr Berlin.

[8] Rundschreiben SenStadt VI A Nr. 14/2004: *Verwendungsverbote und Verwendungsbeschränkungen von Baustoffen*—hier: *Bauteile aus Tropenholz* of 2004. Senatsverwaltung für Stadtentwicklung Berlin.

On 1 January 2013 the Administrative Circular on Procurement and Environment[9] became the effective regulation for public procurement in Berlin. It is to be applied for all procurements of deliveries, construction works and services from an approximated order value of €10,000 and must be used by the senate departments and their subordinate authorities, the district administrations and the federal authorities, establishments and foundations governed by public law. The principle of this circular is to guide the sustainable contribution of Berlin towards a minimisation of the environmental burden. The city's public sector shall move further towards environmental protection in preferring environmentally friendly products and services. Procurement should not only be practiced in a way that favours the financial interests of the employer but must furthermore involve ecological considerations. Within section I, para. 4 no. 13 of the circular the procurement restrictions are listed. Regarding tropical timber, the ban involves wood and wood products not demonstrably originating from sustainably managed forests. The proof has to be adopted by the bidder via a certificate from FSC or equivalent. Equivalent schemes are accepted if proof for the application of FSC corresponding criteria can be provided.

13.3.2 Brandenburg

The guidelines for environmentally friendly procurement in the federal state of Brandenburg demonstrate possibilities that are, however, predominantly not tangible. Also the new Brandenburg Procurement Act[10] of 21 September 2011 does not include bans or restrictions of products for ecological reasons. It does not even mention environmental concerns in procurement procedures. In such a case, a reference to the federal Act on Life-Cycle Management (the former Waste Management Act)[11] and the Brandenburg Waste Management and Soil Protection Act[12] may help, as in the waste related regulations it is stated that preference is to be given to products that are made in a low-waste and resource saving manner, that are durable, easy to repair and reusable and can be subject to high quality recycling

[9] Administrative Circular on Procurement and Environment (*Verwaltungsvorschrift für die Anwendung von Umweltschutzanforderungen bei der Beschaffung von Liefer-, Bau- und Dienstleistungen – VwVBU*) of 23 October 2012. Senatsverwaltung für Stadtentwicklung und Umwelt Berlin; available under: http://www.bsr.de/assets/downloads/VwVBU.pdf, last accessed, 29 August 2013.

[10] Brandenburg Procurement Act (*Brandenburgisches Gesetz über Mindestanforderungen für die Vergabe von öffentlichen Aufträgen – BbgVergG*) of 21 September 2011, Brandenburg Law Gazette I, 11, no. 19.

[11] Of 24 February 2012, Fed. Law Gazette I, p. 212, amended by the Act from 22 May 2013, Fed. Law Gazette I, p. 1324.

[12] Of 06 June 1997, Law Gazette I/97, p. 40, last amended by Law of 15 July 2010, Law Gazette I/10, p. 1.

(Art. 23 Act on Life-Cycle Management). But this obligation needs specification in ordinances which are not prepared for larger sectors, as yet. It is added that these characteristics are to be mentioned in the solicitation of public orders. Finally environmental considerations in the process of procurement shall be within an acceptable financial frame. Profitability is, however, to be judged in a broad sense that includes financial aspects but also environmental and energy efficiency factors (Hermann and Acker 2011).

Furthermore, all agencies under the Ministry of Environment, Health and Consumer Protection are to use the Procurement and Contracting Handbook for Guidance in Planning and Implementation of Construction Works.[13] A former version of the federal "Joint Instruction on the Procurement of Wood Products" from 17 January 2007[14] is part of this handbook and, therefore, applies to Brandenburg. Accordingly, all timber products that are used on behalf of the public sector have to be certified by FSC, PEFC or equivalent (Hermann and Acker 2011).

13.3.3 North Rhine-Westphalia

As in the other federal states, the Waste Act of North Rhine-Westphalia[15] comprises an article that generally outlines the importance and manner of considering environmental factors in public procurement (Hermann and Acker 2011), but also here a specification is missing. Exact provisions on the dealings with wood are given in the Circular of the Ministry of Economy, Trade and Energy (*Ministerium für Wirtschaft, Mittelstand und Energie*) on the "Consideration of Aspects of Environmental Protection and the Energy Efficiency in Public Procurement",[16] released in 2010. The three main statements given in Sect. 2.3.3 of the circular are that (1) timber products must originate from verifiably legal and sustainable forestry, (2) certificates from FSC or PEFC, or analogous schemes are accepted and (3) any scheme other than FSC or PEFC has to be verified by the Johann Heinrich von Thünen Federal Research Institute in Hamburg.

[13] German Procurement and Contracting Handbook for Guidance in Planning and Implementation of Construction Works (*Vergabe- und Vertragshandbuch für die Baumaßnahmen des Bundes—VHB*) from 2008, last amended by circular of 19 September 2012, no B 15-8164.2/2.

[14] Joint Instruction on the Procurement of Wood Products (*Gemeinsamer Erlass zur Beschaffung von Holzprodukten*) from 17 January 2007, Joint Ministerial Gazette, p. 67.

[15] Of 21 June 1988, Law Gazette, p. 250, last amended by Act from 21 March 2013, Law Gazette, p. 148.

[16] Circular on the Consideration of Aspects of Environmental Protection and the Energy Efficiency in Public Procurement in North Rhine-Westphalia (*Runderlass zur Berücksichtigung von Aspekten des Umweltschutzes und der Energieeffizienz bei der Vergabe öffentlicher Aufträge*) of 12 April 2010, Ministerial Gazette, no. 14 of 2010 from 3 May 2010.

13.3.4 Main Differences Between the Federal States' Regulations

The main difference between the federal states is the concreteness with which regulations are formulated. While some federal states use exact and legally tangible formulations, others have rather implemented guidelines for action. Even though it is to be appreciated that ecological factors are, at least to some extent, included in all federal states' procurement policies, lacking concreteness might hinder practical implementation. For example, within section 4.3 of the Guidelines for Procurement of Goods and Services by the Saarland Administration,[17] it is stated that within the selection process for wood products priority shall lie with products from FSC certified production. This formulation lacks any basis for mandatory compliance and leaves room for individual decision making. In case of shortages of financial or technical means, decisions are likely to fall contra certified timber.

Bremen, Hessen, Mecklenburg Western-Pomerania, Lower Saxony, Rhineland-Palatinate, Saxony-Anhalt, Saxony and Thuringia are the federal states in which no separate guidelines or regulations directly regarding the procurement and application of timber exist. Saxony has been chosen as a reference state for further investigation on the legal situation. Therefore, an inquiry was issued towards the Public Enterprise on Property and Construction Management Saxony (*Staatsbetrieb Sächsisches Immobilien- und Baumanagement – SIB*) as it is the relevant authority for all construction projects in Saxony. The first issue to be clarified was how timber procurement is regulated in the public sector, as there are no explicit provisions on this matter in the Saxon legislation. It has been verified that Saxony is, although not separately implemented, still exposed to the federal Joint Instruction on the Procurement of Wood Products of 2011 that requires certified timber from FSC or PEFC.[18]

13.4 Inconsistencies Between Legal Requirements

13.4.1 Timber Trade Regulations

The regulations on timber trade are to be divided into two branches. The most important provisions are visualised in Fig. 13.2. The first branch links to the international requirements of FLEGT and the EU Timber Regulation which create

[17] Guidelines for Procurement of Goods and Services by the Saarland Administration (*Richtlinien für die Vergabe von Aufträgen über Lieferungen und Leistungen durch die saarländische Landesverwaltung*) of 16 September 2008, Law Gazette, p. 1683, last amended by guideline from 28.12.2010, Law Gazette 2011 II, p. 3.

[18] Written statement of the Saxon Public Enterprise on Property and Construction Management of 31 January 2013.

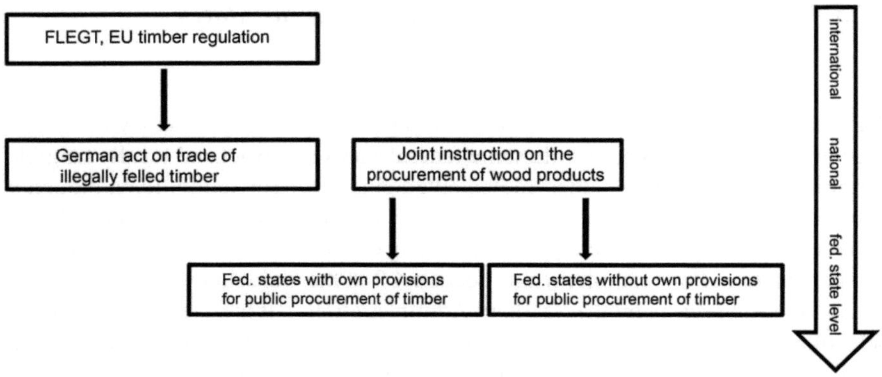

Fig. 13.2 Representation of the legal background on tropical timber trade

the foundation for trade with legally felled timber in the European market. The German Act on Trade of Illegally Felled Timber[19] is the national law that demands for examination and control of FLEGT licenses with timber imports from FLEGT partner countries. The law has recently been amended in order to additionally conform to the EU Timber Regulation. From March 2013 onwards, any imported timber will have to carry a license which proves compliance with the due diligence system (see Sect. 12.2.2 in Chap. 12). Both the license from FLEGT and the one under the due diligence system are to ensure legal origin of all imported timber.

The second branch of regulations for timber use in Germany regards the sector of public procurement. As explained in the former sections, public orders add up to a major part of the total timber demand. Therefore, governments shall adopt responsibility for the regions of timber origin. The Joint Instruction on the Procurement of Wood Products provides the unique usage of timber certified by FSC or PEFC which shall encourage timber producing countries to commit to sustainable dealing with the forests.

Critique has to be issued for insufficient cooperation between the two legislative branches at the national level. To date, both legal sectors do not recognise the verification of the other, i.e. a certificate from FSC or PEFC cannot function as the required proof of legal origin. Vice versa, the required proof of origin is not regarded as sufficient evidence for sustainable production. Hence, for the utilisation of timber in the public sector the different regulations have to be fulfilled separately which implies that more than one certificate is needed. This induces high administrative burden and the potential for confusion.

A hint towards this issue is given in the Explanatory Notes Regarding the Procurement of Wood Products (BMELV Ministry section 3.2) where it is stated that "a review will be conducted in 2013 in order to ascertain if and how wood and

---

[19] Of 11 July 2011, Fed. Law Gazette I, p. 1345, amended by Act of 3 May 2013, Fed. Law Gazette I, p. 1104.

wood-based products from countries with which the EU has concluded Voluntary Partnership Agreements (VPA) can be included in the procurement regime". Thus, from the side of the public procurement regime any certificate that has been issued in the frame of VPAs between timber producing countries and the EU might be accepted as a sustainability certificate. Even though this step would slightly simplify compliance with all relevant regulations it would not tackle the entire problem as only one-sided recognition of VPAs is implied. A more suitable step would be to change the concept of the VPAs in a way to make them a reliable proof of legal and sustainable origin of timber. A concept to approach this step has been developed by the Johann Heinrich von Thünen Federal Research Institute. The general problem of certificates not being a reliable source of origin might be tackled with the help of a biotechnological method that identifies trees according to their DNA (Tnaha et al. 2010). In that way the exact origin can be determined and, hence, a certain statement on legality made. At the moment this practice is in the developmental state and still too complex to be practiced on a large scale.

13.4.2 Environmental Protection vs. International Trade Agreements

Further inconsistencies and potential sources of conflict can be detected with the coming together of legislation for environmental protection such as green procurement laws and international agreements for free trade, especially the General Agreement on Tariffs and Trade (GATT) by the World Trade Organisation (WTO).

The focal points of future WTO developments are the elimination of tariffs and non-tariff barriers in all sectors, including the forestry sector and the examination of environmental labelling practices (FAO 2003; WTO, official webpage (a)).

Alternatively there are increasing trade-restricting requirements for public procurement which demand environmental labelling. These actions might internationally be considered as violation of the GATT. Due to the complexity of the existing multi-level governance system, within commercial and environmental law, it is difficult to find consensus within individual cases. Generally the WTO agreements against trade restrictions have high priority and there have been historical cases where nationally implemented environmental regulations have been annulled for being incompatible with existing WTO law.

However, according to the modern legal interpretation of WTO legislation, especially the GATT, there are numerous opportunities for member states to implement regulations that shall have the aim to protect the environment and human health as long as they comply with the given trade regulations. Yet, such restrictions must be according to non-discrimination, transparency, predictability and the prohibition of quantitative restrictions, which are fundamental principles of the WTO (WTO, official webpage (b)).

The following requirements for trade restricting environmental laws are concluded by Hilf (2000) on the basis of the shrimp-turtle case[20]:

- Trade restricting action must conduce to legitimate public interest of protection. This is especially the case when these interests are accepted within international conventions on environmental protection.
- Protection measures must be compliant with national legislation of the importing country which means the legal system has to be consistent.
- Measures must effectively serve protection.
- Measures must be essential, meaning no other measure than the import restriction that would be less burdening may exist. In the shrimp-turtle case this point was not met since the concerned countries were not given the chance to implement other measures for turtle protection first.
- The commandment for international cooperation which is registered in the Rio Declaration, Agenda 21 and several international environmental agreements has to be regarded. Before the implementation of unilateral actions there must be efforts of coming to a joint agreement.
- Discrimination and differentiation among different exporting countries is not permitted.
- Transparency and verifiability are foundation of import ban decisions.
- There from results the inadmissibility of any arbitrary discrimination or concealed trade barrier.

If these principles are adequately regarded for unilateral timber restrictions from the German side there should not be any major obstacles in respect to WTO rules. For public procurement measures, however, it has to be particularly obvious that no further objectives are aimed at. Imaginable here would be the enhancement of market advantage for specific companies or the protection of the domestic market.

13.5 Conclusions and Recommendations

In the course of the fight against deforestation and global warming the legislative background on timber trade and green procurement is steadily improving from the side of the European Union. Regulations for securing legality are very contemporary. Here the tendency to tackle problems from producer and consumer sides is to be highlighted. Illegal logging is an issue that exists mainly in tropical regions and is regarded as a major crime business. UNEP and Interpol estimate that the damage of illegally harvested tropical timber is between US$30 and US$100 billion, or 10–30 % of global wood trade (Nellemann and INTERPOL Environmental Crime Programme 2012, p. 6). Therefore, in the course of the FLEGT regulation, the

[20] United States—Import Prohibition of Certain Shrimp and Shrimp Products, WTO case nos. 58 (and 61), India etc versus US, ruling adopted on 6 November 1998.

partner countries were chosen in a manner that covers the areas in need for support. Presently, a first VPA is formally concluded with Ghana.[21] VPAs with the Republic of Congo[22] and with Cameroon[23] are presently in the ratification process. VPAs with Malaysia, Indonesia and Vietnam are presently negotiated, as well as with the Central African Republic, Liberia, Gabon and the Democratic Republic of Congo. Furthermore VPAs may be concluded in the future with other countries in Central Africa and Southeast Asia, as well as with Latin American states and Russia. The VPAs shall not only ensure the trade of legal timber but also the conversion of capacity building and expertise for sustainable forest management. This bilateral approach shares responsibility among the two parties involved. The EU Timber Regulation is the next step at EU level that will theoretically eliminate any possibility for introducing illegally harvested timber into the internal market of the EU and it includes also Iceland, Liechtenstein and Norway, because of the relevance of the regulation for the European Economic Area (EEA). But, the EU FLEGT-system is not working well, yet, as a study for the EU, carried out by the non-profit organisation Resource Extraction Monitoring (REM) operating as independent monitor of Law Enforcement and Governance, found out between 2010 and 2013 in the Democratic Republic of Congo (REM 2013; Kwasniewski 2013).

On a national level in Germany, green procurement is given high priority when it comes to the use of tropical timber. If regulations for environmentally friendly public procurement of wood are implemented in Europe, the market demand for certified wood is automatically being raised. Therefore, within the last years the certified forest area quickly increased in order to enable the timber suppliers to keep up with current market demands. However, from the total area of certified forests only 5 % are to be detected in tropical countries (UN 2009) and, according to Thang (2003), there were no significant reductions of tropical deforestation after the implementation of certification.

From the European side, especially for the analysed public procurement regulations, it seems to be disregarded that there is a second part to this evolving environmental awareness: Developed countries cannot demand sustainably produced timber without contributing the relevant transfer of knowledge, technology and capacity towards the suppliers. Sustainable certification schemes for forest management would have to be developed under the guidance of participants with the necessary expertise (Ozinga 2004). Otherwise, they will never tackle the problem of deforestation profoundly. Cooperation between the producer and the

[21] Voluntary Partnership Agreement between the European Community and the Republic of Ghana on forest law enforcement, governance and trade in timber products into the Community, OJ L 70, p. 3.

[22] Voluntary Partnership Agreement between the European Union and the Republic of the Congo on forest law enforcement, governance and trade in timber and derived products to the European Union (FLEGT), OJ L 92, p. 127.

[23] Voluntary Partnership Agreement between the European Union and the Republic of Cameroon on forest law enforcement, governance and trade in timber and derived products to the European Union (FLEGT), OJ L 92, p. 4.

consumer side is the most important issue to be improved. All stakeholders that are involved in the industry must be motivated to support sustainable forest management that results in certification. The preliminary investments on that account are to be done by forest owners. The motivation and means towards these investments must, however, at least partially, be given by the customer countries (Muthoo 2012). Strengthening the national and local governance of the tropical countries is an important step that needs to be attempted.

As discussed in Sect. 13.4, European legislation is divided into two branches regarding tropical timber. The first one aims at proving legal origin while the second demands for sustainable forest management which does not necessarily include exact information on origin. However, as legality and sustainability are strongly interconnected, they are to be reached jointly. At the moment, the voluntary certification schemes are considered insufficient to cover the aspect of legality. Here the need for profound restructuring arises as legal origin is one of the most basic demands that shall be fulfilled by certificates from voluntary sustainability standards. From the legal perspective there are attempts for better mutual fulfilment and recognition of the two aspects, however, as the issue is still under development the results are yet to be seen.

References

Beeko C, Arts B (2010) The EU–Ghana VPA: a comprehensive policy analysis of its design. Int Forestry Rev 12(3):221–230

Blaser J, Sarre A, Poore D, Johnson S (2011) Status of tropical forest management 2011. ITTO Technical Series No. 38. International Tropical Timber Organization, Yokohama

FAO – Food and Agriculture Organization of the United Nations (2003) Forestry trade issues in the WTO. FAO Fact Sheets. Input for the WTO Ministerial Meeting, Cancun

FAO – Food and Agriculture Organization of the United Nations (2012) State of the World's Forests 2012. Rome. http://www.fao.org/docrep/016/i3010e/i3010e00.htm. Last accessed 13 Sept 2013

Greenpeace (2002) Hamburgs Senat tritt Regenwald mit Fuessen – Trotz Beschluss soll Tropenholz aus Mosambik im Congress Centrum verlegt werden (Senate of Hamburg tramples on rain forest – despite a decision tropic timber shall from Mozambique shall be used in the Congress Centre). Press release, Hamburg. http://www.pressrelations.de/new/standard/result_main.cfm?pfach=1&n_firmanr_=101150&sektor=pm&detail=1&r=100548&sid=&aktion=jour_pm&quelle=0. Last accessed 13 Sept 2013

Hermann A, Acker H (2011) Regelungen der Bundesländer auf dem Gebiet der umweltfreundlichen Beschaffung (Regulation of the federal states on environmental friendly procurement). Öko-Institut e.V. commissioned by Umweltbundesamt. Texte 52/2011, Freiburg

Hilf M (2000) Freiheit des Welthandels contra Umweltschutz (Freedom of international trade vs. environmental protection). In: Documentation to the 23rd Scientific Conference of the Gesellschaft für Umweltrecht e.V. (German Society for Environmental Law), Berlin, 1999

Kwasniewski N (2013) So kommt illegales Tropenholz nach Deutschland (How illegal tropic timbers comes to Germany), Spiegel-online, 13 August 2013. http://www.spiegel.de/wirtschaft/wie-illegales-tropenholz-auf-den-deutschen-markt-kommt-a-915804.html. Last accessed 13 Sept 2013

Muthoo MK (2012) Forest certification and the green economy. Unasylva – Int J Forestry Forest Industries 63(239):17–23

Nellemann C, INTERPOL Environmental Crime Programme (eds) (2012) Green Carbon, Black Trade: Illegal Logging, Tax Fraud and Laundering in the Worlds Tropical Forests. A Rapid Response Assessment. United Nations Environment Programme, GRIDArendal. http://www.unep.org/pdf/RRAlogging_english_scr.pdf. Last accessed 13 Sept 2013

Ozinga S (2004) Time to measure the impacts of certification on sustainable forest management. Unasylva – Int J Forestry Forest Industries 55(219):33–38

Purbawiyatna A, Simula M (2008) Developing forest certification: towards increasing the comparability and acceptance of forest certification systems worldwide. International Tropical Timber Organization, Technical Series No. 29, May 2008

REM – Resource Extraction Monitoring (2013) Independent Monitoring of Forest Law Enforcement and Governance (IM-FLEG) in the Democratic Republic of Congo – Final Report. http://www.observation-rdc.info/documents/REM_IMFLEG_2013_report_DRC.pdf. Last accessed 13 Sept 2013

Robin Wood (2002) Offener Brief von Greenpeace, ROBIN WOOD und WWF Deutschland zum "Runden Tisch" am 11. Dezember 2002 zur Diskussion der Eignung des malaysischen Zertifizierungssystems MTCC als Nachweis für nachhaltige Waldbewirtschaftung (open letter of Greenpeace, ROBIN WOOD and WWF Germany on the 'round table meeting' from 11 December 2002 on discussing the ability of the Malayan certification system MTCC as proof for sustainable forest management). Hamburg & Frankfurt (Main). http://www.robinwood.de/german/trowa/offenerbrief.htm. Last accessed 13 Sept 2013

Thang HC (2003) Current Perspectives of Sustainable Forest Management and Timber Certification. Paper submitted to the XII World Forestry Congress, Quebec City, 2003. http://www.fao.org/docrep/ARTICLE/WFC/XII/MS10-E.HTM. Last accessed 13 Sept 2013

Tnaha LH, Leea SL, Nga KKS, Faridahb Q-Z, Faridah-Hanumb I (2010) Forensic DNA profiling of tropical timber species in Peninsular Malaysia. Forest Ecol Manage 259(8):1436–1446

UN – United Nations (2009) The forest sector in the green economy. Timber and Forest Discussion Paper 54, United Nations Economic Commission for Europe and Food and Agriculture Organization of the United Nations Timber Section, Geneva

UN – United Nations (2011) Forest products – annual market review 2010–2011. United Nations Economic Commission for Europe/Food and Agriculture Organization of the United Nations, Timber and Forest Study Paper 27, Geneva

WTO official webpage (a): Eliminating trade barriers on environmental goods and services. http://www.wto.org/english/tratop_e/envir_e/envir_neg_serv_e.htm. Last accessed 13 Sept 2013

WTO official webpage (b): An introduction to trade and environment in the WTO. http://www.wto.org/english/tratop_e/envir_e/envt_intro_e.htm. Last accessed 13 Sept 2013

Chapter 14
The Role of Sustainability Standards in the Energetic Use of Palm Oil Plantation Residues: Case Study of Cameroon

Michael Schmidt, Berthold Hansmann, and Pia Dewitz

14.1 Introduction

This chapter aims to discuss the sustainability aspects of using residual wood from plantations for the production of electricity and heat. There are continuous debates about the potential of biomass feedstock to help solve the numerous environmental problems related to the growth of global energy consumption we face nowadays. Political aims and commitments of individual firms are putting pressure on the limited biomass resources of the European Union. As a consequence, various options for importing biomass from countries outside the EU are being assessed for their sustainability and impacts on the climate.

For several years, the majority of politicians, economists and the public have considered bioenergy a solution to the climate crisis and the need to meet the increasing global energy demand. However, biomass as a renewable energy source has been recently discussed controversially. The many facets of biomass as an energy source have caused misleading assumptions and wrong judgments about its sustainability and its impact on the global carbon household. The ongoing debates

M. Schmidt
Department of Environmental Planning, Brandenburg University of Technology (BTU) Cottbus, P.O. Box 101344, 03013 Cottbus, Germany
e-mail: michael.schmidt@tu-cottbus.de

B. Hansmann
Programme for Social and Environmental Standards, Deutsche Gesellschaft für Internationale Zusammenarbeit (GIZ) GmbH, Eschborn, Germany
e-mail: berthold.hansmann@giz.de

P. Dewitz (✉)
Department of Environmental Planning, Brandenburg University of Technology (BTU) Cottbus, P.O. Box 101344, 03013 Cottbus, Germany

Programme for Social and Environmental Standards, Deutsche Gesellschaft für Internationale Zusammenarbeit (GIZ) GmbH, Eschborn, Germany
e-mail: pia.dewitz@hotmail.de

C. Schmitz-Hoffmann et al. (eds.), *Voluntary Standard Systems*, Natural Resource Management in Transition 1, DOI 10.1007/978-3-642-35716-9_14,

about the negative impacts of biofuels produced from palm oil, soy and energy plants on the climate, local economies and international trade have lead to a review of the concept and strategy among politicians and society.

For biofuels and liquid energy carriers, the EU Renewable Energy Directive (RED) (2009/28/EG) provides criteria and requirements for the proof of a product's origin. For the use of solid and gaseous biomass for electricity generation and heating, in contrast, the European Commission has published recommendations and guidelines for sustainability criteria which are voluntary and have not yet been implemented into national law by any of the member states (IFEU 2011).

The purpose of residual biomass is its processing to wood chips or pellets and subsequent combustion in existing power plants, so-called *co-firing*. This technology is the process of replacing a share of fossil fuel supplied to a power station with a renewable alternative, such as solid biomass (Kaltschmitt et al. 2009).

The main focus of this chapter is the discussion of relevant criteria for the assessment of the sustainable production of wood chips or pellets and how these criteria are addressed by voluntary sustainability standards.

With this view, the chapter has the following objectives:

- Providing an overview of the available resources suitable for the production of solid biofuels on the basis of the Cameroon case;
- Identifying relevant criteria for sustainable energy production from woody organic residues; and
- Discussing examples of standards and certification schemes applicable to the specific case of imported energy wood.

In this context, this chapter supplements and continues the studies on voluntary sustainability standards in the palm oil and forestry sector by proposing their extension to cover aspects which are significant for the responsible use of plantation residues. The chapter starts by presenting the current situation of the palm oil industry and the related biomass accumulation in Cameroon. Section 14.3 discusses various sustainability aspects among which are ecological impacts and socio-economic considerations attributed to the energetic use of plantation residues. As part of the challenge of balancing greenhouse gas (GHG) emissions, the carbon dioxide emission reduction potential will be explained.

The chapter ends with conclusions and recommendations in Sect. 14.4, highlighting the most important criteria which need to be considered for possible future binding guidelines on the political and corporate level in order to ensure sustainable provision and use of solid biofuels.

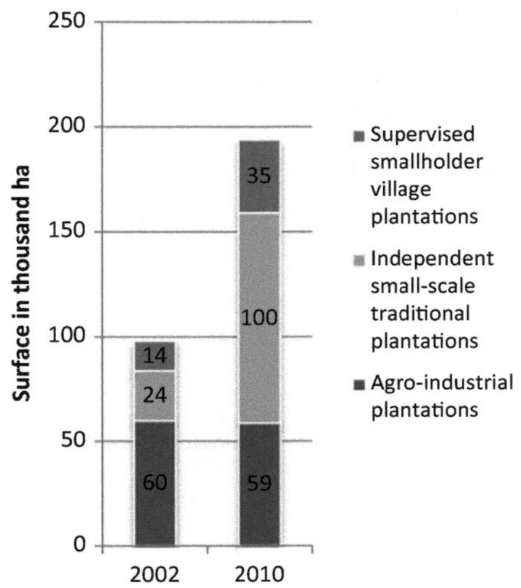

Fig. 14.1 Development of area of palm oil plantations in Cameroon (adapted from Hoyle and Levang 2012; World Rainforest Movement 2006)

14.2 Oil Palm Residues in Cameroon and Their Energetic Potential

The apparent trend of increasing palm oil production in many African countries, e.g. Cameroon (Hoyle and Levang 2012), leads to a proportionately increasing amount of biomass on palm oil plantations. The following figures provide data on the surface extension (Fig. 14.1) and the development of oil palm production of three different plantation types (Fig. 14.2) which is necessary to estimate the amount of the available biomass and its energetic value.

The data show that the surface, and consequently the production of palm oil, has increased significantly over the past 10 years. Today, 194,000 ha of land in Cameroon are dedicated to oil palm cultivation (Hoyle and Levang 2012). While the total surface of palm oil plantations in Cameroon has almost doubled since 2002, the raw production of palm oil has risen by around 64 %. The largest area of oil palm plantations is in the hands of independent small-scale traditional plantations, which has also expanded the most. The respective area is now four times as large as it was in 2002. At the same time, the surface area of agro-industrial plantations has remained more or less stable. Despite the strong area expansion of independent small-scale plantations, their production of palm oil has not increased proportionately, implying a decreasing productivity of palm oil per unit area. Simultaneously, a productivity increase of agro-industrial plantations and smallholder village plantations can be concluded from the observation that palm oil production of these two types has increased more than the corresponding area. According to the data, the overall productivity has decreased from 1.43 t/ha in 2002 to 1.19 t/ha in 2012.

Fig. 14.2 Development of palm oil production in Cameroon (adapted from Hoyle and Levang 2012; World Rainforest Movement 2006)

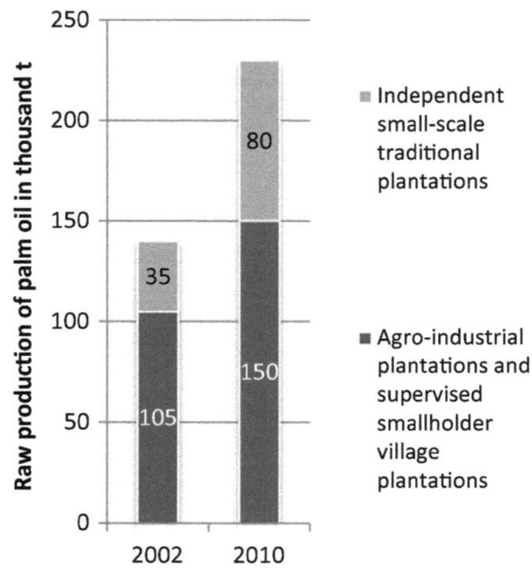

Nevertheless, the trend towards a further expansion of palm oil production is obvious. An increase in production to 300,000 t in 2015 and 450,000 t in 2020 is proposed by the Government of Cameroon's Rural Sector Development Plan. This aim will be achieved mainly by increasing oil palm production, oil extraction yield and the area under production. The latter is more in the government's focus than the palm oil yield or any environmental or biodiversity concern. The Government of Cameroon plans to further promote industrial palm oil production as part of its growth, employment and poverty reduction policies. In order to meet domestic demand and for export purposes, palm oil production is a national priority. According to the 1994 New Agricultural Policy of the Ministry of Agriculture and Rural Development (MINADER), there was a need for increased investment in agro-industry through privatisation of existing public institutions and the creation of new agro-industrial palm oil plantations (Hoyle and Levang 2012).

Increasing amounts of biomass on palm oil plantations imply higher masses of residues, e.g. felled, old plants or palm kernel shells as a residue of the palm oil production process. Besides the availability of biomass, heating values are a decisive factor determining the suitability of organic residues for energetic purposes. The heating value of conventional wood sourced in Europe, for example spruce or beech, is 15 MJ/kg (air dry) (LWF 2011). In comparison, heating values of oil palm trunks (OPT) of around 17 MJ/kg (UNEP 2012) can easily compete with conventional raw material used for biofuel production. Consequently, plantation residues theoretically bear a relatively high potential for energy generation despite their high moisture content. Especially OPT are a suitable source of raw material for the production of wood chips or pellets because they account for almost half of all oil palm compartments available at felling after 25 years (UNEP 2012). Assuming similar values for the case of Cameroon, more than 9 million tonnes of OPT accumulate on the total area of plantations.

Bioenergy produced from palm kernel shells (PKS) could be one of the most promising alternatives because it does not involve costly extraction of biomass from the plantation site since it accumulates during the production of palm oil. This would imply lower potential impacts on soil fertility and consequential changes in fertiliser use. They already find use as boiler feed in palm mills (UNEP 2012) and have become a popular trading commodity and subject of research projects despite their relatively small availability compared to other residual products of the oil palm. Refining of raw oil palm biomass, e.g. pelletisation or torrefaction, can achieve even higher energetic values and thus gives an even stronger competitive advantage regarding the energy potential of organic combustibles.

The area of palm oil plantations in Cameroon is still quite small compared to countries leading in palm oil production, such as Indonesia and Malaysia (Hoyle and Levang 2012). However, the growth trend of the palm oil industry in Cameroon promises an increase in biomass accumulation on plantations leading to a higher availability of residues suitable for energy generation in the future. This trend is supported by increasing regulations preventing the clearing of forests such as the Reduced Emissions from Deforestation and Degradation (REDD) mechanism becoming more active in the leading palm oil producing countries rather than in Africa (Hoyle and Levang 2012). A longer planting history and a higher total hectarage could make OPT available on a continuous basis. Whether such development is desirable from an environmental viewpoint, of course, is questionable.

14.3 Sustainability Aspects of Imported Solid Biofuels

Sustainable sourcing of solid biofuels addresses a wide range of criteria to be taken into account. According to the European Commission, efficiency of energy transformation, social aspects, the protection of ecosystems with a large biodiversity and high carbon stock, as well as the reduction of greenhouse gases, play an important role in this context. The "Report from the Commission to the Council and the European Parliament on sustainability requirements for the use of solid and gaseous biomass sources in electricity, heating and cooling" (further referred to as COM (2010)11), published by the European Commission in 2010, provides recommendations and guidelines for criteria for GHG emission reduction related to the use of solid biofuels. It states that biomass cannot be taken from protected nature reserves or primary forests. With reference to woody biomass, deforestation and extensive extraction of forestry residues shall be avoided. Moreover, there are potential conflicts with the food industry, crop prices, land use rights and national standards of labor. Specific on-site ecological problems concern soil fertility, as determined by nutrient content and biomass turnover in plantations ecosystems.

Furthermore, there is still a high potential of increasing the efficiency of energy transformation, especially for electricity and heat generation with woody biomass. The European Commission therefore recommends the promotion of combined heat and power systems (EC 2010). Consequently, important criteria for the evaluation of the sustainability of an energy carrier also lie in the energy generating technology applied to it, e.g. measured by energy and heat conversion rates. With reference to palm oil plantations, the Roundtable on Sustainable Palm Oil (RSPO) can serve as an example of how energy efficiency and renewable energy carriers can be covered by a sustainability standard (RSPO 2013). However, explicit requirements for the energetic use of plantation residues are still undefined.

As can be seen, independent of the claim to achieve sustainable bioenergy, some of the mentioned criteria can be considered as increasingly significant also for a holistic view on the sustainable management of palm oil plantations. At this point, a clear distinction between sustainability criteria for imported solid biofuels and for palm oil is necessary. However, the strong interrelation between agricultural production systems and raw material supply for renewable energy production, in this case palm oil production and the sourcing of organic residues for wood chip or pellet production, is apparent. Because of this, standard systems need to be assessed and compared from the perspective of both systems. Existing certification schemes for biomass production often overlap or apply to a specific production system, e.g. the Roundtable on Sustainable Biofuel (RSB) or the Better Sugar Cane Initiative (BSI). In forestry in particular, various voluntary certification systems exist for the proof of sustainability, e.g. the Forest Stewardship Council (FSC) and the Programme for the Endorsement of Forest Certification (PEFC) (IFEU 2011). The challenge is to further investigate whether the same standards can, if only partially, be applied to plantations as specific production systems and the residues accumulating on site. In the following, some case examples on sustainability aspects will be discussed in order to show the complexity of applications and interfaces between the different standards. Based on such considerations, the integration of criteria for the sustainable use of woody residues into existing VSS systems, whether in the context of palm oil or biofuel production, shall be achieved in the future.

14.3.1 Greenhouse Gas Emissions

Among many other sustainability aspects, the carbon footprint of importing woody biomass from tropical countries is the most important because it has a direct influence on the global climate. There is no justification for trading biomass for the production of 'renewable' energy if its entire life cycle turns out have a negative impact on the climate. According to the Intergovernmental Panel on Climate Change (IPCC), bioenergy has a significant GHG mitigation potential if resources are developed sustainably and efficient bioenergy systems are used. However, the net positive GHG reduction impact can be lessened by direct loss of carbon stocks and indirect land use change (ILUC) (Chum et al. 2011).

'Carbon neutrality' for bioenergy has been the presumption in many scientific concepts so far, which assumes that end-of-pipe emissions are automatically offset by regrowth of biomass. This characterisation generalises the GHG emission intensity of biomass use and leads to a problematic misrepresentation of the heterogeneity of biomass. The Institute for European Environmental Policy (IEEP) contests that through international trade of wood pellets, especially from the tropics, this form of bioenergy sourcing can no longer be considered as 'climate neutral'. It is thus unavoidable to assess the climate balance of the whole supply chain. Consequently, for every separate type of biomass import, a complete Life Cycle Assessment (LCA), an analysis of all fluxes of energy and material from production until disposal of a product, should be carried out as proposed by the EU (EC 2010). According to the IEEP (2012), however, many of the data used in LCAs are based on flawed life cycle metrics. The various supply chains of biomass exhibit a great variety with regard to their respective climate impact. Characteristics and conversion pathways through which biomass can be utilised lead to considerable variations of the emission intensities associated with the different feedstocks. This is because for biomass sources where carbon neutrality is presumed, the risk of underestimating emissions in accounting exercises can vary considerably. The greatest weaknesses have their origin in the assumptions underlying the life cycle calculations which have been largely adopted by EU bioenergy policies. Some of these errors can be the risk of underestimating or ignoring alternative uses, the need to replace nutrients or potential effects on soil productivity and soil carbon stocks (IEEP 2012).

It is often assumed that emissions associated with the combustion of bioenergy are automatically offset. In reality however, the extraction of biomass from its original source will inevitably lead to a temporary carbon imbalance on site. This amount can only be balanced in favor of the climate if the carbon content added to the total atmospheric carbon stock is compensated by the re-growth of oil palm biomass capturing carbon from the atmosphere. Nevertheless, COM(2010)11 determines that both for the firing of woody biomass in coal power plants within the emission trading scheme of the European Union and for biomass thermal power plants subsidised by the RED, GHG emissions are generally balanced 'zero' (EC 2010).

At the start of the life cycle of a biofuel, the energy consumed for the cultivation of raw biomass and the related GHG emissions are of particular importance. In order to comply with the requirement that no additional energy is consumed during cultivation, it is crucial that oil palm biomass does not lose its residual character as opposed to biomass exclusively cultivated for energetic purposes, such as short rotation coppice or other so-called 'energy-plants'. The advantage for the overall climate balance of residual wood is that calculations of GHG emissions only start at the point of biomass collection and extraction from the original production site, namely the plantation. The energetic use of plantation residues is linked to the already present expansion and intensification of oil palm cultivation, therefore indirect additional GHG emissions due to deforestation and land use change might have to be taken into account in the calculations. A clear definition of what is 'residual' biomass and how it can influence the development of the palm oil industry is thus necessary to be included in sustainability standards.

Moreover, the accounting exercise may ignore or underestimate the temporal dimension of decomposition or existing uses, or a so-called *counterfactual*. This is relevant, for example, if the cut or pruned oil palm biomass was otherwise oxidised immediately, as for site preparation through burning (IEEP 2012). Since there is no proof of effective residue management in many palm oil producing countries, it can be assumed that residues accumulating on palm oil plantations are usually left to decompose or burned on site, meaning a direct release of carbon into the atmosphere in the latter case. Oil palms store about 40 tonnes of carbon per hectare (UNEP 2012) which is released into the atmosphere when burned. Applied to the total area of palm oil plantations in Cameroon, one calculates around 7.8 million tonnes of carbon contained in oil palm biomass.

Against this background, alternative uses of residual biomass such as soil fertiliser (Choong 2012) or local applications of solid biofuels for cooking and heating can be preferable in terms of resource efficiency, climate impact or local socio-economic benefits. Especially in developing countries, the majority of the poorest households and small industries depend on the supply of woody biomass (Chum et al. 2011). Furthermore, forest residues are already being utilised in the wood panel or chemical industry. In many other cases, residues are relied upon in order to maintain carbon stores in managed forests or they are used as bedding or soil improver in agriculture (IEEP 2012). If such uses are not existent however, the exportation of solid biofuels produced from plantation residues can be considered as a means to reduce the overall (indirect) climate impact related to the palm oil industry.

From another perspective, the conventional use of fossil fuels constitutes an alternative to the option of co-firing which is even more harmful to the climate. From this viewpoint it is worth assessing the GHG emission reduction potential of pellets made from plantation residues destined in particular for substituting emission-intensive fossil fuels in existing coal power plants in Germany. Here, the blending share and the type of fossil fuel to be replaced also influence how climate-friendly the energetic use of solid biofuels is. For example, the specific emission reduction potential of substituting lignite is much higher than that of replacing hard coal because power generation from lignite involves much higher emissions (Vogel et al. 2011).

Concerning the origin of solid biofuels, the highest GHG reduction potential lies in the use of forestry residues from the EU in the form of wood chips or pellets; the latter only if wood is used as process fuel for pellet production. With these options, emission savings of more than 90 %, compared to fossil fuels, can be achieved for both electricity and heat generation (EC 2010). However, most national wood resources are limited for energetic applications, especially in Germany, due to the existing demand for wood for material use such as furniture or paper. An increase of energy production from woody biomass will depend on technology development and the access to additional sources of woody biomass (Vogel et al. 2011). Wood pellets imported from the tropics still bear an emission reduction potential of around 30–70 %, also depending on the process fuel. Using natural gas as process fuel reduces the GHG avoidance potential of wood pellets from both the EU and the tropics by a considerable amount. For example, GHG avoidance for wood pellets from the tropics is reduced by more than 50 % if natural gas instead of wood is used as process fuel for pellet production.

It becomes clear that a final judgment can only be made if a fair baseline for comparing the GHG emission intensity of different energy sources or uses of biomass is provided. For the application of sustainability standards to the energetic use of woody biomass, a highly-differentiated scale of emission profiles of alternative energy sources will be required to enable the comparison and sustainability evaluation of a single energy product. As an integral part of an LCA, a sustainability standard for imported solid biofuels should particularly account for the process fuel used in production, the type of fossil fuel which is substituted and the blending share of co-fired biomass since these factors have a major influence on an energy product's global GHG emission reduction potential.

Some voluntary certification schemes such as the Roundtable on Sustainable Biomaterials (RSB, formerly known as the Roundtable on Sustainable Biofuels) describe requirements for the certification of sustainable biofuels with regard to their contribution to climate change mitigation. Principle 3 requires biofuels to significantly reduce lifecycle GHG emissions (by 50 %) as compared to a fossil fuel reference. The RSB lifecycle GHG emissions calculation methodology includes emissions from land use change, carbon stock changes and gives incentives to use residues in a way that GHG emissions of the biofuel are reduced (RSB 2010). Another certification system for biomass from agriculture and forestry is the Green Gold Label (GGL) which covers all process steps beginning with cultivation until the final energetic or material use of biomass. A separate standard (GGLS8) determines aims and calculation methods in compliance with the EU described formula for GHG emission reduction. The GGL also defines an emission reduction by 50 % (GGL 2012).

Both the RSB and the GGL bear a relatively high potential of application to imported woody biofuels. The principles and criteria of the RSB have been adapted to be in agreement with the sustainability requirements of the EU-RED, including the calculation methodology of GHG. In Agriculture, Standard 2 of the GGL (GGLS2) is a widely accepted certificate (IFEU 2011) but it does not cover the criteria of GHG mitigation addressed by GGLS8. In contrast to the RSB, it only uses the EU-RED as orientation and does not consider the variety of indirect effects on the GHG balance of biofuels.

Most existing standards for sustainable biofuels require a clear understanding of the source, transportation and final use of the type of biomass under assessment in their principles. They often do not describe, however, additional factors influencing the overall emission profile, among which are, for example, negative consequences of biomass extraction through increased fertiliser use, which is known to be highly intensive in GHG emissions. The process fuel used for biofuel production or the type of fossil fuel replaced in the case of co-firing are also ignored in most cases. Eventually, the use of plantation residues for energy production can never be completely 'climate neutral'. The natural decay of biomass will always have a better climate balance than its use for co-firing in power plants with additional emissions through fuel production and transport. As a conclusion, a certain type of biofuel should always be certified only as 'climate-friendly' or relative to the available counterfactuals and not as 'sustainable' *per se*.

14.3.2 Ecological Impacts

As another crucial aspect of sustainability, the protection of ecosystems and biodiversity was mentioned. In most tropical countries, biodiversity loss and ecosystem disturbances are attributed to ongoing deforestation. Especially in developing countries, a reason among others is the absence or insufficiency of decision-making structures concerning the sustainable management of forest resources (EC 2010). This allows the continued expansion of palm oil plantations posing a direct threat to biodiversity through land conversion. As an example, recently planned plantation sites in Cameroon lie inside globally recognised biodiversity hotspots between protected areas (Hoyle and Levang 2012).

From an economic viewpoint, the possibility of additional market value for oil palm biomass can give further incentives to expand plantation areas independent of the already lucrative business of palm oil production. On the contrary, meeting the demand for bioenergy with residual wood from plantations instead of (primary) forests might help to combat extensive extraction of biomass from other sources with higher ecological and biological diversity. The energetic use of plantation residues might become an additional criterion for sustainable palm oil production since it can contribute to the reduction of overall GHG emissions within the entire production system.

The present problem of deforestation attributed to the extension of palm oil plantations has already been addressed by many initiatives, e.g. by the Roundtable on Sustainable Palm Oil (RSPO). One of its criteria (Principle 5) is the environmental responsibility and conservation of natural resources and biodiversity. The certification system explicitly requires that new plantations after 2005 shall not replace primary forests (Principle 7) (RSPO 2013).

The International Sustainability and Carbon Certification (ISCC) introduced in 2010 also define principles referring to the ecological impacts of biomass and bioenergy. It demands that biomass is not sourced from species-rich areas and those of high nature protection value and includes a criterion for sustainable agriculture with a special focus on water and soil quality (IFEU 2011).

As another crucial component of the sustainable management of biomass-producing ecosystems, soil quality and nutrient cycling play a major role. In many forests, the extraction of residues has beneficial effects on nutrient stocks, but only to a limited extent. An increasing demand for biomass residues, however, can cause losses to the soil carbon stock if too few residues are left on site (EC 2010). For some primary forestry and agricultural residues it may be better to keep the residual material in situ, helping to maintain soil carbon stock (IEEP 2012). Especially nutrient-poor soils rather benefit if residues are left where they accumulate (Flaig et al. 1999). In Malaysia, one of the major palm oil producing countries, studies on nutrient cycling and residue management have been carried out since 1994. The so-called 'Zero Burning Technique' involves the planting of oil palm seedlings into the rows of old palm biomass residues in order to conserve nutrients. It is described as the "most effective method of biomass management during replanting" because young palms can easily take up the nutrients released by chipped or shredded residues. The major benefits of this method are the conservation of soil fertility through nutrient recycling and moisture conservation, as well as the maintenance of biodiversity of soil fauna and microbial communities (Khalid et al. 2007).

Another positive side-effect also with regard to the energy and GHG emission intensity of palm oil production is the reduction of chemical fertiliser input to 50 % (Khalid et al. 2007). Total estimated emissions related to the life cycle of fertiliser accounts for around 2.5 % of total global GHG emissions (IFA 2009). Against this background, innovative residue management on oil palm plantations does not only reduce spending on fertiliser, but also the environmental impact of fertiliser production and use, without expanding the area under agricultural production in order to maintain yields. The returning of nutrient-rich ashes after combustion is another alternative method to maintain soil fertility. This idea, however, seems too unrealistic for long-distance biotrade as in the case of Cameroon which makes the returning of ashes economically inefficient and thus undesirable.

Many standard systems are limited with regard to the effect of energetic use of residues on plantation ecosystems. As stated before, a holistic understanding of sustainable palm oil production should integrate the sustainable use of residual biomass not only with regard to its climate impact and its energetic potential, but also it's potential as soil fertiliser. The extraction of woody biomass for energetic use can compete with the sustainable management of plantation residues from an environmental viewpoint. Therefore, the local use of plantation residues as a nutrient source should be outweighed against its energetic use in terms of costs, energy consumption and GHG emissions along the entire life cycle of the product.

To give an example, one principle of the Roundtable on Sustainable Biomaterials (RSB) is the requirement that biofuel operations shall apply practices that help to reverse soil degradation and protect soil quality (IFEU 2011). The intensive extraction of residual biomass on palm oil plantations could severely lower the soil's nutrient content. If the production of woody biofuels from plantation residues can be defined or interpreted as such operation, there is no hindrance for the application of such criteria to this specific case.

With regard to palm oil plantations, the RSPO has formulated a principle for appropriate best practices by growers and millers (Principle 4). Criterion 4.2 requires practices which maintain soil fertility "at a level that ensures optimal and sustained yield" (RSPO 2013). It further explicitly states that nutrient recycling strategies should include empty fruit bunches (EFB), palm oil mill effluent (POME) and palm residues after replanting and any of its use for energy production (RSPO 2013). To which type of energy conversion pathway should priority be given however, is not defined. Nevertheless, it becomes clear that the RSPO acknowledges the nutritional and energetic potential of oil palm residues and requires the avoidance of fire and residue burning for land preparation (Criterion 5.5) as it goes along with unnecessary environmental pollution (RSPO 2013). The recognition of utilising by-products for added value will be essential for certifying the sustainability of palm oil biomass energy products and at the same time benefit the sustainability of palm oil production (NL Agency 2013).

14.3.3 Socio-Economic Aspects

In many regions of this planet, biomass potentials are left unused or used inefficiently. In Africa, local uses of biomass as an important traditional source of energy in the rural areas are prevailing, but there is still a surplus. In 2009, 20 % of biogenic solid fuels are used in Africa and the majority (80 %) is used by the poorer population for meeting daily energy demands, e.g. heating and cooking (Kaltschmitt et al. 2009).

Due to the lack of access to modern technologies, however, the combustion efficiency of traditional open fires and simple stoves is usually low. A more efficient energetic use of woody biomass by the local population can be achieved through improved cookstoves (ICS) and other advanced biomass systems for cooking (Chum et al. 2011). The local use of refined biomass as fuel for households in the country where biomass is produced, e.g. in the form of wood chips or pellets, presents a promising alternative to biomass exports. It can help meet the energy demand of the local population in a more efficient way while reducing the dependence on fossil fuel imports and adds value to the local and national economy. Individual local added value can further lead to an improvement of the financial situation of individual plantation owners. From a microeconomic perspective, the sales of palm oil plantation residues provide the opportunity for an extra income for local oil palm cultivators. Small-holders might especially benefit from a higher profit margin than large agribusinesses, directly passing on additional incomes to their families. In general, a stronger relationship between owner and property, e.g. plantation residues as a potential source of bioenergy, yields a better understanding of production processes and consequently a stronger interest to protect and sustain them. In this way, uncontrolled and inefficient uses of biomass by burning or decomposition can be avoided. According to the principle that ownership entails obligation, the chances of illegal use of residual biomass can be reduced. Positive impacts of introducing pellet production in Cameroon in general are additional employment, revenue to the state and infrastructure expansion. In the long run, a sub-industry of residue processing associated with oil palm cultivation and palm oil production can develop over time, achieving economic development and labor opportunities in the region.

For future developments of bioenergy systems, trade-offs between environmental and socio-economic criteria need to be taken into account. With regard to the climate impact of imported solid biofuels, the optimal scenario would be the local use of bioenergy, given that logistics are a major emitter of GHG along the supply chain. However, this option has been shown to be less promising than international biomass transports despite energy consumption and emissions caused by sea transport over large distances. The comparison of various bioenergy systems has proven that international compared to local utilisation of biomass has a higher overall reduction potential of GHG emissions (Uasuf 2010). As a consequence, a judgement on the sustainability of the energetic use of plantation residues depends on whether the focus lies on its climate impact or on socio-economic benefits to local economies. Eventually, those aspects prioritised by a specific sustainability standard will determine whether the type of biomass under assessment should rather be used locally or traded internationally.

Closely linked to the criteria related to the social and economic impacts of the palm oil industry and bioenergy production in developing countries, sustainable residue management and its processing should undergo similar treatment by the respective standards. A current issue, for example, is the fact that large agribusinesses usually seek large tracts of land and do not involve smallholders in their projects (Hoyle and Levang 2012). It is likely that residue extraction will take place most efficiently in areas where large-scale oil palm cultivation takes place thanks to existing management structures and economies of scale. Government support of industrial palm oil production, as in the case of Cameroon, could strengthen the position of agribusinesses investing in the trade or energetic use of residual biomass. As an important criterion for both sustainable bioenergy and palm oil production, the treatment of plantation residues as an integrated part of agricultural management needs to be carried out in a way that is smallholder-friendly and maximises economic efficiency with special focus on local added value. Social aspects referring to land use rights and living conditions of the local population have been adopted partly by the RED, but more comprehensively by the RSB, SAN and RSPO (IFEU 2011).

The RSPO is continuously developing detailed guidance for the application of its principles and criteria by smallholders which supports them in the certification procedure. However, it does not identify residue management and bioenergy production as a crucial socio-economic factor for rural development, an aspect which might have to be included in the future in order to further strengthen the position of smallholders (RSPO 2013). The Sustainable Agriculture Network Standard (SAN), linked to the Rainforest Alliance in contrast includes as a critical criterion that farm management must "implement policies and procedures for identifying and considering the interest of local populations and community interest groups" (SAN 2010). As part of Principle 7 on community relations, the standard requires the certified farm to collaborate with the development of the local economy and infrastructure (SAN 2010). Given that oil palm biomass is a resource with increasing economic value, the extraction of plantation residues in this context could be considered a new farm activity with a potential to have an impact on employment and local resource economy, as described in the principle. At this point, a clearer definition of such activities would help to identify the energetic use of residual biomass among other measures which influence the social and economic wellbeing of local communities.

14.4 Conclusions and Recommendations

Given the increasing amounts of solid residues on oil palm plantations and the high energetic potential of the accumulating organic matter, the production of solid biofuels from such raw materials becomes increasingly relevant for environmental, social and economic considerations. As for the case of palm oil production, for credible certification of sustainable production of biofuel from solid organic matter, all aspects of sustainability need to be addressed. Especially the GHG emission balance of exported solid biofuels, impacts on biodiversity through direct and indirect land use change as well as soil fertility deserve particular attention. When it comes to socio-economic considerations, sustainability standards should take into account the smallholder-friendliness of its application. Stable and reliable energetic use of solid biomass on this scale appears to be of higher significance to traditional agricultural structures, especially for the poorer share of the population.

Concerning the sustainability of biofuels produced from plantation residues, its local use is recommendable in order to reduce the climate impact attributed to the logistics phase of imported biofuels. Nevertheless, the co-firing of imported wood pellets produced from plantation residues in developing countries reveals a much more climate-friendly emission profile compared to the business-as-usual combustion of fossil fuels. In general, the international aim of GHG emission reduction should always be viewed from a global perspective rather than a local one because climate warming affects the whole planet. However, local efforts can help achieve this global aim, and bioenergy systems applied close to the sites where raw material is supplied should be given priority provided that they show a better climate balance relative to international trade of biofuels. Against this background, one of the most important principles of sustainable biofuels will be a clearly defined method of GHG life cycle calculations. The calculation methods described by most standards addressing the reduction of GHG emissions, including RSB, ISCC and GGL, refer to the RED by the European Commission. For the correct allocation of GHG emissions, all of the mentioned standards can thus be referred to when the certification of imported biofuels is at stake. However, as emphasised before, remaining errors in the method of calculation and estimations proposed by the EU should be removed, taking into account further parameters such as the carbon intensity of existing uses, long-term soil carbon stock and the temporal dimensions of biomass production and biofuel use.

The trend of expanding palm oil plantations in developing countries such as Cameroon is apparent and implies ecological problems related to indirect land use change and biodiversity loss through deforestation. The impact of biomass extraction on soil quality and thus long-term productivity of palm oil plantations is still vague or considered irrelevant. Woody biomass loses its residual character if its energetic use competes with local, innovative residue management techniques which aim to maintain the nutrient balance in soils. The efficient use of residues for the purpose of soil fertility conservation should thus have priority over any other alternative usage and should be a component of any standard system certifying sustainable palm oil production such as proposed, to a limited extent, by the RSPO. A measure against the ongoing deforestation of primary rainforests and deteriorating soil health could be in the form of maximum extraction limits of residual woody biomass for any purpose other than soil fertiliser. The reduction of soil carbon through extensive use of woody residues is dealt with for example by the RSB, but it does not consider the nutrients contained in the residues (IFEU 2011). Consequently, a high potential of developing the component of soil protection for the standardisation of sustainable biofuels made from plantation residues remains. From an environmentalist perspective, the extraction of biomass from the site of its initial growth should be avoided for the simple reason that all human interference in plant ecosystems can disturb its carbon and nutrient balance in soils. The term *sustainability* of solid biofuels should thus always be used in a relative and cautious manner.

From a socio-economic viewpoint, many principles on the valorisation of labour and land use rights, business transparency, standards of living and the interests of the local population are included in many existing standards, e.g. the RSB, ISCC, FSC, SAN and RSPO (IFEU 2011). They do not require further extension in order to be applicable to the production of bioenergy from organic residues. However, the scope of application of the sustainability criteria suggested by COM(2010)11 for standard systems should also cover small-scale users of biomass when applied to developing countries despite the additional administrative burden. The production and use of certified solid biofuels could add value to local economic entities and encourage higher performance and efficiency for those who depend on local biomass sources the most. From another perspective, the certification of biofuels imported from developing countries should also involve an assessment of competing, economically more sustainable uses of biofuels on smaller local scales.

For any sustainability standard addressing the use of woody biomass as a source of energy, a transparent, non-confusing and open communication to consumers of which aspects are covered by a single certificate is of superior necessity. One needs not forget that there is a vast range of existing certification schemes which go far beyond those mentioned in this chapter. It is therefore recommendable that new subjects affected by the international discussion about the sustainable production of biofuel, respectively palm oil, are included in highly-developed and established standard systems already addressing similar products or production systems. With regard to the issue of deforestation, a final challenge will be the social acceptance of palm oil in general and bioenergy imports especially from tropical countries, which is independent of a specific sustainability standard. Eventually, the certification of imported energy carriers produced from woody residues as well as products containing sustainable palm oil requires, as in many similar cases, accurate and factual information accessible to the end user in order to achieve acceptance and consequently market access.

References

Choong MY (2012) Waste not the palm oil biomass. http://www.thestar.com.my/Lifestyle/Features/2012/03/27/Waste-not-the-palm-oil-biomass/. Accessed 23 Mar 2014

Chum H, Faaij A, Moreira J, Berndes G, Dhamija P, Dong H, Gabrielle B, Goss Eng A, Lucht W, Mapako M, Masera Cerutti O, McIntyre T, Minowa T, Pingoud K (2011) Bioenergy. In: Edenhofer O, Pichs-Madruga R, Sokona Y, Seyboth K, Matschoss P, Kadner S, Zwickel T, Eickemeier P, Hansen G, Schlomer S, von Stechow C (eds) IPCC special report on renewable energy sources and climate change mitigation. Cambridge University Press, Cambridge

EC – European Commission (2010) Report from the Commission to the Council and the European Parliament on sustainability requirements for the use of solid and gaseous biomass sources in electricity, heating and cooling. http://eur-lex.europa.eu/LexUriServ/LexUriServ.do?uri=COM:2010:0011:FIN:EN:PDF. Last accessed 27 Aug 2013

Flaig DH, Leuchtweis C, von Lüneburg E, Ortmaier E, Seeger C (1999) Biomasse – nachwachsende Energie. Potentiale – Technik – Kosten. Expert Verlag, Renningen-Malmsheim

GGL – Green Gold Label (2012) GGLS8 – Greenhouse gasses and energy balance calculation standard. http://www.greengoldcertified.org/site/pagina.php?id=11. Last accessed 27 Aug 2013

Hoyle D, Levang P (2012) Oil palm development in Cameroon. http://awsassets.panda.org/downloads/palmoildevelopmentcameroon_english.pdf. Last accessed 27 Aug 2013

IEEP – Institute for European Environmental Policy (2012) Does bioenergy have a role to play in reducing Europe's GHG emissions? http://www.ieep.eu/assets/1008/IEEP_-_The_GHG_Emissions_Intensity_of_Bioenergy_-_October_2012.pdf. Last accessed 28 Aug 2013

IFEU – Institut für Energie- und Umweltforschung Heidelberg GmbH (2011) Kriterien zur nachhaltigen Beschaffung holzartiger Biomasse für die Strom- und Wärmegewinnung im Land Berlin. http://www.ifeu.de/nachhaltigkeit/pdf/IFEU%20nachhaltiges%20Holz%20VattenfallSenGUV%2016-03-11_FINAL.pdf. Last accessed 28 Aug 2013

IFA (International Fertilizer Industry Association) (2009) Fertilizers and climate change. Enhancing agricultural productivity and reducing emissions. http://www.fertilizer.org/ItemDetail?iProductCode=8863Pdf&Category=ENV&WebsiteKey=411e9724-4bda-422f-abfc-8152ed74f306. Accessed 23 Mar 2014

Kaltschmitt M, Hartmann H, Hofbauer H (2009) Energie aus Biomasse. Grundlagen, Techniken und Verfahren, 2nd edn. Springer, Heidelberg

Khalid H, Chan KW, Ahmad TM (2007) Nutrient cycling and residue management during oil palm replanting in Malaysia. Malaysian Palm Oil Board. http://portal.fedepalma.org/conferencia2009/presentaciones/M1_13_Khalid_Haron.pdf. Last accessed 28 Aug 2013

LWF – Bayrische Landesanstalt für Wald und Forstwirtschaft (2011) Merkblatt 12. Der Energiegehalt von Holz. http://www.lwf.bayern.de/veroeffentlichungen/lwf-merkblaetter/mb-12-energiegehalt-holz.pdf. Last accessed 28 Aug 2013

NL Agency – Ministry of Economic Affairs, Agriculture and Innovation (2013) Valorization of palm oil (mill) residues. Identifying and solving the challenges. http://english.rvo.nl/news/palm-oil-mill-residues-wrongly-considered-waste. Accessed 23 Mar 2014

RSB – Roundtable on Sustainable Biofuels (2010) RSB principles & criteria for sustainable biofuel production. http://rsb.org/pdfs/global/11-03-08-RSB-PCs-Version2.pdf. Last accessed 28 Aug 2013

RSPO (Roundtable on Sustainable Palm Oil) (2013) RSPO-principles and criteria for sustainable palm oil production. http://www.rspo.org/en/document_principle_and_criteria_certification. Accessed 23 Mar 2014

SAN (Sustainable Agriculture Network) (2010) Sustainable Agriculture Standard, July 2010 (version 3). http://sanstandards.org/sitio/subsections/display/9. Accessed 23 Mar 2014

Uasuf A (2010) Economic and environmental assessment of an international wood pellets supply chain: a case study of wood pellets export from northeast Argentina to Europe. http://www.freidok.uni-freiburg.de/volltexte/7905/. Last accessed 17 Mar 2013

UNEP – United Nations Environment Programme (2012) Converting waste oil palm trees into a resource. http://www.unep.org/ietc/Portals/136/News/Waste%20Palm%20Tree%20study%20report%20publication/Converting%20Waste%20Oil%20Palm%20into%20a%20Resource_FINAL%20REPORT.pdf. Last accessed 27 Aug 2013

Vogel C, Herr M, Edel M, Seidl H (2011) Die Mitverbrennung holzartiger Biomasse in Kohlekraftwerken. Ein Beitrag zur Energiewende und zum Klimaschutz? Deutsche Energie-Agentur GmbH (DENA). http://www.dena.de/fileadmin/user_upload/Presse/studien_umfragen/Holzmitverbrennung/Endbericht_Biomassenutzung_in_Kohlekraftwerken_final.pdf. Last accessed 17 Mar 2013

World Rainforest Movement (2006) Cameroon: oil palm plantations fostered by new biofuel market harm local livelihoods. http://www.wrm.org.uy/oldsite/bulletin/112/Cameroon.html. Accessed 23 Mar 2014

Chapter 15
The Adoption and Impact of Forest Stewardship Council Standards in the Congo Basin Forestry Sector

Mercy Nambu Diangha and Gerhard Wiegleb

15.1 Introduction

The forestry sectors of the respective countries within the Congo Basin region are aware of the essential roles of forest to their communities and are hence making efforts to manage it. Unfortunately, forests are being degraded and reduced in size at an alarming rate, mainly due to unsustainable management practices (Akinwande 2012). Voluntary Standard Systems (VSS) may help to overcome the situation. In an effort towards more responsible forest management, the Forest Stewardship Council (FSC) standards of certification were adopted in the region to promote social, economic and environmental balance in forest management. Since FSC's adoption, over 5 million ha of forest have received certification for this increasingly recommended approach (Hakizumwami 2011). Also the number of companies seeking certification has increased. Nevertheless, a larger area of the Congo Basin forest and many logging industries in the region has remained uncertified. The impact created since the emergence of FSC in the region is very important for the sustainable management of the region's forest but has not been fully reported.

This chapter therefore seeks to present the impact experienced since the adoption of FSC standards in the Congo Basin region and focuses on the economic, social and environmental aspects of its application. The main objectives are to appraise benefits provided by the standing forest in the Congo Basin, impacts created since the adoption and implementation of FSC standards in the region, and limitations of FSC certification in the region. In addition, the information found in scattered publications and internet sources will be presented in a systematic way.

In this chapter the Congo Basin forest region is described in Sect. 15.2, regarding the biophysical characteristics (15.2.1), benefits offered by standing forests (15.2.2) and current threats faced by forest in the region (15.2.3). The FSC philosophy is

M.N. Diangha (✉) • G. Wiegleb
Department of General Ecology, Brandenburg University of Technology (BTU) Cottbus, P.O. Box 101344, 03013 Cottbus, Germany
e-mail: diangmer@tu-cottbus.de; wiegleb@tu-cottbus.de

C. Schmitz-Hoffmann et al. (eds.), *Voluntary Standard Systems*, Natural Resource Management in Transition 1, DOI 10.1007/978-3-642-35716-9_15,

outlined in Sect. 15.3, focusing on the principles of the FSC (15.3.1); the emergence and adoption of FSC standards (15.3.2); the economic, environmental, and social impact from its application (15.3.3); and some limitations experienced since FSC adoption in the region (15.3.4). The chapter concludes in Sect. 15.4, highlighting the need for an increase in efforts to ensure more tangible economic, social and environmental protection in the Congo Basin Region.

15.2 The Congo Basin Forest Region

15.2.1 Biophysical Characteristics

The Congo Basin forest in Africa remains one of the most important wilderness areas left on Earth (WWF 2013a). It contains the world's second- largest and intact humid tropical forest, after the Amazon Basin (Sonwa et al. 2009). While nine countries have part of their territory in the Congo Basin (Butler 2013), the Congo Basin forest is conventionally associated with the rainforest that spreads across six countries (see Fig. 15.1) including Cameroon, Central African Republic (CAR), Gabon, Equatorial Guinea, the Republic of Congo and the Democratic Republic of Congo—DRC (CBFP 2006; Sonwa et al. 2009; WWF 2013b). The area of forest attributed to these six countries vary in available literature, with estimations including 180 million ha (RFF 2011; WWF 2013b), approximately 202 million ha (WWF 2013a), 251 million ha (CBFP 2013) and approximately 388.5 million ha (Sonwa et al. 2009). These discrepancies have raised considerable uncertainty about the true extent of the Congo Basin forest.

This forest has a global reputation for its rich biodiversity, comprising over 400 mammal species, more than 1,000 bird species and over 10,000 plant species of which about 3,000 are endemic (WWF 2013b). Peaks of endemism are found in the region particularly in Cameroon, DRC and Gabon (RFF 2011). Approximately 30 million people from 150 ethnic groups depend and benefit from a wide range of ecosystem services delivered by this forest (CBFP 2006).

15.2.2 Benefits Provided by the Standing Forest in the Congo Basin

The forest of the Congo Basin, offers provisioning, regulating, cultural and supporting services to the region (Nlom 2011). It remains a vital economic resource and provides a significant source of domestic export for the Congo Basin countries (Ndoye and Tieguhong 2004; Karsenty 2007), where millions of hectares of the forest has been allocated for commercial timber exploitation (Alemagi and Nukpezah 2012) for export. A wide range of provisioning services including

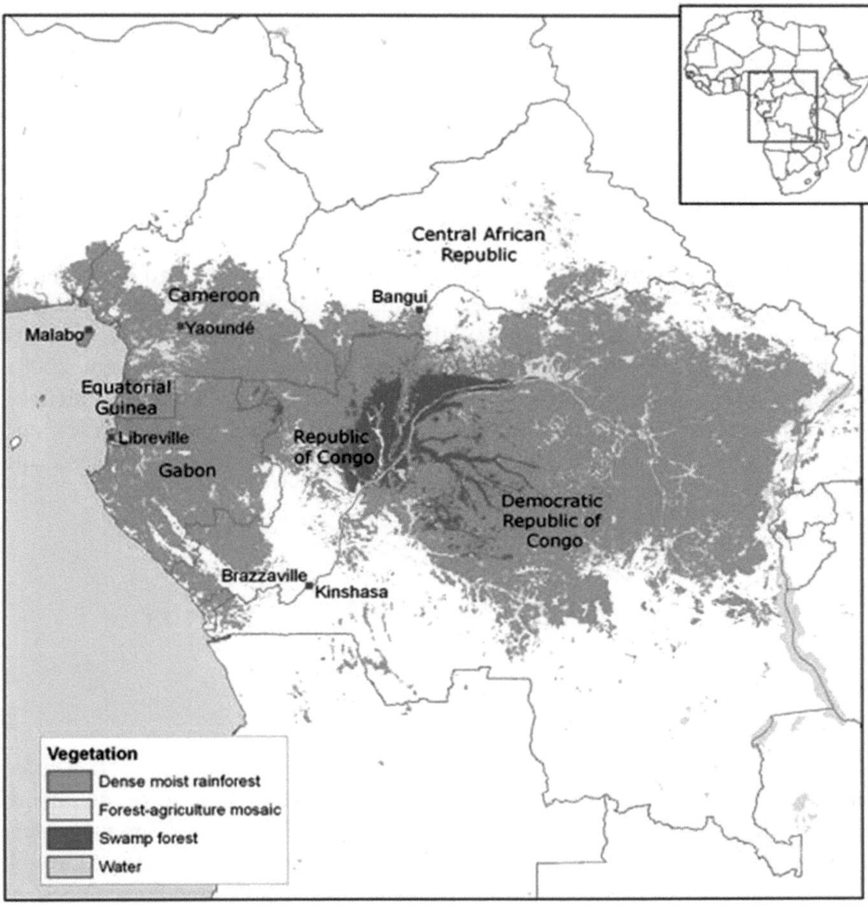

Fig. 15.1 Forest of the Congo Basin (*Source*: Austin et al. 2010)

medicinal and genetic resources; food resources from hunting, gathering and fishing as well as timber; fuel-wood and other tangible goods (Ndoye and Tieguhong 2004; Austin et al. 2010) are enjoyed by the dependent population.

The Congo Basin forest also offers extensive regulating services including watershed protection, carbon sequestration, microclimate and global rainfall regulation (Nlom 2011).

Recent international efforts to reduce emissions from deforestation and degradation (REDD+) hope to direct payments through funding of projects to the Congo Basin region, which will potentially improve maintenance and conservation of biodiversity in addition to improving human well-being (Brown et al. 2010). Culturally diverse ethnic groups inhabit the Congo Basin region where local cultures and livelihoods are fully and intimately entwined with the forest (Nlom 2011).

Additionally, the rich and unique biodiversity and ecosystems in the region has advanced recreational activities, which now serve as sources of revenue and employment for local people.

15.2.3 Problems and Threats to the Congo Basin Forest

Despite the socioeconomic and biophysical importance of the Congo Basin forest, logging, agricultural expansion, bush meat hunting, oil and mineral extraction by the ever-growing and demanding population are exerting pressure on biodiversity and threatening the region's forest ecosystems (Usongo and Nagahuedi 2008). Lacking compliance with national laws and international agreements, undefined land tenure and land use rights, little perceived respect of indigenous peoples' and worker's rights, no management plans, sparse monitoring schemes, and unsustainable plantations may increase the problems.

15.3 Forest Stewardship Council (FSC) and Forestry

15.3.1 Principles of FSC

FSC was founded in 1993, following a recommendation from the 1992 United Nation Conference on Environment and Development (UNCED) at Rio de Janeiro (FSC 2012a) to deal with the environmental impact of poor forest management (Guillery 2011; Green3Dhome 2013). FSC has since then been working through programs and services to advance its voluntary sustainable standards (VSS) of forest certification, empowering organizations, timber extraction companies and communities to support forest management that meets social, economic and ecological needs of the present and future generations (FSC 2012b). FSC is working towards increasing the level of communication and the demand for forest products in the marketplace through its labeling scheme (FSC 2010). It is also developing innovative forest management solutions through stakeholder collaboration and is supporting stakeholder engagement in the certification process.

The requirements set forth by the FSC for adoption are based on ten principles and 56 criteria that describe the essential elements or rules of environmentally appropriate, socially beneficial and economically viable forest management (FSC 2013a). Table 15.1 presents the ten core FSC principles that must be adopted to invite FSC certification along with authors' views on potential problems that may limit their application. Each principle is supported by criteria that help in judging if the principle has been met in practice (FSC 2013a). As for all voluntary standard systems, formulations are vague and leave room for broad interpretations.

Table 15.1 FSC guiding principles and anticipated problems (adapted from FSC 2013b)

Principle	Anticipated problems
Compliance with all applicable laws and international treaties	Incentives missing
Demonstrated and uncontested, clearly defined, long-term land tenure and use rights	Often not guaranteed
Recognition and respect of indigenous peoples' rights	Lack of measures to verify if consent exist, is informed and given freely
Community relations and worker's rights—to maintain or enhance forest workers' and local communities' social and economic well-being	Not enough commitment
Benefits from the forest—to maintain or enhance long term economic, social and environmental benefits from the forest	Often not guaranteed
Environmental impact—to maintain or restore the ecosystem, its biodiversity, resources and landscapes	Inadequate prior information about environmental values and impact that operations could bring on them
Management plan—to have a management plan, implemented, monitored and documented	Often incomplete with limited considerations
Monitoring and assessment—to demonstrate progress towards management objectives	Often expensive. Basis of conflict of interest among stakeholders
Maintenance of high conservation value forests—to maintain or enhance the attributes, which define such forests	Sensitive forest not put off limit to logging
Plantations—to plan and manage plantations in accordance with FSC principles and criteria	Missing incentives

15.3.2 Emergence and Adoption of FSC in the Congo Basin Region

Since 1996, Congo Basin countries have pursued FSC certification, through adherence to the prescribed FSC standards of Sustainable Forest Management (SFM). The first ever FSC certificate in the Congo Basin region was issued to Leroy Gabon in 1996 (Hakizumwami 2011) but was shortly after withdrawn as a result of non-compliance to FSC standards and particularly, due to lack of a complete management plan, inadequate consideration of stakeholders and the partial overlapping of forest units onto the Lopé protected reserve (Atyi 2004). In January 2006, 10 years after the withdrawal of the first certificate, over 41,000 ha of forest concession managed by Wijma Cameroun was certified (Hakizumwami 2011). Since then, FSC standards adoption has made a significant leap forward in the Congo Basin (Alemagi and Nukpezah 2012), where FSC certified forest area has increased from nothing to 5.2 million ha in 2010, while about 5 million ha awaits certification in the near future (Hakizumwami 2011). Figure 15.2 presents the proportion of the total FSC certified forest area by region in Congo Basin in 2010. Republic of Congo is leading with 47 %, followed by Gabon (37 %) and

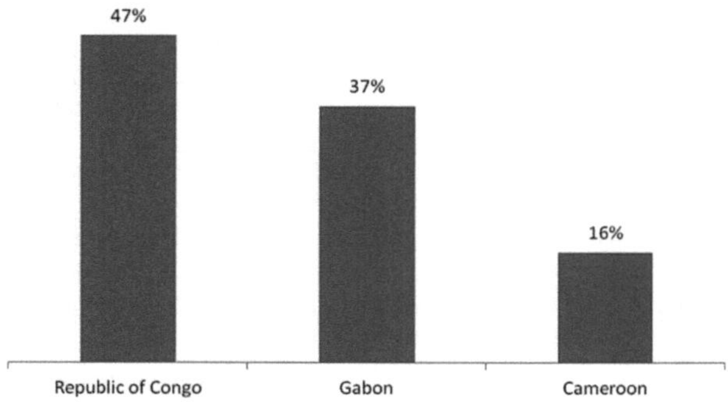

Fig. 15.2 Proportions of the FSC certified area in the Congo Basin region in 2010 [adapted from Cerutti et al. (2011) and Hakizumwami (2011)]

Cameroon (16 %). No forest area had been FSC certified in DRC, Equatorial Guinea and CAR by the time of recording these figures.

The number of companies awarded FSC certificates is on the rise. Companies including Wijma, Transformation Reef Cameroun (TRC), Pallisco and Société d'Exploitations Forestières et Agricoles du Cameroun (SEFAC) in Cameroon, Congolaise Industrielle des Bois (CIB) in Republic of Congo, Compagnie Equatoriale des Bois (CEB) and Rougier-Gabon in Gabon are now holders of FSC certificates, and are responsible for managing over 5 million ha of forest area (Karsenty 2007; Hakizumwami 2011) in the Congo Basin region. Chain of Custody (CoC) certificates have been issued to Pallisco in Cameroon, Industrie Forestière d'Ouesso (IFO) and Olam in the Republic of Congo, and to SODEFOR and SIFORCO in DRC for all their processed timber products (Alemagi and Nukpezah 2012; Karsenty 2007). In addition, Cameroon, Republic of Congo, DRC, and Gabon are holders of FSC controlled wood (CW) certificates that support the production of FSC mixed sources labeled products.

Beside FSC certification, CEB and Rougier-Gabon of Gabon were awarded forest management (FM) certificates by Keurhout and according to ISO 14001 to manage additional forest areas in Gabon. Table 15.2 shows the distribution of FSC certification in the Region. With only approximately 800,000 ha of forest certified in Cameroon (Table 15.2), it holds the greatest proportion of FSC Forest Management (10) and Chain of Custody certificates (16) in the region. Despite the spread of FSC certification in the Congo Basin region, the forestry sectors in Equatorial Guinea, CAR and DRC are yet to reach this milestone.

15.3.3 Impact of FSC in the Congo Basin Region

The environmental, social and economic impacts resulting since the adoption of FSC standards are of the utmost importance to the sustainable management of forest in the Congo Basin region. FSC certification is gaining momentum in this region and is evident by the area of forest certified, and the number and type of FSC certificates issued since its adoption.

Table 15.2 Distribution of FSC's SFM/CoC, controlled wood (CW) and chain of custody (CoC) certificates in the Congo Basin region (adapted from Cerutti et al. 2011 and Hakizumwami 2011)

Country	Total area certified (ha)	SFM/CoC	CW	CoC
Cameroon	824,730	10	1	16
Republic of Congo	2,430,996	3	2	2
Gabon	1,873,505	3	4	15
Democratic Republic of Congo (DRC)	0	0	1	1
Equatorial Guinea	0	0	0	0
Central African Republic (CAR)	0	0	0	0
Total	5,129,231	16	8	34

Environmental Benefits as a Result of FSC Certification

Though the proportion of certified forest area (over 5 million ha) is relatively low in comparison to the total area of forest cover in the Congo Basin—180 million ha (RFF 2011; Hakizumwami 2011), FSC certification has nevertheless made its greatest impact through its influence on the implementation of sound management practices in the certified forest, which has permitted the forest ecosystems to continue to deliver goods and services.

FSC certified companies are now placing serious considerations on environmental concerns through the adoption of reduced impact logging approaches and on ensuring proper planning of forest management operations (Karsenty et al. 2008). Measures for the conservation of resources, including regulations against hunting of bush-meat particularly in concession areas are being considered in management planning. Silvicultural approaches in certified concessions are being improved (e.g. TRC planted about 7,000 trees from four different indigenous species within one of its forest management units) and monitored for good practice.

Companies are increasingly willing to partner with environmental and conservation organizations to ensure environmentally sustainable logging. CIB in the Republic of Congo, for example is involved in a long term process with conservation NGOs to improve the understanding of the ecosystem from which it is harvesting and is working towards developing procedures to minimize inevitable impacts of timber extraction on the forest as well as on the communities living around the forest (Lewis and Nelson 2006).

Through partnership, international NGOs in the Congo Basin, in support of FSC objectives are now offering technical guidance in the field. WWF international and WCS for example are guiding Pallisco in the Eastern region of Cameroon and CIB in the North of Congo respectively, towards a technically sound management of forest (De Blas and Ruiz-Perez 2008).

The presence of the FSC in the Congo Basin has intensified international and national commitment, towards recognizing the environmental value of the forest. For example, a voluntary partnership agreement (VPA) has been signed between the European Union and three Congo Basin countries (Cameroon, CAR, Republic of Congo) so far, to tackle illegal logging through the Forest Law Enforcement Governance and Trade (FLEGT) process (EU and MINFOF 2010; EU and MSDFE

2010; EU et al. 2011). Also a national consortium of Congo Basin countries under the auspices of the Commission des Forêts d'Afrique Centrale' (COMIFAC) and REDD+ schemes are now engaged in national and international discussions to manage the Congo Basin forest in a sustainable manner (Sonwa et al. 2009).

Economic Benefits Resulting from FSC Certification

The economic benefits from FSC certification in the Congo Basin region have been felt by government, communities and certified companies operating in the region. Once certified, logging companies gained market access and sold selected timber species to Europe for up to 30 % more per cubic meter (Kollert and Lagan 2007). Unfortunately, this additional revenue could only cover the cost of certification with no additional profit from the venture. Certified companies are yet to experience big increases in premium returns from this initiative (Rougier and Clément 2012).

Certified companies are beginning to experience a competitive advantage over uncertified companies in the market place where FSC label attracts premium (Rougier and Clément 2012). While other companies are experiencing a decline in demand for their products, Rougier and Clément (2012) reported a greater demand and higher value for FSC labeled hardwood and plywood produced by Rougier-Gabon in the northern European market that has helped maintain its market share. Timber extraction companies through FSC certification have also benefited from additional funding, from multiple stakeholders in support of FSC objectives. The Tropical Forest Trust (TFT) for instance, has helped CIB in the Republic of Congo to secure World Bank funding for its project (Lewis and Nelson 2006).

Governments of the respective countries have experienced substantial improvement in tax revenue from the export of responsibly managed timber. For example, the government of Cameroon now collects about 39.7 million Euros of forest related taxes every year (Atyi 2009) since the adoption of FSC and the start of verification of legality in the country. Certified companies contribute annually large sums in forest royalties to the government and local communities, distributed as follows: 50 % for the state, 40 % for municipality and special council support fund and 10 % for the local population (Pallisco and CIFEM 2013; Oyono et al. 2005).

Consideration of benefits sharing from sound forest management between certified logging companies and local communities has increased in the region (FSC 2013b). The companies are making efforts to comply with the guidelines '*Cahier des charges*' of the respective countries. For example, they pay approximately 1.5 Euros for every cubic meter of timber exploited, to local communities (Oyono et al. 2005). Communities have also earned income as compensation from logging companies through the development of micro-projects. Nana (2005) reported an estimated 76,000 Euros (50 million CFA) sawmill was donated to a local community in Cameroon by Wijma. Forest workers from nearby communities and the general public have benefited from employment and remuneration, which at times, is higher than the salaries of civil servants employed in comparable positions within the public sector.

Social Benefits from FSC Certification

Certification has increased communication and collaboration between stakeholders (logging companies, government, local communities, corporate bodies, NGOs and external consumers) in the forestry sector at both local and national scales and at policy fora (Atyi 2004; Hakizumwami 2011). Pallisco Cameroon for example has promoted open dialogue, to institute Peasants Forest Committees (CPF) in 2006 in approximately 80 villages, surrounding their six forest concessions to participate in forest management (Pallisco and CIFEM 2013). CIB in the Republic of Congo is also in the process of developing a conflict resolution mechanism acceptable to all stakeholders. External NGOs have agreed to monitor this mechanism to ensure that conflicts are resolved with the free, prior and informed consent of disputants (Lewis and Nelson 2006). Communication and dialogue has in effect promoted legal and responsible management in forestry concession in the Congo Basin region (Hakizumwami 2011).

Respect for the rights of indigenous people is increasingly being considered. CIB in the Republic of Congo for instance has formally recognized the rights of indigenous peoples to their traditional territory and has agreed to establish processes to ensure that timber harvesting takes place only after obtaining their free, prior and informed consent (Lewis and Nelson 2006). Workers rights to good housing and medical facilities are being considered. In addition to financial benefits obtained through employment and support for community proposed micro-projects, local communities have benefited from infrastructural development. Roads, portable pipe borne water points and health units have been provided to a small number of communities by some certified companies (Wood Hole Research Center 2007; Wijma 2008; Pallisco and CIFEM 2013).

15.3.4 Limitations in the Application of FSC Standards in the Congo Basin

FSC adoption and implementation in the Congo Basin region has been limited by several factors. First, the principles and criteria of FSC were too complicated for the region and, only recently in 2012, were considerations made to develop regional FSC standards (FSC 2013c). The dysfunctional society characterized by under-resourced government agencies in the Congo Basin region, has limited the implementation of effective consultative processes required for certification (Barume et al. 2012).

The forestry sector is gradually embracing FSC standards but is more skeptical about the status of government legalized companies and legally logged timber from the region in the marketplace, in the face of an increasingly recommended FSC approach of SFM.

Some certified companies have failed to fully abide by the stipulated principles and criteria of the FSC (Greenpeace 2013) and this has limited the success of FSC in the region. Poor implementation of silvicultural practices, harvesting and consumption of bush meat, and non-compliance with legal allowable cuts and diameter limits of specified timber species in forest concessions in the region are some of the failures recorded (Cerutti et al. 2008; Alemagi and Nukpezah 2012).

Though much has been done for the rights of local communities in the Congo Basin region since FSC adoption, they've remained marginalized (Greenpeace 2013). In most cases, they have been marginalized in important decision-making steps. Skilled employment opportunities are limited within forest communities around concessions and according to Alemagi and Nukpezah (2012), local communities who have witnessed timber exploitation within forests in their localities are still poor and faced with a variety of health problems. Guidelines (*cahier des charges*) of their involvement in forest management in most cases have not been fully observed by operators in the region (Alemagi and Nukpezah 2012). The cost of certification and insignificant market premium achieved in the region so far, and the fear of adopting a less flexible SFM approach has dissuaded many small holders from adopting the process (Atyi 2004).

15.4 Conclusions and Recommendations

The Congo Basin forest is important for its wood products and the myriad of ecosystem benefits it provides. Regrettably, the forest ecosystem has continued to decline as a result of unsustainable management practices. SFM has advanced since the adoption of FSC's approach in the region. Positive environmental, social and economic impacts may indicate that forestry stakeholders have adhered to the process of promoting responsible forestry in the region. The dysfunctional society, inadequate skills of stakeholders, uncertainty of market premiums, failure to fully respect standards and high cost of certification, are still the main limitations. However, there is an increasing interest for FSC certification by extraction companies in the region (Durst et al. 2006).

Though advances have been made, critical reports already presented by Greenpeace (Barume et al. 2012; Greenpeace 2013), on the credibility of the process in the region highlight the remaining challenges to be overcome. More efforts are required therefore to improve communication, participation and to develop skills needed for the implementation and monitoring of FSC standards and criteria in the Congo Basin region. FSC certification so far remains the only credible system of SFM in the Congo Basin (De Blas and Ruiz-Perez 2008) that if properly applied, will ensure a more tangible and long-term economic, social and environmental benefit to the Congo Basin region. Thus, the progress of the VSS approach can be regarded as at least partly successful at this juncture.

References

Akinwande B (2012) Under threat: deforestation pressure on Congo Basin forests increasing: Forest News. http://blog.cifor.org/11433/under-threat-deforestation-pressure-on-congo-basin-forests-increasing/#.Ua8j3tJkM08. Last accessed 28 Aug 2013

Alemagi D, Nukpezah D (2012) Assessing the performance of large-scale logging: companies in countries of the Congo Basin. Environ Nat Resour Res 2(2):38–47

Atyi RE (2004) Forest certification in Gabon. Internet adress: http://yale.edu/forestcertification/symposium/pdfs/gabon_symposium.pdf. Accessed 25 Mar 2014

Atyi RE (2009) Study on the development and progress in Timber procurement policy: case study: Cameroon. http://www.itto.int/direct/topics/topics_pdf_download/topics_id=2304&no=0&disp=inline. Last accessed 28 Aug 2013

Austin K, Stolle F, Elmore S (2010) Preparing for REDD in the Republic of Congo. World Resources Institute. http://www.wri.org/stories/2010/08/preparing-redd-republic-congo. Last accessed 28 Aug 2013

Barume A, Heuse E, Ozinga S (2012) Formal complaint by Greenpeace against SODEFOR's association with FSC. http://www.fsc.org/download.complaints-panel-evaluation-report-jan-2012.346.htm. Last accessed 28 Aug 2013

Brown P, Somorin A, Nkem N, Sonwa D (2010) Reduced Emissions from Deforestation and Forest Degradation (REDD): Institutional Perspectives on Opportunities and Challenges in the Congo Basin. 18th Commonwealth Conference. http://www.cfc2010.org/papers/session14/brown-s14.pdf. Last accessed 28 Aug 2013

Butler R (2013) The Congo Basin and its forest ecosystems. http://rainforests.mongabay.com/congo/#.UhqAXD8uHIU. Last accessed 28 Aug 2013

Cerutti P, Nasi R, Taconi L (2008) Sustainable forest management in Cameroon needs more than approved forest management plans. Ecol Soc 13:36–48

Cerutti P, Taconi L, Nasi R, Lescuyer G (2011) Legal vs certified timber: preliminary impact of forest certification in Cameroon. Forest Policy Econ 13(3):184–190

CBFP – Congo Basin Forest Partnership (2006) The forest of the Congo Basin. http://www.giz.de/Themen/de/dokumente/en-state-of-forests-congo-basin-2006.pdf. Last accessed 28 Aug 2013

CBFP – Congo Basin Forest Partnership (2013) The Congo Basin and its forest ecosystems. http://ccr-rac.pfbc-cbfp.org/Stateoftheforest.html. Last accessed 28 Aug 2013

De Blas D, Ruiz-Perez M (2008) Prospects for reduced impact logging in Central African logging concessions. Forest Ecol Manage 256:1509–1516

Durst P, Mckenzie P, Brown C, Appanah S (2006) Challenges facing certification and eco-labelling of forest products in developing countries. Int Forestry Rev 8(2):193–200

EU – European Union and MINFOF – Cameroon Ministry of Forest and Wildlife (2010) FLEGT Voluntary Partnership Agreement between Cameroon and the European Union: Briefing Notes. http://www.illegallogging.info/sites/default/files/uploads/AW3602EFICameroonBriefingNoteEngvisual.pdf. Last accessed 26 Aug 2013

EU – European Union and MSDFE – Ministry for Sustainable Development, Forest and Environment (2010) FLEGT Voluntary Partnership Agreement between the Republic of Congo and the European Union: Briefing Notes. http://www.euflegt.efi.int/files/attachments/euflegt/eccongobriefingnoteenglish.pdf. Last accessed 26 Aug 2013

EU – European Union, Zanga A, M'Bangolo D (2011) FLEGT Voluntary Partnership Agreement between the Central African Republic and the European Union: Briefing Notes.

http://capacity4dev.ec.europa.eu/system/files/file/20/04/2012_-_1318/car_vpa_-_bn_2010.pdf. Last accessed 29 Aug 2013

FSC – Forest Stewardship Council (2010) Market potential of FSC products coming from smallholder forestry operations in Europe. CeFCo Project. http://www.cefcoproject.org/fileadmin/cefco/pdf/6-Analysis_of_market_demand_of_FSC_products-EN.pdf. Last accessed 26 Aug 2013

FSC – Forest Stewardship Council (2012a) History: an innovative idea takes root. https://ic.fsc.org/our-history.17.htm. Last accessed 26 Aug 2013

FSC – Forest Stewardship Council (2012b) Factsheet about FSC. http://ic.fsc.org/force-download.php?file_connector=779. Last accessed 26 Aug 2013

FSC – Forest Stewardship Council (2013a) FSC principles and criteria: international guidelines to forest management. https://ic.fsc.org/download.revised-fsc-pc-v-5-0-high-resolution.a-1780.pdf. Last accessed 26 Aug 2013

FSC – Forest Stewardship Council (2013b) The 10 principles: ten rules for responsible forest management. https://ic.fsc.org/the-ten-principles.103.htm. Last accessed 29 Aug 2013

FSC – Forest Stewardship Council (2013c) Forest stewardship Standard for the Congo Basin Region: Norme FSC pour la certification des forêts du Bassin du Congo. https://ic.fsc.org/download.fsc-std-cb-01-2012-regional-standard.a-1724.pdf. Last accessed 29 Aug 2013

Green3Dhome (2013) Forest certification and labeling. http://www.green3dhome.com/GoGreen/FSCCertificationandLabeling.aspx. Last accessed 29 Aug 2013

Greenpeace (2013) FSC at risk: a joint 4-step action plan to strengthen and restore credibility. http://www.greenpeace.org/international/Global/international/briefings/forests/2013/FSC-at-risk.pdf. Last accessed 29 Aug 2013

Guillery P (2011) Traceability and the Forest Stewardship Council. In: Presentation, International Seminar on Forest Certification, Forest Stewardship Council. http://www.tracefood.org/images/4/4b/8%29_Phil_Guillery,_Forest_Stewardship_Council_%28FSC%29.pdf. Last accessed 29 Aug 2013

Hakizumwami E (2011) Progress of FSC certification in the Congo Basin. In: Presentation, International Seminar on Forest Certification, Forest Stewardship Council. http://awsassets.wwf.es/downloads/elie_hakizumwami_overview_of_fsc_certification_in_the_congo_basin.pdf. Last accessed 29 Aug 2013

Karsenty A (2007) Overview of industrial forest concessions and concession-based industry in Central and West Africa and considerations of alternatives. http://pfbc-cbfp.org/tl_files/archive/thematique/Forest_Concessions_and_Concession_Industry_Central.pdf. Last accessed 29 Aug 2013

Karsenty A, Drigo GI, Piketty M, Singer B (2008) Regulating industrial forest concessions in Central Africa and South America. Forest Ecol Manage 256(7):1498–1508

Kollert W, Lagan P (2007) Do certified tropical logs fetch a market premium? A comparative price analysis from Sabah, Malaysia. Forest Policy Econ 9(7):862–868

Lewis J, Nelson J (2006) Logging in the Congo Basin. What hope for indigenous peoples' resources and their environments? http://data.cameroun-foret.com/system/files/18_39_09.pdf. Last accessed 29 Aug 2013

Nana W (2005) Wijma Sawmill launched in Tinto. http://www.postnewsline.com/2005/05/strongwijma_mob.html. Last accessed 29 Aug 2013

Ndoye O, Tieguhong JC (2004) Forest resources and rural livelihoods: the conflict between timber and non-timber forest products in the Congo Basin. Scand J Forest Res 19:36–44

Nlom JH (2011) The economic value of Congo Basin protected areas goods and services. J Sustain Dev 4(1):130–142

Oyono PR, Kouna C, Mala W (2005) Benefits of forest in Cameroon: global structure issues involving access and decision making hiccoughs. Forest Policy Econ 7(3):357–368

Pallisco and CIFEM (2013) A platform for social dialogue. http://ang.pallisco-cifm.com/social-20082010163535.asp. Last accessed 29 Aug 2013

RFF – Rain Forest Foundation (2011) About Congo Basin Region. http://www.rainforestfoundationuk.org/Congo%20Basin%20Region. Last accessed 29 Aug 2013

Rougier F, Clément M (2012) Preliminary feedback on FSC certification from an operator's point of view. Proparco Magazine 14:22–24

Sonwa D, Bele Y, Somorin O, Jum C, Nkem J (2009) Adaptation for forests and communities in the Congo Basin. ETFRN NEWS 50:93–100. http://r4d.dfid.gov.uk/PDF/Outputs/ClimateChange/ETFRN_50_Forests_and_Climate_Change93-100.pdf. Last accessed 29 Aug 2013

Usongo L, Nagahuedi J (2008) Participatory land-use planning for priority landscapes of the Congo Basin. Unasylva 59(230):17–22

Wijma (2008) Étude d'impact environmental et sociale de l'exploitation forestiére de l'UFA 09-024. http://wijma.prosygma-cm.com/download/DocumentsDeTravail/EIE_UFA_09024.pdf. Last accessed 29 Aug 2013

WHRC – Woods Hole Research Center (2007) Impacts of Industrial Logging In Central Africa Studied. ScienceDaily. http://www.sciencedaily.com/releases/2007/06/070608093811.htm. Last accessed 29 Aug 2013

WWF – World Wide Fund for Nature (2013a) Congo Basin. http://worldwildlife.org/places/congo-basin. Last accessed 29 Aug 2013

WWF – World Wide Fund for Nature (2013b) Congo Basin Forest. http://swp.gmu.edu/silvacarbon/sites/default/files/DOCS/DRC/congo_forest_cc_final_13nov07.pdf. Last accessed 29 Aug 2013

Chapter 16
Issues and Opportunities for Implementation of VSS in China

Ni An and Eberhard Schaller

16.1 Introduction

This chapter aims to discuss the current situation and the implementation of VSS in China where it is still not widely recognised. China is one of the fastest developing countries in the world and plays a very important role in the worlds' economy. However, along with its rapid development, a series of consequences, both economic and environmental, have drawn massive attention from the public and media. Chinese industries have an urgent need of tools to enhance their reputation internationally. Since the mid-1990s when voluntary sustainability standards (VSS) were first introduced in mainland China, many companies especially SMEs (small and medium enterprises) were reluctant to engage with VSS. It is therefore important to study the reasons for these setbacks and possible solutions while taking into account the political background, the specific conditions and the institutional realities in China.

The main focus of this chapter is to discuss the possible benefits from implementing VSS and to analyse the possible difficulties for both manufacturers and customers. With this view, the chapter has the following objectives:

- To analyse the current situation of China's economy and the development of VSS;
- To explain the VSS scheme and the potential benefits;
- To define the reasons behind difficulties in VSS implementation;
- To analyse feasible procedures for future VSS development.

Section 16.2 is a short description of both the past rapid economic growth in China and the damage and problems that come along with it, which makes the sustainable development of economy the new direction for China's future.

N. An (✉) • E. Schaller
Department of Environmental Meteorology, Brandenburg University of Technology (BTU) Cottbus, P.O. Box 101344, 03013 Cottbus, Germany
e-mail: anni@tu-cottbus.de; schaller@tu-cottbus.de

C. Schmitz-Hoffmann et al. (eds.), *Voluntary Standard Systems*, Natural Resource Management in Transition 1, DOI 10.1007/978-3-642-35716-9_16,

Section 16.3 introduces VSS as a useful tool on the China's road towards sustainability. The principle of the VSS will be explained and the necessities and benefits for companies in adopting such a system will be analysed. The current situation for most companies is a lack of knowledge of VSS, and therefore lack of enthusiasm for participation. A few companies that are willing to engage with the system suffer from obstacles such as ambiguous policy and low quality advisory services. Section 16.4 discusses challenges in implementation of VSS and the reasons for the difficulties experienced by suppliers and customers under the specific Chinese societal structure. The chapter ends in Sect. 16.5 with conclusions and some recommendations for future development of VSS.

16.2 Economic Development in China and Its Consequences

In the past two decades, China's economy has experienced rapid development. According to the latest report on the Chinese economy, in 2011, China had the second highest total GDP of 8 trillion USD with a growth rate of 9.2 % from the previous year, ranked after the USA and followed closely by Japan and Germany. In 2011, China's foreign trade import and export value was 3.64 trillion USD with an increase of 22.5 % compared to the same period in 2010, which is a new national Chinese record (Qingmin 2012; Xinhua 2012). There is no doubt that economically, China has become one of the strongest countries in the world.

16.2.1 Environmental Damages and the Chinese Sustainability Strategy

Although China's rapid development has certain positive effects on the life of its citizens, it induces severe environmental consequences as well. The urbanisation-induced pollution of croplands prevents cultivation that is needed to feed an estimated 65 million of the Chinese population (Edmonds 1999). The Chinese Ministry of Civil Affairs issued a 2011 Annual Natural Disaster Evaluation Report, which shows that in 2011, China's natural disasters mainly consisted of floods, drought, hail, freezing, snowstorms and earthquakes. Forest and grassland fires, pests, diseases and other disasters had varying degrees of occurrence. According to the report, natural disaster caused a total of 1,126 deaths and many more injuries. According to estimations by the World Bank, by 2020, China will have to spend 390 billion USD annually or 13 % of the national GDP to treat diseases caused by pollution (Lam 2008). The list of serious problems is growing.

This situation has raised attention in the media as well as within the government. In the 11th Five-Year plan (for 2006–2011), the Chinese government turns away from growth driven policies towards sustainable development, which demonstrates that China's leaders are ready to address the country's environmental problems. This shift towards sustainability in China is still facing many obstacles, including inadequate legal protection, lack of standard metrics and local non-compliance. With a growing recognition of an environmental crisis, China will be pushed towards sustainable development.

16.2.2 Economic Consequences and Barriers

In economic development, efforts to maximise profit alone lead not only to environmental damage but also to unforeseen national consequences in addition to potential losses on the global market. Domestically, Chinese people have encountered several high profile incidents related to food safety issues and working condition issues. The 2008 Chinese milk scandal in China for example, resulted in 53,000 victims with four infant fatalities from kidney stones and similar renal complications. This was later traced to deliberate melamine addition to milk in order to alter the protein content to circumvent quality testing (Ministry of Health 2008). The consequences of this scandal continue to the present day, which makes Chinese consumers distrustful of milk products on domestic markets and instead purchase elsewhere (i.e. Europe). On an international level, Chinese brands are suffering from global recognition problems. 'Made in China' has become synonymous with a low-quality product. Chinese products tend to give an impression of being environmentally unfriendly and of poor quality, resulting in enormous economic losses.

Green barriers are trade protection measures that importing countries develop, promulgate and implement. They consist of environmental regulation of technical standards. These standards include green tariffs (customs), market entry verification system, and green anti-dumping and other sustainability standards. For example, in recent years, China's aquatic products frequently suffered from the impact of international green barriers. In 2001, imported shrimp from China were tested and contained 0.2–5 ppb of chloramphenicol which exceeded the EU standard. As a result, the EU commission decided to ban this product imported from China since 31 January 2002 for human consumption and animal feed. Such incidents have occurred frequently in the past several years.

To be successful in the game, one must be familiar with the rules. Chinese enterprises must address these frictions actively, understand the system, and improve their production to avoid further losses.

16.3 Opportunities of Implementing VSS in China

If China fails to achieve global recognition on their commitment to sustainable development, it will aggravate the negative aspects of the China brand, which in turn, will limit China's economic success domestically and internationally.

The Chinese business community must in the first place obey the international and national laws of those countries in which their businesses operate. However this is not sufficient to strengthen the brand competitiveness. For further success in global markets, they must reach compliance with other frameworks and sustainability-related standards evolved from expectations and interests of consumers, employees and others, such as ISO 14000 Environmental Standards and ISO 26000 Social Responsibility Guidelines (Long et al. 2009). VSS is a new generation of sustainability standards that covers a multitude of issues in different fields.

16.3.1 VSS: One Powerful Tool

VSS provide a set of criteria which is embedded into a governance structure to govern production using proven techniques for goods and services with the goal of a sustainable economy. They are mainly developed and governed jointly by businesses, civil society, trade unions and the government (Dietmar et al. 2011). The system covers a wide range of environmental (e.g. compliance with prohibited chemical list), social (e.g. International Labour Organization) and economic issues (e.g. product quality and supply chain management). Thousands of sustainability standards encompass hundreds of sectors. For example, almost 180 million ha of global forests are certified by the Forest Stewardship Council (FSC), as of May 2013 (FSC 2013). VSS include both technical and structure-related elements. Technical elements establish the content and technical functionality of the system and the latter provides dynamism to the system and the regulated processes. The high effectiveness of the standard system requires the successful interaction between both elements. VSS have some remarkable advantages for compliance:

- *Certification*: the system provides a buyer-recognised certificate that proves the compliance of the participants.
- *Simplicity*: the standards are a fixed set of practical criteria defining sustainability.
- *Visibility*: allows for permission to use a label after certification.
- *Improvement*: compliance is assured through periodic audits instead of a one-time audit.

Different sections within society are in demand of VSS for many reasons. Firstly the consumers want their products to be safe and to be of high-quality. With a visible certification label, the consumers can be assured of the quality of a product as it is easily indicated to them that the product conforms to particular quality standards, giving a sense of self-satisfaction to the consumer that they have made an informed decision. Secondly, the brands need the certification to differentiate themselves from competing products. Last but not least, the public and society benefit from such standards due to the improved social environmental conditions, enhanced reputation and newly created values. The producing companies can also access significant benefits through VSS. Engaging with VSS can enhance their reputation and image with consumers, attracting new customers whilst bolstering confidence with existing customers, and help open new routes for business into new markets with potential sales increases being realised. By modifying for efficiency and productivity through efficient use of natural resources, the final goal of reducing costs and increasing revenue can be achieved.

16.3.2 Status of VSS Systems in China

Despite the major potential benefits and the urgent need for such systems, VSS in China are generally not well recognized and therefore underappreciated. There are many reasons for the reluctance of VSS uptake by Chinese companies. The most significant is lack of familiarity with VSS and proper guidance from decision makers. In China, as in many developing and emerging market countries, some decision makers still have doubts regarding VSS and treat them as a restriction in international trading, rather than an advantageous tool towards increasing competitiveness. There are few regional governments in China which actively promote VSS and these activities concentrate heavily on the development of domestic standards rather than on participating in the development of international VSS (Dietmar et al. 2011).

Nevertheless, there are already some successful case studies of VSS implementation within China. For example, the most prominent VSS is the Chinese Social Compliance Management System CSC9000T. The China National Textile and Apparel Council (CNTAC) launched CSC9000T in response to international standard initiatives such as the Fair Labour Association (FLA), Social Accountability International (SAI) and the Ethical Trading Initiative (ETI). CSC9000T underlines the Chinese approach in the textile and apparel sector, which is to “engage with existing international initiatives, rather than trying to shape existing standards” (Long et al. 2009). Another successful example is the Forest Stewardship Council (FSC), which is the most important forestry certification scheme in China; with approximately 6.8 million ha certified thus far (FSC 2013). Further examples of VSS and their status can be found in Table 16.1.

Table 16.1 Scale and status of some sustainability standards (*Source*: Long et al. 2009, p. 42)

Standard	Scale	Chinese trade
Fairtrade Labelling Organisation: non-profit, multi-stake-holder association involving 23 member organisations, traders and external experts	75 % UK residents recognise 'fair trade' logo. As of 2005, Fairtrade scales increased 37 % annually, accounting for 1.1 billion Euros in sales worldwide	China has minimal association in Fairtrade coffee and tea; some speciality goods sold in foreign supermarkets, collaborating with WWF (World Wide Fund for Nature)
Marine Stewardship Council: the world's leading certification and eco-labelling program for sustainable seafood	Certifying 10 % of global wild caught fish. Until 2008, 589 companies from 36 countries have been certified	China is world's largest producer. 16–17 million tons of wild capture; 13 million fishermen. 2011, first fishery in Dalian, China was under the accreditation stage
Principles for Responsible Investing: 2006 UN initiative to make environmental and social governances part of investment analysis	444 signatories and 14 trillion Euros under management	China Investment Corp. and China Banking Corp. became signatories
Extractive Industries Transparency Initiative	24 candidate countries with high foreign investment in various sectors; 40 companies signed up	No Chinese companies

16.4 Challenges and Difficulties

Although there are several successful examples of VSS, many companies, especially SMEs, are reluctant to engage with the systems. There are several challenges to their development:

16.4.1 Limited Awareness and Lack of Understanding of VSS

The key reason of the under-developed situation of VSS is that both public and private sectors have limited understanding of available standards, characteristics and benefits. Most of the companies are uncertain about the certification systems and therefore unwilling to participate. According to a study carried out by a Sino-German CSR project for 2010 in Guangdong Province, fewer than half of the respondents indicated that factory audits were useful in bringing direct economic or social benefits (Fig. 16.1). Among 50 respondents to the question "whether or not the CSR factory audit and approval is helpful?", most of them (65 %) agree that

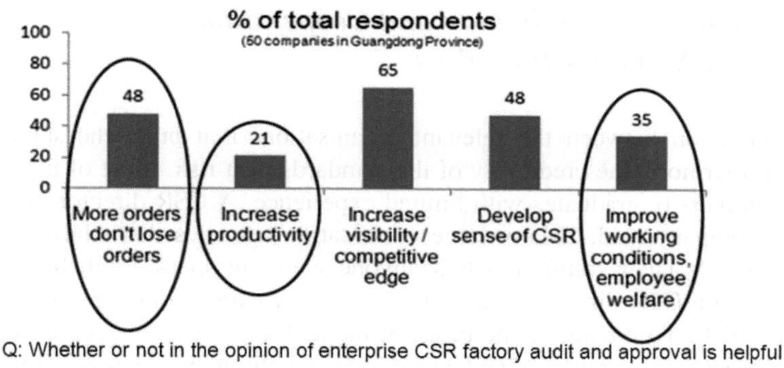

Fig. 16.1 Baseline study on VSS in Guangdong Province, China 2010 (*Source*: Taras 2010)

they will improve competitively, but few of them realize the direct benefits, such as bringing more orders, increased productivity and socially improved working conditions.

16.4.2 Ambiguous Policy Environment

As of yet there have been no clear indications regarding government policy for voluntary standards. Unlike most western countries, China has a distinctive economic and social system, where the government has a strong influence upon the developing trends of domestic economy. Especially for state-owned companies, the challenge regarding VSS is the lack of explicit guidance from Chinese governmental institutions and from other important stakeholders. Some decision-makers are still sceptical of VSS with the perception that VSS are entry barriers to global markets. They are most of the time forced to obey formal laws rather than choose to adopt those standards as a result from a strategic business decision.

16.4.3 Low Capacity and Scale of Advisory Services

Due VSS unpopularity, there are few advisory services in the market. It is not only hard to find a company that provides high quality advice services but also expensive, which makes companies unwilling to pay and for some small companies, unaffordable. For example, the costs of FSC certification depend on several factors: area of the forest, location and distribution of the area, experts in the auditing team, profits of the auditing company and other factors. Generally speaking, the cost of FSC auditing is approximately 6,000–20,000 Euros (FSC China). Furthermore, if using advisory companies, there are more costs. It is certainly a large amount of money and is very difficult for some small companies to afford.

16.4.4 Lack of Coordination and Cooperation Among the Standard Initiatives

The cooperation between the relevant organisations that build the standards is weak. Furthermore, the credibility of the standards is at risk. Most of the auditors are new university graduates with limited experience. A CSR director of a multinational company said, "Many large certification agencies don't have qualified auditors at all. Their solution is that anyone who can speak English can be an auditor" said a CSR director of a multinational company from experience (Sino-German 2010). To maximize the profit of the auditing company, some avoidable auditing processes are repeated and the aim of the auditing becomes an effort to find mistakes rather than solutions. Fraud and corruption in advisory businesses is very common which largely reduces the credibility of VSS. Most advisory services focus on how to pass audits rather than on the improvement of production processes.

The Chinese industries have urgent need of proper guidance from the government or a clear policy on support, such as tax reduction or advisory support guidelines. For those suppliers and companies that are already willing to engage with the VSS, proper help and advice are strongly needed.

16.5 Conclusions and Recommendations

VSS will continue to be an important tool in global trade in the future. As for the current situation of the Chinese business community suffering from recognition problem, they are in urgent need to enhance competitiveness using international sustainability standards. Therefore, it is very important for China to get involved and promote the development of VSS. Due to the special economy and social system; Chinese government should be supportive and play a positive role in promoting VSS, by framing specific policy to encourage the engagement for enterprises in the development and application of sustainability standards. For example, a preferential income tax policy for companies embedding sustainability strategies can be an incentive while making investment decisions. Operation of some international investment funds or suitable public insurance products can be an encouragement of international investment (Long et al. 2009).

Chinese enterprises must develop the knowledge behind the true meaning and rules of VSS. They must be aware of the true economic benefits of VSS and choose to engage with VSS from the perspective of it being a sound business decision. Meanwhile, high-quality advisory services facilitating the auditing and management of VSS are badly needed. With the aim of promoting VSS, the costs to participants must be reduced. Other measures such as increased transparency of auditing processes and coordination between various standards will also be fundamental to VSS uptake.

For a healthier future development of the country, Chinese government should treat sustainable trade as an integral element when forming governmental strategy. China has a long way to go before truly achieving the goal of sustainable development and it requires cooperation within government departments and the wider business communities.

References

Chinese Ministry of Civil Affairs (2012) 2011 Annual Natural Disaster Response Work: Summary Assessment Report

CSC9000T – China Social Compliance 9000 for Textile & Apparel Industry (2005) Launched in 2005 by the China National Textile and Apparel Council and is an industry specific management system for social compliance of China's textile and apparel sector. www.csc9000.org.cn. Last accessed 16 Sept 2013

Dietmar R, Shepherd I, Taras D, Thomas-Dirla P (2011) A brief introduction to voluntary standard systems and related trends in China. An overview prepared in the Framework of the Sino-German Corporate Social Responsibility (CSR) Project. German Society for International Cooperation (GIZ) GmbH, Beijing. http://www.chinacsrproject.org/Resources/Resource_List_EN.asp?LstFlt_D1=4. Last accessed 16 Sept 2013

Edmonds RL (1999) The environment in the People's Republic of China 50 years on. China Q 159:640–649

FSC – Forest Stewardship Council (2013) Facts and figures 2013. https://ic.fsc.org/facts-figures.19.htm. Last accessed 16 Sept 2013

Lam D (2008) The reality of environmental sustainability in China. City 12(2):245–254

Long G, Zadek S, Wickerham J (2009) Advancing sustainable competitiveness of China's transnational corporations. AccountAbility, London

Ministry of Health (2008) Ministry of Health: 1.2 million infants hospitalized after consuming milk powder, press release by the Xinhua, 21 September 2008. http://news.xinhuanet.com/newscenter/2008-09/21/content_10088082.htm. Last accessed 16 Sept 2013

Qingmin Y (2012) Promote China's factoring business to better support the real economy: speech by Assistant Chairman of CBRC at the 44th FCI Annual Meeting on 4 June 2012. http://www.china-cba.net/bencandy.php?fid=43&id=9596. Last accessed 16 Sept 2013

Sino-German (2010) Baseline study on voluntary social standards in China, a study commissioned by the Sino-German Corporate Social Responsibility (CSR) Project. German Society for International Cooperation (GIZ) GmbH, Beijing

Taras D (2010) Challenges in implementing voluntary sustainability standards in China. Presentation given at the Hong Kong Quality Assurance Agency's Guangzhou Office, Guangzhou, 22 July 2010. http://www.chinacsrproject.org/Resources/Resource_List_EN.asp?LstFlt_D1=3. Last accessed 16 Sept 2013

Xinhua – Xinhua News Agency (2012) China calls for efforts to promote more balanced, sustainable foreign trade. Press release on the NAM News Network, Beijing, 1 May 2012. http://www.namnewsnetwork.org/v3/read.php?id=MTkyNzEz. Last accessed 16 Sept 2013

Chapter 17
A Feasibility Study of Utilising Voluntary Sustainability Standard (VSS) Systems in Paper-Making Enterprises in Liaoning Province, China

Xiaoying Gu and Gerhard Wiegleb

17.1 Introduction

This chapter aims to discuss the feasibility of utilising certifications from Voluntary Sustainability Standard (VSS) systems in paper-making enterprises in Liaoning Province, China. The Chinese paper-making industry produces many pollutants which are regarded as harmful both to the living environment and public health. Regulating and controlling the paper-making industry is an urgent task for the Liaoning provincial government. Besides strengthening the enforcement of environmental laws and related regulations, new methods of environmental management are to be implemented, some of which are already well established in many others countries. It may be worth considering applying VSS systems in the Liaoning Province.

In this context, the present chapter continues and supplements previous studies on the feasibility of using VSS systems in paper-making industry in Liaoning Province. The objective of the chapter is to compare existing VSS systems and to outline criteria for the selection of the most feasible approach the paper-making industry in Liaoning Province. The chapter begins by introducing the information about of the paper-making enterprises in Liaoning Province in Sect. 17.2. Section 17.3 will highlight the comparison of different VSS systems related to paper-making enterprises in China. Section 17.4 elucidates which VSS system is the most appropriate for Liaoning Province. Section 17.5 contains the conclusions and recommendations.

X. Gu (✉) • G. Wiegleb
Department of General Ecology, Brandenburg University of Technology (BTU) Cottbus, P.O. Box 101344, 03013 Cottbus, Germany
e-mail: guxiaoying1210@gmail.com; wiegleb@tu-cottbus.de

C. Schmitz-Hoffmann et al. (eds.), *Voluntary Standard Systems*, Natural Resource Management in Transition 1, DOI 10.1007/978-3-642-35716-9_17,

17.2 The Introduction of Paper-Making Enterprises in Liaoning Province

17.2.1 The Operating Situation of Paper-Making Enterprises in Liaoning Province

In 2007, 417 paper-making and related enterprises were operating in Liaoning Province. The main paper products were toilet paper, corrugated fibreboard and cardboard paper (Zhu 2010). The paper-making industry in Liaoning Province has the following characteristics:

1. Most of the paper-making enterprises operate on a small scale. 225 of the 417 paper making enterprises have an annual production of less than 34,000 tonnes (Zhu 2010). These enterprises distribute their products in a rather uncoordinated and dispersed manner without concentrating on a specific marketing strategy. All products are produced primarily for the domestic market.
2. The average economic output attributed to the paper-making industry in Liaoning Province is relatively low. Most of the production facilities are out of date as these enterprises were established long before the implementation of environmental management systems. The production technology is not well developed and the products are far from being sophisticated. The number of products is relatively limited. This may be due to the fact that paper products do not face much competition on the domestic market (Zhu 2010).
3. Pollution caused by the paper-making industry in Liaoning Province is still very serious. Almost every city in Liaoning Province has at least one paper-making plant. The small-scale enterprises do not have any supporting facilities to treat the emitted pollutants. Pollution is strongly affecting the surface waters, which is harmful to the local living conditions as well. The industrial pollution caused by the paper-making industry in Liaoning Province is 41 % of the total industrial pollution, but the industrial value produced by paper-making industries makes up only 0.44 % (Zhu 2010).

The Liaoning government has already realised the unbalanced situation between the high pollution and low industrial productivity of the local paper-making industry. Since April 2008, Liaoning government started closing down those paper-making enterprises which do not reach the existing legal standard. Until June 15th 2010, 285 out of 417 paper making enterprises were closed down, 132 enterprises are under reconstruction. While small enterprises are closed down, most of larger enterprises using more environmental friendly technology and equipment are encouraged to improve their environmental performance (Zhu 2010).

17.2.2 The Policy Considering the Paper-Making Industry

According to the 12th Five-year Plan in China, the main objective of the Chinese paper industry is to continue and complete the policy of energy saving and pollution reduction as well as developing an effective certification of forest products (Chinese Government 2011):

1. The government will reward enterprises which performed well in environmental protection, such as providing more emission targets or approving new construction projects applied for by certain enterprises.
2. The government is also committed to promote environmental certification systems, environmental labelling systems, and evaluation of the clean production process.
3. The public is encouraged to be a part of the monitoring efforts for environmental protection performance of enterprises. For example, public awareness of saving paper as well as encouraging consumers to buy environmentally friendly paper is increased.

In such a situation, more paper making enterprises are motivated to get a certification from a VSS system.

17.3 The Introduction of VSS Systems Related to Paper-Making Enterprises in China

17.3.1 What Is a VSS System?

In China, paper-making enterprises can be certified by two different kinds of VSS systems. One system is the *international voluntary sustainability standard system*, e.g. certification by Forest Stewardship Council (FSC) or Programme for the Endorsement of Forest Certification (PEFC). Certification from the FSC is a relatively popular international award in China. It is a globally operating non-profit organisation dedicated to the promotion of responsible forest management worldwide. The other available system is the *Chinese national voluntary sustainability standard system*, offered by the Chinese Forest Certificate Committee (CFCC). Since 2001, China has started to build up a national forest certification system according to its own national conditions. CFCC was established by the State Forestry Administration of China (SFA) and the National Certification and Accreditation Administration of China (CNCA) in 2008 (ICFCC 2012).

If any enterprise reaches the certification standard set by these organisations, it can receive certification and use special labels on their products. The labels enable both businesses and consumers to make informed choices about which forest products they buy and to create positive change by using the power of market dynamics (WWA 2012).

17.3.2 Differences and Similarities Between the Chinese National VSS System and International VSS Systems

A comparison between the FSC standard and the CFCC standard reveals two big differences between the respective approaches of the international and national organisations:

Setting of Voluntary Sustainability Standards

The FSC sets the standards for forest management and chain of custody certification, defining all procedures, which certification bodies should follow in their own certification assessments. All FSC standards and policies are developed by the "Policy and Standards Unit" based at the FSC International Centre in Bonn (5S 2012).

The CFCC standard is set by the SFA. The SFA is a Chinese government department, while the FSC is a non-government organisation. The CFCC can use two different standards, namely Forest Management (FM) and Chain of Custody (CoC). The FM certification is awarded to forest managers or forest owners whose management practices meet the requirements of the FSC principles and criteria. CoC certification applies to manufactures, processors and traders of FSC certified forest products. It verifies FSC certified materials and products along the production chain (TFSC 2012). The paper making industry needs to apply for the CoC. Despite the formal independence, CFCC certification is linked to internationally accepted criteria in all cases.

Certification Bodies

The FSC does not issue their own certificates, rather independent certification bodies carry out the forest management and chain of custody assessments that lead to FSC certification. Only FSC accredited certification bodies are authorised to issue FSC certificates. Additionally, Accreditation Services International (ASI) is responsible for checking and accrediting the certification body. All FSC accredited certification bodies must meet the FSC accreditation requirements (AP 2012).

Compared to the FSC, the CFCC system has only one certification body, namely Beijing Zhonglin Tianhe Forest Certification Centre (ZTFC), which was built up by the SFA and the CNCA in April 2009. It is an independent enterprise, but it is still under the purview of the SFA (ICFCC 2012). Since SFA is a department of the

Chinese government, the ZTFC is not as independent as the certification bodies of the FSC.

The procedures which enterprises must undergo in order to attain certification, from either an international system or a national system, are very similar. The stepwise procedure to get a certification from the FSC can be summarised as follows (5S 2012):

1. The enterprise contacts one or several FSC accredited certification bodies. As a first estimate regarding the financial cost and time required, the certification body will need some basic information (e.g. the resource of the raw material, the type of the product, the producing technology and equipment). The certification body provides the enterprise with information about the mandatory requirements for FSC certification.
2. The enterprise decides which certification body it would like to work with and signs an agreement with the certification body.
3. A certification audit takes place to assess the enterprise's qualifications for certification.
4. The data collected at the audit are the basis of the audit report based on which the certification body makes the final certification decision.
5. If the certification decision is positive, the enterprise will receive a FSC certificate. If the audit reveals that the operations are not yet in full compliance with FSC requirements, the enterprise can apply for further audits after having implemented the changes suggested in the certification report.

To receive certification from the CFCC, the ZTFC enterprises must follow the same procedure as described in steps 3–5, with the exception that the enterprises cannot choose the certification body (step 2).

17.4 Which VSS System Should the Paper-Making Enterprises in Liaoning Province Choose?

All paper-making enterprises in Liaoning Province are focused on local markets (Zhu 2010). Due to the small scale and insufficient financial abilities of most enterprises, it is still very hard for them to utilise the VSS system. Only if the public has higher expectations as to their living habitat and is more motivated to join the cause for environmental protection, the implementation of a VSS system would become much easier. From a long-term perspective, introducing the VSS System in Liaoning Province has the potential to bring many benefits. As already mentioned, the Liaoning government is already starting to reorganise the paper-making industry. They have closed down small-scale enterprises, which can never reach any environmental standard, thus, enabling the remaining enterprises to apply for the necessary VSS certification.

17.4.1 Benevolent Policy and Domestic Acceptance

The Chinese government is the main policy maker in all fields including the environment. The paper-producing enterprises can reach towards triggering a more benevolent policy from the Chinese government by using eco-labelling and other certification methods that have been encouraged by the Chinese government in the past. This is also true for the CFCC certification system, monitored by the SFA.

The Chinese national certification is expected to receive wide acceptance by the Chinese public as paper products are focused on the domestic market. In contrast, enterprises in the coastal areas of China have more international connections with European and USA markets. For them, an international VSS certification could be the so-called 'entrance ticket' to a wider international market. Public awareness of international eco-labels in China is still limited. Enterprises focused on the domestic market are not motivated to achieve certification.

17.4.2 The Risk to Lose the Certification from the CFCC

The Chinese Forestry Department is already initiating cooperation and mutual authentication with many international institutions, for example, World Wide Fund for Nature, Rainforest Alliance, and The Nature Conservancy. The Chinese Forest Department actively promotes mutual recognition and cooperates with the American Forest & Paper Association, Australian Forestry Standard Company, the Malaysian Timber Certification Committee, and the Indonesian Eco-labeling Institute of the National Forest Certification Agency (Shi 2011).

By 2011, FSC certification covered nearly 2.7 million ha of forest in China, with over 2,100 Chinese companies having gained FSC CoC certification. Seven out of the ten major paper companies in China were FSC certified. Furthermore, 12 out of 23 'vice president level' members of the Chinese National Forest Product Industry Association are FSC certified.

Cooperating with international VSS systems may become a more risky choice for enterprises in the future. According to the laws and regulations of 'Chinese Certification and Accreditation Ordinance', international VSS agencies that want to do business in China, requires a mutual recognition with the 'Chinese National Forest Certification System' first, and additionally should not oppose national security and public interests. However, FSC policy implies that an enterprise is not to carry out any mutual recognition acts with any other national forest certification system worldwide.

As a rule, the FSC selects certification standards and certification bodies by its own standards. Due to this, since 2012, the FSC has been in danger of being considered as an illegal certification instrument. Perhaps it will be excluded from the Chinese market of forest certification by the Chinese government. If the FSC is to become illegal in China, all the certifications from the FSC will become invalid as well. In the worst case, certification bodies will have to pay a fine for auditing for an illegal certification system (Zhang 2012).

17.5 Conclusions

In the short term, it will remain difficult for the paper-making enterprises in Liaoning Province to apply for a certification from any VSS system. Nevertheless, in the long run, it will be possible to realise this aim with the guidance and monitoring from the provincial government. The government must play the most important role in this process. The paper-making enterprises that cannot reach the basic legal requirements will be shut down by the Liaoning government in the future. Benevolent policy on the one hand and strict laws on the other hand will be used at the same time to regulate and motivate the enterprises towards VSS. In this way, VSS systems can be established even in relatively undeveloped areas such as Liaoning Province. Given the small size and the precarious financial situation of many paper-making enterprises, applying for certification from the CFCC is feasible and less risky.

References

5S – 5 Steps (2012) 5 Steps towards FSC certification, helping you along the path to becoming certified. http://ic.fsc.org/5-steps-to-certification.36.htm. Last accessed 29 May 2013

AP – Accreditation Program (2012) Ensuring excellence in certification. http://ic.fsc.org/accreditation.28.htm. Last accessed 29 May 2013

Chinese Government (2011) The 12th five year plan of China considering paper-making industry (in Chinese). http://www.miit.gov.cn/n11293472/n11295176/n11298883/n14422053.files/n14422454.pdf. Last accessed 29 May 2013

ICFCC – Introduction of Chinese Forest Certification Centre (2012). http://www.cfcc.org.cn/gsjj.asp. Last accessed 16 Oct 2012

Shi F (2011) The development of forest certification in China at the beginning of the 12th five year plan (in Chinese). Green Times of China of 16.07.2011. http://www.cfcc-ztfc.com/zxdtinfo.asp?id=802&nid=428. Last accesses 29 May 2013

TFSC – Types of FSC certificates (2012) From the forest through the supply chain. http://ic.fsc.org/types-of-certification.35.htm. Last accessed 29 May 2013

WWA – Who We Are (2012) Global, multi-stakeholder, membership organization. http://ic.fsc.org/about-us.1.htm. Last accessed 29 May 2013

Zhang K (2012) Invisible forestry war in Chinese Forest Certification Market (in Chinese). First Financial Daily of China of 04.07.2012. http://www.yicai.com/news/2012/07/1866657.html. Last accessed 29 May 2013

Zhu J (2010) Current situation and further development of paper-making in Liaoning Province (in Chinese). J Environ Prot Circular Econ 9. http://www.cnki.com.cn/Article/CJFDTOTAL-LNCX201009027.htm. Last accessed 29 May 2013

Chapter 18
Voluntary Environmental Certification in Ukraine: Experience and Issues

Dmitry Palekhov and Michael Schmidt

18.1 Introduction

As is well known, the 'green' segment of the world economy accounts for less than 2 %; and in post-Soviet countries its share is a mere fraction of 1 % (SP Ecological Union 2012). In this regard, great expectations are placed on voluntary eco-certification, which should become an instrument for harmonising business interests with sustainable development targets, i.e. by promoting the economic growth while raising the level of environmental safety and improving the state of the environment (Darnall and Sides 2008).

In Ukraine the development of voluntary eco-certification is a relatively new direction of the national and regional environmental policy, which is of major significance for fulfilling Ukraine's international commitments as a member of the World Trade Organisation (WTO) and in the framework of the EU–Ukraine Association Agenda (EC 2013; EU–Ukraine Cooperation Council 2013). Ukraine has already made certain efforts for the realisation of the indicated policy. In 2004 Ukrainian non-governmental organisation Living Planet became a full member of the Global Ecolabelling Network (GEN), and in 2011 its Ecolabelling Programme was accredited through the GEN peer review process—GEN Internationally Coordinated Ecolabelling System (GENICES) (MENR 2012; GEN 2012). Furthermore, in the last 2 years Ukrainian government has been conducting a number of dynamic reforms aimed at reducing the involvement of the state in technical regulation. As a result, the scope of compulsory certification has been drastically decreased; this process was in particular supported by the adoption of a series of legislative acts regulating the voluntary eco-certification and ecolabelling.

These developments offered Ukrainian producers the opportunity of gaining access to globalised production and supply chains. However, to take advantage of

D. Palekhov (✉) • M. Schmidt
Department of Environmental Planning, Brandenburg University of Technology (BTU) Cottbus, P.O. Box 101344, 03013 Cottbus, Germany
e-mail: dmitry.pal@gmail.com; palekdmy@tu-cottbus.de; michael.schmidt@tu-cottbus.de

C. Schmitz-Hoffmann et al. (eds.), *Voluntary Standard Systems*, Natural Resource Management in Transition 1, DOI 10.1007/978-3-642-35716-9_18,

this opportunity, the producers have to comply with certain requirements, e.g. prove the conformity to general or industry-specific environmental standards. At the same time, the necessary organisational restructuring could not sufficiently support and keep up a pace with this new policy of voluntary environmental certification. As a consequence, Ukrainian business had so far a very limited use of voluntary eco-certification and ecolabelling as an instrument for improving its competitiveness in domestic and international markets (see Chap. 11 for the detailed discussion on gaining and sustaining the competitive advantage based on voluntary environmental certification in the post-Soviet states).

The objective of this study is, therefore, to analyse problems and suggest possible solutions for the activation of voluntary eco-certification in Ukraine. The chapter starts with the discussion of distinctive features of voluntary environmental certification in Ukraine in Sect. 18.2. The chapter then continues with the review of issues related to Ukrainian procedure of voluntary eco-certification in Sect. 18.3. Section 18.4 provides an overview of Ukraine's experience with ecolabelling, with a special emphasis put on the discussion of barriers hindering its development and extended use. The chapter ends with conclusions and recommendations in Sect. 18.5 highlighting which measures could activate the development of voluntary eco-certification and labelling in Ukraine.

18.2 Distinctive Features of Voluntary Environmental Certification in Ukraine

Until recently, the Ukrainian system of certification remained cumbersome and inefficient. This was a significant limiting factor for the integration of Ukraine into the European economic space. In this context, the EU–Ukraine Association Agenda formulated an objective of harmonising the principles of technical regulation with the EU *acquis communautaire* (EU–Ukraine Cooperation Council 2013). The following actions were determined as significant for the achievement of this objective: (1) limit the scope of compulsory certification as much as possible, and promote the voluntary use of European and international standards; (2) simplify the certification procedure and start the transition to the module-based conformity assessment procedures as formulated in the EU's 'New Legislative Framework' for marketing of products[1]; (3) conduct administrative reforms with respect to the state regulation of certification procedure.

[1] The new regulatory Framework (often referred to as the New Legislative Framework—NLF) is a recent effort to strengthen the effectiveness of the EU's legislation on product safety, its implementation mechanisms, and ensure a greater consistency throughout all the different economic sectors (EC 2010). The NLF is formed by two distinct regulations: Regulation (EC) No 765/2008 of the Parliament and the Council setting out the requirements for accreditation and market surveillance relating to the marketing of products (OJEU L 218/30 of 13.8.2008), and Decision No. 768/2008/EC of the European Parliament and of the Council on a common framework for the

In an effort to comply with these obligations to the EU, Ukraine has undertaken a complete reorganisation of its certification system. Over the last 2 years, the list of products subject to compulsory certification has decreased by almost 65 %. More specifically, the compulsory certification was abolished for 50 groups of industrial products and 92 groups of other products/services (CMU 2013). As of February 2013 the certification is compulsory only for 19 product groups including approximately 200 product types. At the same time, more than 40 technical regulations on industrial products and conformity assessment procedures were introduced (e.g. Technical Regulation on Ecolabelling), which were developed in full accordance with the New Legislative Framework of the EU (Ukrmetrteststandard 2013b).

The Law of Ukraine "On main principles (strategy) of the state environmental policy of Ukraine for the period up to 2020" declared voluntary environmental certification as basis for the integrated environmental governance. And still, the organisational format of voluntary environmental certification in Ukraine has substantial differences from the European or international practice. This is caused by several reasons.

First of all, in Ukraine voluntary environmental certification is realised almost exclusively as certification of management systems on the basis of formal cross-sectoral ISO standards of the 14000 family, which were adopted in 1997 as national standards (i.e. as DSTU—State Standards of Ukraine[2]). Furthermore, as a rule, environmental standards of the ISO 14000 family are implemented as an add-on or improvement to the quality management systems. For example, as of 1 April 2013 in Ukraine there existed 3,064 management systems certified for conformity with the DSTU ISO 900 standard, 120 management systems among them were certified for conformity with the DSTU ISO 14001 standard (UkrSREC 2013).

With regard to the international sector-specific voluntary sustainability standards, such as Forest Stewardship Council (FSC), Programme for the Endorsement of Forest Certification (PEFC), Marine Stewardship Council (MSC), Sustainable Agriculture Network (SAN), etc., there are no principles, criteria and indictors adapted to the local, country-specific situation. And, as the result, their application is still very limited in Ukraine. For example, only a few Ukrainian companies have certificates for conformity with the FSC standard: five wood processing companies certified their products, and one juice producing company uses Tetra Pak packages with the FSC label (Eco-Live 2012; also see Fig. 18.1).

---

marketing of products (OJEU L 218/82 of 13.8.2008). Based on principles of proportionality and effectiveness, the NLW provides for a set of common conformity assessment procedures for products, referred to as '*Modules*'. The conformity assessment procedures are divided into eight basic modules (A–H), ranging from a manufacturer's self-declaration without reference to any independent third-party (Module A) through to full quality assurance leading to the issuing of a certificate of conformity (Modules F and G) (Ecorys 2011, pp. 151–156).

[2] DSTU is the anglicised acronym for the State standards of Ukraine (DSTU—"*derzhavni standarty Ukrainy*").

Fig. 18.1 A juice package used by the Ukrainian producer with a variety of labels on it, including the FSC label and a claim that the cardboard used for this package comes from sustainable sources (photograph by: Eco-Live 2012)

This tendency can, to a certain extent, be explained by the fact that voluntary environmental certification in the majority of cases is conducted through the Ukrainian State Certification System (hereinafter referred to as the UkrSEPRO). UkrSEPRO is a set of rules and procedures for proving the conformity of products and services that are being certified to the legal requirements, technical regulations and standards of Ukraine, as well as any other national or international standards that are adopted in Ukraine (DSTU 3410-96, Art. 3).

It should be emphasised that the UkrSEPRO is strictly regulated by the state standards of the DSTU 3410–3420 series (11 standards in total), through which it is tuned for the compulsory conformity assessment; however, the UkrSEPRO system is also open for voluntary environmental certification, mainly in accordance with the ISO 14000 family. The compulsory conformity assessment is conducted for all products listed in the regularly amended official "List of products that are subjected to the mandatory certification in Ukraine".[3] Voluntary certification through the UkrSEPRO system is carried out on the basis of an agreement between the applicant (manufacturer or supplier) and certification body. Certification of the imported products and domestic products is performed in accordance with the

[3] The List is approved by the Order of the State Committee of Ukraine on Technical Regulation and Consumer Policy No 28 of 01.02.2005, last amended on 29.12.2012. The State Committee of Ukraine on Technical Regulation and Consumer Policy was abolished in 2011; its functions were succeeded by the State Inspectorate of Ukraine on the Protection of Consumer Rights and the Ministry of Economic Development and Trade of Ukraine. The List is now amended by the Ministry of Economic Development and Trade.

same procedure. The right to carry out the certification is granted to certification bodies, testing laboratories and centres, as well as to officially-recognised auditors, who are included within the UkrSEPRO register (Ukrmetrteststandard 2013a).

After a number of regulatory bodies became defunct in the process of administrative reforms, the functions of a national certification body were delegated to the structural unit of the Ministry of Economic Development and Trade of Ukraine—the Department of Standardisation and Conformity Assessment.[4] The competence scope of this body includes the assistance available to Ukraine in joining the various international certification systems. In reality, however, its organisational and methodological activities are limited to the scope of the UkrSEPRO system.

There is no doubt that the UkrSEPRO has a number of advantages. This system is well organised at the nationwide level, its certification procedures are unified and can be applied to any company or enterprise. The UkrSEPRO register contains the complete information on certified companies and products, authorised certification bodies, and ensures the legal validity of the certification and accreditation results. Information contained in the UkrSEPRO register is freely available to any interested persons. It is also important to note that the UkrSEPRO certificate is recognised by all countries of the Commonwealth of Independent States (CIS), i.e. this certificate allows the export of products to other markets.

In the broader context, the access of Ukrainian producers to other voluntary systems of eco-certification is extremely limited. At the same time, according to the data from the UkrSEPRO register (UkrSREC 2013), in 2012 the proportion of environmental certificates in the total number of valid certificates remained at the level of 2006 and constituted less than 4 %. This trend is also reflected by the stagnating dynamics in the development of environmental certification (see Table 18.1).

The low flexibility of the UkrSEPRO system combined with a narrow scope of formal environmental standards of the ISO 14000 family, with which it mainly works, remains a formidable barrier to the realisation of the voluntary environmental certification concept in Ukraine. Furthermore, the voluntary environmental certification is sensitive to deficiencies in other instruments of environmental management, including environmental regulation and planning, environmental assessment, monitoring, public participation, etc.[5]

The latter factor became particularly apparent in the course of a project series on national adaptation of the FSC standard that were initiated in 2003 with the support of the World Bank and the World Wide Fund for Nature (WWF), and later, in

---

[4] The Ministry of Economic Development and Trade of Ukraine was formed in 2011 by the Order of the President of Ukraine No 634/2011 of 31.05.2011. It is a legal successor of a number of functions previously fulfilled by other ministries and government bodies, including: Ministry of Economy, Ministry of Industrial Policy, State Committee of Ukraine on Regulatory Policy and Entrepreneurship, State Service on Technical Regulation.

[5] See Palekhov et al. (2008) for the detailed analysis of strengths and weaknesses in Ukrainian practice of environmental regulation (i.e. setting environmental standards and thresholds) in the context of applying environmental assessment procedures.

Table 18.1 The dynamics of voluntary environmental certification to ISO 14001 in 2012 as compared to 2006 (*Source*: UkrSREC 2013)

	Number of valid certificates at the beginning of the period	Number of issued certificates	Number of cancelled or expired certificates	Number of valid certificates at the end of the period
Quarters of the year 2012				
I	92	4	3	93
II	93	7	2	98
III	98	14	4	108
IV	108	16	5	119
Total in 2012	92	41	14	119
Quarters of the year 2006				
I	23	2	–	25
II	25	2	–	27
III	27	4	1	30
IV	30	15	13	32
Total in 2006	23	23	14	32

2006–2008, supported by IKEA grant (ENPI FLEG 2012). The project encountered a number of organisational, legal and methodological difficulties. For example, it was found that there are discrepancies between Ukrainian norms and regulations on timber production and the FSC principles and criteria, that there are differences in interpretation of the 'sustainable forest management' concept. Another critical issue is that there is no adapted methodology for the assessment and monitoring of actual impact of certification on the environmental performance of certified companies (see Chap. 9 for the related discussion on measuring the impact of voluntary sustainability standards). Also questions of voluntary environmental certification are not sufficiently addressed by Ukrainian authorities in their plans and programmes (e.g. regional development programmes).

Therefore, following the discussion above, it is possible to conclude that there is a need in adapting the national system UkrSEPRO to make it compatible with various programmes of voluntary environmental certification. Furthermore, it might be necessary to introduce and develop other certification systems that would be specifically designed for the work with international voluntary sustainability standards in accordance with the latest developments and trends in this field.

18.3 The Procedure of Voluntary Environmental Certification and Its Challenges

In Ukraine, voluntary environmental certification is regulated in Art. 17 of the Law of Ukraine "On conformity assessment" (Law No. 2406-III of 17.05.2001). This Law stipulates that voluntary certification is conducted on a contract basis between the applicant (a manufacturer or a supplier) and the certification body. Voluntary environmental certification can be carried out for products and environmental management systems that are not subject to compulsory certification in the UkrSEPRO system. During the voluntary certification process, a certification body verifies the conformity of any claims with particular voluntary requirements, e.g. standards.

In accordance with the state standard DSTU 3410-96 "Certification System UkrSEPRO. Main principles", the following items are subject to compulsory environmental certification:

- Products and services included in the official List of products that are subjected to mandatory certification in the UkrSEPRO system;
- Activities and objects prone to causing higher environmental risks (approved by the Cabinet of Ministers Decree No. 554 of 27.07.1995, with amendments);
- Production and consumption waste, including objects of transboundary movement, and waste management activities;
- Environmental management systems at enterprises producing environmentally dangerous products (in accordance with standards developed by the technical committee ISO/TC 207—Environmental management, in which Ukraine participates as an observer member);
- Natural resources and environmental components (or their parts) that fall under the scope of the Convention on International Trade in Endangered Species of Wild Fauna and Flora, including those harvested in the open sea by vessels under a Ukrainian flag.

The decrease in the scope of compulsory certification has brought the issue of bodies performing environmental certification higher on the agenda. In Ukraine currently there are no domestic certification bodies with broad competence in voluntary environmental certification. In practice, all certification bodies are accredited by the National Accreditation Agency of Ukraine (NAAU) in accordance with the Law of Ukraine "On accreditation of conformity assessment bodies" (Law No. 2407-III), and consequently gain the right to provide certification services on the basis of formal standards, i.e. state standards of Ukraine DSTU, ISO 9000 and 14000 family standards, OHSAS standards, etc. As of 2 August 2013 there were 117 valid accreditation certificates issued by the NAAU for certification bodies, including: 32 accreditation certificates in accordance with ISO/IEC 17021:2011—Bodies providing audit and certification of management systems; 79 accreditation certificates in accordance with DSTU EN 45011-2001—Bodies for certification of products, processes and services; 6 accreditation certificates in accordance with

ISO/IEC 17024:2003—Bodies operating certification of persons (NAAU 2013). As a consequence, companies willing to obtain a certificate in accordance with any other type of voluntary sustainability standards have to search for an appropriate certification body in other countries.

In addition, the analysis of services performed by Ukrainian certification bodies revealed the limited variation in scope of environmental certification. The majority of companies, which sought voluntary certification, certified their environmental management systems—ca. 60 %. Approximately 30 % of companies certified food products, consumer goods, and industrial products. Around 20 % of certifications account for various services, e.g. wholesale and retail trade, maintenance of power supply units, accommodation services, travel services, etc. Since recently, the eco-certification and labelling unit of the non-governmental organisation 'Living Planet' offers the Ukrainian market a new certification service, 'Green Office', a voluntary certification of office facilities on the basis of the private standard SOU OEM.08.036.67 "Administrative services (offices). Environmental criteria" that was developed in accordance with the ISO 14000 standards family (Living Planet 2012).

Unfortunately, the engagement with voluntary environmental certification of products in Ukraine is significantly less active than in the EU states, or even in Russia. According to information provided by Living Planet, at the moment, only 60 Ukrainian manufacturers certified their own-produced products, in total 230 brands. The procedural difficulties with voluntary certification are one of the reasons for this situation. According to the Law of Ukraine "On conformity assessment" (Art. 17), the rules for voluntary certification are established by the certification body. As the result, during the voluntary certification in the majority of cases the bodies apply procedures of compulsory certification in accordance with schemes provided by the state standard DSTU 3413-96. Table 18.2 characterises the voluntary certification procedure that is conducted in accordance with the formalised approach of compulsory certification in the UkrSEPRO system.

As can be seen from Table 18.2, the procedure of voluntary environmental certification for products conducted in accordance with the state standard DSTU 3413-96 can be rather complicated and expensive. However, the validity period of certificates varies only between 1 and 5 years. During such a short time it is difficult to start gaining market benefits from the certified products, and as a consequence some of the companies refuse carrying out the repeated certification.

The interest in voluntary environmental certification can be also decreased by certain difficulties in fulfilling the requirements of standards. For example, in case of the ISO 14001 standard, the auditors are often asked questions such as: How to implement the process of "continuous improvement" and how to measure its results? How to measure economic efficiency and other real benefits arising from the certification to ISO 14001? What should be done, if the environmental performance is not improved?

At the same time, the auditors are mainly involved in certification and controlling the state of enterprises. They do not have sufficient time, and sometimes also knowledge, to suggest solutions for potential problems in the post-certification

Table 18.2 Voluntary environmental certification schemes for products in accordance with the state standard DSTU 3413-96

Certification schemes	Manufacturing plant inspection	Attestation of manufacturing	Quality management certification	Tests during certification	Technical supervision	Type/validity of certificates
Certification of individual products	Is not required	Is not required	Is not required	Conducted for every product unit	Is not required	Conformity certificate for every product unit
Certification of product batches	Is not required	Is not required	Is not required	Conducted for product samples from the batch in accordance with the appropriate procedure	Is not required	Conformity certificate for product batch with specification of the batch size
Certification of product lines	Is not required	Is not required	Is not required	Conducted for product samples in accordance with the appropriate procedure	Conducted through control tests or manufacturing plant inspection	Conformity certificate with a validity of not more than 1 year
	Is required	Is not required	Is not required	Conducted for product samples in accordance with the appropriate procedure	Conducted through control tests or manufacturing plant inspection	Conformity certificate with a validity of not more than 2 years
	Is not required	Is required	Is not required	Conducted for product samples in accordance with the appropriate procedure	Conducted through control tests or manufacturing plant inspection	Conformity certificate with a validity of not more than 3 years, taking into account validity of the manufacturing attestation certificate
	Is not required	Is not required	Carried out by the body for certification of quality management systems	Conducted for product samples in accordance with the appropriate procedure	Conducted in accordance with the certification methodology and quality management system	Conformity certificate with a validity of not more than 5 years, taking into account validity of the quality management system certificate

period, as well as methods for overcoming developmental stagnation. The role of auditors and capacity building in promoting voluntary environmental certification should be arguably increased. The companies should have an opportunity for getting assistance with the selection of appropriate marketing strategies based on the use of voluntary environmental certification. Also, in addition to ISO 14000 family standards, companies should get easier access to certification with a broader range of voluntary sustainability standards, e.g. FSC, MSC, SAN, the Initiative for Responsible Mining Assurance (IRMA), and many others.

18.4 The Experience with Voluntary Ecolabelling

The contemporary principles of voluntary ecolabelling in Ukraine were de facto established in 2003. At that time, Ukrainian non-governmental organisation 'Living Planet' with support of the government initiated the national programme "Development of sustainable (balanced) production and consumption in Ukraine", in the framework of which an environmental certification and labelling programme was developed and implemented in full accordance with the national standard DSTU ISO 14024:2002 "Environmental labels and declarations. Type I environmental labelling. Principles and procedures".[6] The best practice and experience from such voluntary eco-certification schemes as the German 'Blue Angel' (Der Blaue Engel, launched in 1978), and the EU 'Flower' Ecolabel (introduced in 1992) provided a strong basis for its development. The new ecolabelling programme was first presented during the Fifth Ministerial Conference "Environment for Europe" that was conducted from 21 to 23 May 2003 in Kiev. In 2004 Ukraine with its ecolabelling programme became a full member of the Global Ecolabelling Network (GEN). And in 2011 it was the first voluntary certification scheme in the post-Soviet space to be accredited through the GEN Internationally Coordinated Ecolabelling System (GENICES).

It should be noted that Ukrainian eco-certification and labelling programme is a result of the unique non-governmental initiative that established a voluntary and independent scheme for certification of products allowing the use of ecolabels. It is based on numerous criteria and considers the environmental advantages of products within a particular product category based on the results of their life-cycle assessment (DSTU ISO 14024:2002).

Ukrainian eco-certification and labelling programme includes a label—a crane bird stylised as a green sprout that is depicted on a background resembling the Earth and symbolising life on our planet (see Fig. 18.2).

[6] In its turn, the state standard DSTU ISO 14024:2002 was adopted as a harmonised ISO 14024:1999 standard, which provides guidance on developing programmes that verify the environmental attributes of a product via a seal or a label, i.e. specify the procedures and principles that third-party certifiers, or eco-labelers, must follow (Green Seal 2013; IISD 2013).

Fig. 18.2 Ukrainian ecolabel 'Green Crane' (first from the left) in comparison with other ecolabels according to the ISO 14024 standard (*Source*: MENR 2012)

Accreditation of the Ukrainian eco-certification and labelling programme through the GENICES provides Ukrainian producers with an opportunity to expand the market for products and services that were certified in Ukraine, as well as to obtain a license for using the ecolabels from other countries participating in the GENICES programme through a simplified procedure.

The regulatory basis of ecolabelling was strengthened by the adoption of the Technical Regulation on Ecolabelling (approved by the Cabinet of Ministers Decree No. 529 of 18.05.2011) that was drafted with due consideration of provisions of the Regulation (EC) No. 66/2010 of the European Parliament and of the Council of 25.11.2009 on the EU Ecolabel.

The Technical Regulation on Ecolabelling stipulates the requirements for the conferment and use of voluntary ecolabels in Ukraine, and specifies the procedure for developing and revising the environmental criteria. Ecolabels can be applied only to products for which conformity to the environmental criteria, established for the particular product category, has been assessed. The Technical Regulation on Ecolabelling specifically prohibits the producers and retailers from using claims, such as 'environmentally friendly', 'environmentally safe', 'non-polluting', 'ozone friendly', 'green', etc., as long as these claims are not duly verified. Before the adoption of the Technical Regulation on Ecolabelling, such claims were widely used on the product packaging and in advertising without proper verification of their credibility. Also the marking 'free from . . .' is only allowed if the content of the particular substance in the product does not exceed the background levels.

The label 'Green Crane' can be seen increasingly often on the packages of Ukrainian consumer and food products, in advertising and company materials, in offices. At the same time, ecolabelling in Ukraine still cannot be considered as a powerful instrument of sustainable consumption, as for example in the EU member states. The following issues hinder the development and extended use of ecolabels:

- environmental criteria defining the environmental advantages are still missing for numerous categories of products and services;
- low level of awareness about ecolabelling among the producers;
- the consumers are often not aware of and confused by various ecolabels and seals;
- insufficient state support and promotion of the use of ecolabels.

18.5 Conclusions and Recommendations

Ukrainian schemes of voluntary environmental certification and labelling are currently going through major reforms. The main directions of change include: rapid decrease in the scope of compulsory certification; transition to the conformity assessment procedures based on technical regulations; administrative reforms in the governmental regulation of certification procedures. At the same time, voluntary environmental certification and ecolabelling in Ukraine cannot be characterised by a sustainable development dynamics. The main reason for such situation is a small number of alternative options for the selection of certification programmes, which are mainly limited to schemes offered by the UkrSEPRO system. This system has a number of obvious benefits: it is well organised throughout the country, its certification procedures are uniform and are open for any interested company or enterprise. However, it is also not devoid of certain disadvantages. In particular, the UkrSEPRO system is tuned primarily for the compulsory conformity assessment, and voluntary certification within this system is limited to formal standards of the ISO 14000 family. And difficulties with measuring the real benefits from certification to ISO 14000 standards decrease the interest of business in voluntary environmental certification. Another major issue is the readiness of certification bodies to the work with other contemporary voluntary standards systems, such as FSC, MSC, IRMA, Fairtrade or SAN, etc.

A series of measures may be necessary to encourage the development of voluntary environmental certification, including the following:

- First, the voluntary certification segment of the national certification system UkrSEPRO should be gradually adapted to work with various internationally recognised voluntary sustainability standard systems. The potential and resources of the UkrSEPRO could facilitate an increasing recognition and use of modern voluntary sustainability standards in Ukraine. However, this approach may encounter certain difficulties caused by the excessive formalisation of the UkrSEPRO system, and, as a consequence, integration of new voluntary certification schemes might be complicated.
- Second, the voluntary environmental certification in Ukraine could greatly benefit from specific partnership programmes with leading international standard-setting organisations and voluntary standards systems. For example, newly established national offices and focal points could work on: adaptation of standards to local conditions, including harmonisation of principles, criteria and indictors; development of country-specific guidelines; capacity building programmes and consultations for the business and public, etc.
- Third, the development of voluntary certification could be facilitated by a general organisational and economic support from the government. Green procurement programmes, providing economic benefits and tax reliefs, integration of voluntary certification with other existing environmental management instruments may serve as an example of such support measures.

- Finally, voluntary certification will require support from the national certification body, i.e. the Department of Standardisation and Conformity Assessment of the Ministry of Economic Development and Trade of Ukraine. Such support could be: the training of national auditors; development of training methodologies and guides; assistance for certification bodies with their preparations to the accreditation; maintaining the register of accredited certification bodies and auditors, etc.

References

Cabinet of Ministers Decree No. 554 of 27.07.1995 "On the List of activities and objects prone to causing higher environmental risks". Government courier (Uriadovy kurier) No. 131 of 02.09.1995, last amended by the Cabinet of Ministers Decree No. 630 of 06.06.2011, Official gazette of Ukraine (Ofitsiyny visnyk Ukrainy) No. 45 of 24.06.2011, p. 102, st. 1847

Cabinet of Ministers Decree No. 529 of 18.05.2011 "On approval of the Technical regulation on ecolabelling". Official gazette of Ukraine (Ofitsiyny visnyk Ukrainy) No. 39 of 03.06.2011, p. 131, st. 1599

CMU – Cabinet of Ministers of Ukraine (2013) Information on the Implementation of the EU–Ukraine Association agenda in 2011–2013. http://www.kmu.gov.ua/control/uk/publish/article?art_id=243281941&cat_id=223345338. Last accessed 31 July 2013

Darnall N, Sides S (2008) Assessing the performance of voluntary environmental programs: does certification matter? Policy Stud J 36(1):95–117

DSTU 3410-96. Certification System UkrSEPRO. Main principles. State Standard of Ukraine. State Committee of Ukraine on Technical Regulation and Consumer Policy, Kiev, 1997

DSTU ISO 14024:2002. Environmental labels and declarations. Type I environmental labelling. Principles and procedures. State Committee of Ukraine on Technical Regulation and Consumer Policy, Kiev, 2004

EC – European Commission (2010) State of the Implementation of the New Legislative Framework (NLF). The EC's Directorate-General for Enterprise and Industry, Brussels, 14 September 2010

EC – European Commission (2013) EU–Ukraine: deep and comprehensive free trade area. http://trade.ec.europa.eu/doclib/docs/2013/april/tradoc_150981.pdf. Last accessed 31 July 2013

Eco-Live – Ukrainian Information Eco-Portal "Eco-Live" (2012) FSC in Ukraine (in Ukrainian). http://www.eco-live.com.ua/category/kluchovi-slova/fsc. Last accessed 31 July 2013

Ecorys – Ecorys Nederland BV (2011) Security regulation, conformity assessment and certification. Final report – volume I: Main report, October 2011. http://ec.europa.eu/enterprise/policies/security/files/doc/secerca_final_report_volume__1_main_report_en.pdf. Last accessed 29 July 2013

EU–Ukraine Cooperation Council (2013) EU–Ukraine Association agenda to prepare and facilitate the implementation of the association agreement, as endorsed by the EU–Ukraine Cooperation Council (Luxembourg, 24 June 2013). http://eeas.europa.eu/ukraine/docs/eu_ukr_ass_agenda_24jun2013.pdf. Last accessed 31 July 2013

ENPI FLEG – European Neighbourhood Policy Instrument – Forest Law Enforcement and Governance Programme (2012) Regional Workshop on Voluntary Forest Certification and Sustainable Forest Management Standards in the Southern Caucasus: Workshop Report, 23–24 May 2012, Batumi, Georgia

GEN – Global Ecolabelling Network (2012) GEN Annual Report 2012. The GEN Secretariat, Ottawa, Ontario

Green Seal (2013) International standards for eco-labeling. http://www.greenseal.org/GreenBusiness/Standards/HowGreenSealDevelopsStandards/InternationalStandardsforEcoLabeling.aspx. Last accessed 15 Aug 2013

IISD – International Institute for Sustainable Development (2013) The ISO 14020 series. http://www.iisd.org/business/markets/eco_label_iso14020.aspx. Last accessed 15 Aug 2013

Living Planet (2012) Environmental certification of organisations providing office services according to the programme "Green Office" (in Ukrainian). http://www.ecolabel.org.ua/ekologichna-sertifikatsiya-ta-markuvannya/ekologichna-sertifikatsiya-ta-markuvannya-za-programoyu-zeleniy-ofis.html. Last accessed 10 Aug 2013

MENR – Ministry of the Environment and Natural Resources of Ukraine (2012) Ukrainian programme for voluntary certification and labelling (in Ukrainian). http://www.menr.gov.ua/content/article/11651. Last accessed 29 July 2013

NAAU – National Accreditation Agency of Ukraine (2013) NAAU Register of the accredited conformity assessment bodies (as of 02.08.2013). http://www.naau.org.ua/ua/members.html. Last accessed 05 Aug 2013

Palekhov D, Schmidt M, Pivnyak G (2008) Standards and thresholds for EA in highly polluted areas – the approach of Ukraine. In: Schmidt M, Glasson J, Emmelin L, Helbron H (eds) Standards and thresholds for impact assessment, vol 3, Environmental protection in the European Union. Springer, Berlin, pp 33–48

SP Ecological Union – Saint-Petersburg Ecological Union (2012) Report of representatives of the Russian ecolabel "Vitality leaf" on the status quo of eco-certification in Russia at the Global Ecolabelling Network (GEN) Summit 2012 (in Russian). http://www.ecounion.ru/ru/site.php?content=detailcontent.php&blockType=196&blockID=1444. Last accessed 31 July 2013

Ukrmetrteststandard – State Enterprise "All-Ukrainian Research and Production Center of Standardization, Metrology, Certification and Consumers' Rights Protection" (2013a) Certification System UkrSEPRO. http://www.ukrcsm.kiev.ua/index.php/en/2009-02-12-13-16-36/engineering/110-2009-02-16-14-15-53. Last accessed 01.08.2013

Ukrmetrteststandard – State Enterprise "All-Ukrainian Research and Production Center of Standardization, Metrology, Certification and Consumers' Rights Protection" (2013b) Register of technical regulations, updated on 26.03.2013. http://csm.kiev.ua/index.php?option=com_content&view=article&id=111&Itemid=66&lang=ru. Last accessed 01.08.2013

UkrSREC – State Enterprise "Ukrainian Scientific-Research and Educational Center for Standardization, Certification and Quality" (2013) Monitoring data of the certified management systems (in Ukrainian). http://www.ukrndnc.org.ua/index.php?option=com_content&task=category&sectionid=8&id=46&Itemid=54. Last accessed 31 July 2013

Chapter 19
Comparative Analysis of Environmental and Social Impacts of Cocoa Production: Case Study Cameroon

Eric Ambe Asoh, Ambe Emmanuel Cheo, Michael Schmidt, and Hans-Jürgen Voigt

19.1 Introduction

During the late 1980s, the Cameroonian cocoa sector was controlled by the government in their attempt to stabilise farm income by setting an average purchase price for raw cocoa. This system collapsed around 1987, when there was a drastic fall in the world price of cocoa and the government could not stabilise the market using its traditional support systems of setting an annual purchase price. In 1994/1995 the Cameroon cocoa trade was fully liberalised (Bisseleua 2007). According to the Ministry of Agriculture and Rural Development, the Sixth International Cocoa Agreement which came into force in 2003, inspired the adoption of Law No. 2004/025 on 30th December 2004 that liberalised the country's cocoa sector (UNEP 2009). In this law, issues such as product quality and their derivatives, the collection, analysis and dissemination of data were addressed. Additionally, compliance with international standards for future production was also ensured by the law.

Cocoa production in Cameroon is primarily conducted on small-scale farms. The increased number of smallholder plots has led to fragmentation of many habitats, resulting in a significant decrease in the number of endemic species, especially birds (UNEP 2009). There is an important relationship between cocoa production, local environment, biodiversity and the social character of communities that are involved in cocoa production. The production of cocoa is very labour intensive and requires a substantial portion of available manpower. The lack of this substantial

E.A. Asoh (✉) • M. Schmidt
Department of Environmental Planning, Brandenburg University of Technology (BTU) Cottbus, P.O. Box 101344, 03013 Cottbus, Germany
e-mail: erickarel_a@yahoo.com; asoheric@tu-cottbus.de; michael.schmidt@tu-cottbus.de

A.E. Cheo • H.-J. Voigt
Department of Environmental Geology, Brandenburg University of Technology (BTU) Cottbus, P.O. Box 101344, 03013 Cottbus, Germany
e-mail: cheoambe@tu-cottbus.de; voigt@tu-cottbus.de

C. Schmitz-Hoffmann et al. (eds.), *Voluntary Standard Systems*, Natural Resource Management in Transition 1, DOI 10.1007/978-3-642-35716-9_19,

labour force has led to the employment of children. There is also gender disparity in the acquisition of farm land. Furthermore, multiple studies (Zapfack et al. 2002; Bobo et al. 2006; Sonwa et al. 2007; Sonwa 2004) have identified numerous negative environmental and social impacts resulting from cocoa production in Cameroon. In the wake of such impacts, the voluntary standard systems were introduced so as to curb the aforementioned impacts.

According to Carey and Guttenstein (2008), "*voluntary standard systems are cross-sectoral partner-ships created with a rule-setting purpose, to design and steward standards for the regulation of market and non-market actors*". The identified voluntary standard systems operating in the cocoa sector of Cameroon includes Fairtrade and International Federation of Organic Agriculture Movements (IFOAM). These systems function through farmers organising themselves under a certification body.

Despite the growing popularity of voluntary standard systems, there seems to be little or no consensus that voluntary standard systems are positive for farmers or not. Some consider the standard systems as an adequate management tool in achieving the objectives of the Sustainable Development Goals (SDGs). According to KPMG (2012), some actors in the sector are less optimistic about the net benefits that certification offers at farm level. Furthermore, they highlighted the burden that certification could bring in terms of required investments.

This chapter will continue in Sect. 19.2 by identifying the cocoa production areas and quantifying the potential of the cocoa sector in regard to the whole Cameroonian economy. Following on in Sect. 19.3, the different cultivation methods are characterised and their operational arrangements are summarised.

Drawn from findings, Sect. 19.4 gives a comparative analysis for the socio-environmental impacts of cocoa production under conventional methods and production under voluntary standard systems. Several indicators are identified to focus the analysis. However some of the indicators are not quantified; rather they provide an improved understanding of key issues related to sustainability and biodiversity in Cameroon. In conclusion, Sect. 19.5 consists of recommendations for further research on measuring and quantifying the socio-environmental impacts of the different production systems so that with their result, it is easier to evaluate their performance.

19.2 Potential of the Cocoa Sector in Cameroon

Historically, cocoa production has been a major source for export revenue and internal revenue through taxes and worker salaries/income in Cameroon. Statistics show that about 420,000 ha of land is used for cocoa cultivation in Cameroon (ICCO 2006). According to Sonwa et al. in Bisseleua (2007), 450,000 rural households depend on cocoa for their livelihood (more than a third of the total number of rural households). Production is predominantly done on small family farms with farm size ranging between 2 and 10 ha. There are about 500,000 cocoa

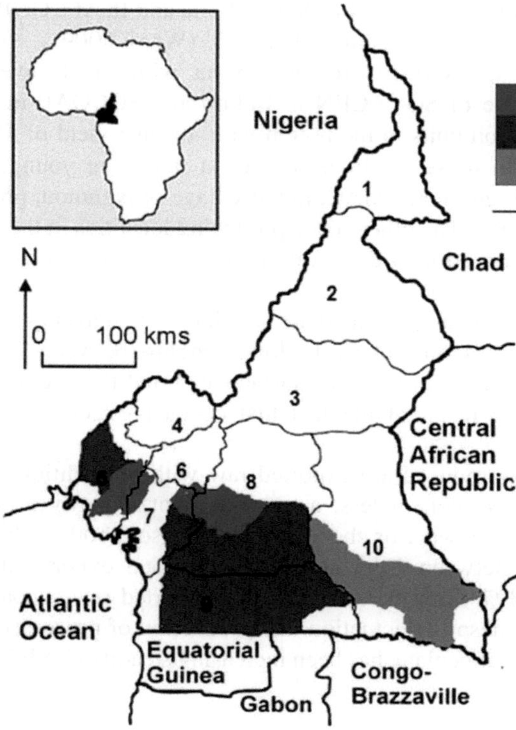

Fig. 19.1 Cocoa production zones in Cameroon (*Source*: adapted from NCCB 2008). *Note*: Regions; 1 = Far North, 2 = North, 3 = Adamawa, 4 = North West, 5 = South West, 6 = West, 7 = Littoral, 8 = Centre, 9 = South, 10 = East

farmers in Cameroon (Righarts 2009). The production of cocoa is predominantly done in two zones shown in Fig. 19.1 as the centre/southern regions and the south-west region.

Cameroon ranks fifth in terms of world cocoa production. In 2002, the production of cocoa was estimated to be at approximately 134,000 tons per annum. By 2009 production had increased to approximately 200,000 tons per annum. Between 2010 and 2011, total cocoa production was estimated at 236,000 tons per annum (Drum Commodities 2012).

The sizes and characteristics of the cocoa fields in the production zones vary according to the regions. The region of the South West holds the largest production surface. This region is characterised by a strong dynamic of cocoa farming, of relatively young cocoa farms with an average age of 35 years. Cocoa cultivation in some localities of this region corresponds to a semi-intensive farming system. The average size of cocoa field per farmer in the South West amounts to 3 ha. The South West has the highest output averaging more than 425 kg ha^{-1}. It is the most important region of production, accounting for about half of the country's output

(Wanji 2009). The Cooperatives of BAFIAFCOOP in Bafia and IKAFCOOP in the village of Ikata (South West) have yields of 500 kg ha^{-1} (Wanji 2009).

In the Centre, the average surface area is 2.5 ha with yields averaging 360 kg ha^{-1}. The cooperative of SOCACEN in Bokito and SOCOAP in Okola are identified as key production units in the region with average yield of 450 and 460 kg ha^{-1} respectively. In these localities the cocoa fields are young, about 40 years and the yields are high. SOCAMAC, in the village of Ngumou, produces an average yield of 420 kg ha^{-1}. The other cocoa producing localities in the Centre region are generally characterised by a decrease in yield caused by the advanced age of the cocoa fields (Wanji 2009).

In the Southern region, the areas cultivated are smaller with an average surface area of 2.3 ha and the yields are relatively low, ranging between 200 and 300 kg ha^{-1}. The Littoral region has the least surface area and the average yield is 350 kg ha^{-1}. This is as a result of the fact that cocoa production is newly introduced to this region.

Most of the cocoa beans produced are exported raw with no additional value through the processing stages. Nonetheless, a small amount is partly or wholly transformed into paste, butter, powder or chocolate which is sold both in domestic and international markets. Between 1992 and 2007, the total export value for transformed cocoa was 24.3 % from the total value of exported raw cocoa beans (UNEP 2009). Furthermore, despite fluctuation, the production of processed products, such as cocoa butter and chocolate, has been increasing steadily (UNEP 2009).

19.3 Characteristics of Cocoa Cultivation in Cameroon

In a broader sense, the production of cocoa can be classified into two major groups; namely the conventional method of production and the production of cocoa under a voluntary standard system. To better understand the socio-environmental implication of these methods, their specific differences are summarised in the subsequent paragraphs.

Conventional methods of production are cultivation methods that are not done under the guidance of any certification scheme. These are the predominant and traditional methods of cocoa production in Cameroon. Production under such methods does not necessarily mean they are environmentally unsustainable. Many farmers under such schemes sometimes use little or no fungicides and insecticides. Some farmers use organic manure as their source of fertiliser. Therefore, there exists a large group of farmers using non certified techniques which are not necessarily destructive to the environment (such as organic manure farming, no fungicide and insecticides farming, no use of synthetic fertiliser, etc.) but their products are not accredited by any certification body. Farmers in KONAFCOOP, SOCAMAC and GROUPEX SC are examples of farmers that have had contact with extension workers and trained to produce sustainably. Farmers from these cooperatives have benefited from training offered to them through Farmer Field School

(FFS). The farmers were trained in using integrated pest management techniques to fight against cocoa diseases and parasites. This scheme was proposed by the International Institute of Tropical Agriculture (IITA), considered under the banner of the Sustainable Tree Crop Program (STCP). The aim was a long-term increase of productivity, reduction of production costs, increase in output of marketable produce, increase of on-farm storage opportunities, improved socio-economic situation of farmers and improved efficiency in the marketing chain (Wanji 2009).

Under conventional methods of production, the cocoa crop is predominantly cultivated in multi-product and multi-strata agroforests (Leplaideur 1985; Bisseleua 2007). This multi-product feature of the system was developed by farmers to diversify production and minimise risk. According to Kanmegne (2004), about 75 % of cocoa farms in Cameroon are cocoa dominating mixed with few trees (fruits, timber, etc.) while 25 % of cocoa farms are cocoa mixed with food crops (such as bananas, plantain, cocoyam, cassava, etc.). When the cocoa tree and other components grow to maturity, the agroforest becomes a more diverse and structurally complex, closed-canopy multi-strata system that resembles natural forest (Duguma et al. 2001). An economics analysis revealed that the agroforest system of agriculture involving cocoa tree has been profitable (UNEP 2009).

The main inputs involved in conventional farming are fungicides, insecticides and hired labour (which can be occasional or permanent), and these components constituent the most important variable costs for the farm. The most severe problem faced by cocoa farmers in Cameroon is the occurrence of pests and diseases. According to Bakala and Kone (1998), about 50–80 % of yield loss in Cameroon is due to pests and diseases. Most often farmers increase their production through the intensification of production or by the expansion of their farm land. The farmers depend greatly on their past experience for decision making and the management of their farms.

Even though similarities exist in the major goals and certification procedures of most standards, there are some significant differences in their emphasis on environmental, social or economic issues. Standards obliged producers to detail all aspects of production before their products are certified by a third party. Quality aspects such as soil management, crop rotation, biological inputs, pest controls, post-harvest techniques, labour force, storage, handling and document tracking, etc., are important parameters considered in standard schemes.

In Cameroon, producers are organised in groups (called co-operatives), in order to participate in a standard scheme, with compliance auditing conducted annually. Co-operatives that were identified to be under standard schemes include MACEFCOOP; KONAFCOOP for Fairtrade (Ebong 2008) and Biotropical for IFAOM. Other companies such as Export Agro and EXODOM companies are involved in organic cocoa production but are certified by complying with the EEC Regulation 2091/92 (FAO 2001). There is no national standard for cocoa producers in Cameroon. Therefore, the operation of voluntary standards systems and other standards is understandable since the target market for the cocoa product is European and North American markets which adhere to very strict agricultural policies.

Voluntary standard systems advocate offering benefits to small producers who have limited access to world markets and a weak bargaining position. They also advocate environmental safety and social benefits for the producers. Furthermore, farmers also benefit through extension workers who train them in integrated pest management to prevent and reduce the incidence and severity of cocoa disease and parasite infestation. However, the major constraint towards achieving the objectives of voluntary standard systems is the high level of illiteracy of the farmers who are unable to document their production process or improve their poor understanding of the standards scheme.

19.4 Analysis of the Socio-Environmental Performance of Cocoa Production

In Cameroon, detailed study about the environmental performance of cocoa production has not been done. The social benefits are only realised when cocoa production remains economically attractive to the smallholder farmer. Sections 19.4.1 and 19.4.2 below summarise some of the identified environmental and social impacts related to cocoa production.

19.4.1 Environmental Impacts

Table 19.1 summarises the environmental impacts for both conventional and voluntary standard systems that are related to sustainability and biodiversity.

Farmers under voluntary standard systems are taught to control pests and diseases using biological methods. Their shade and farm canopy manipulate the environment in a sustainable manner so as to increase productivity of the farms. According to Jaffee (2008) and Krain et al. (2011), farmers under certification schemes adopt environmentally sound measures such as planting shade trees, producing compost and applying this to fields, building live plant barriers, establishing terraces or setting aside certain areas for wild life protection. However, stringency of the requirements for the conservation of soil, flora and habitats differs per scheme (KPMG 2012).

Data for a quantitative environmental impact assessment was not available for cultivation under voluntary standard systems. Nonetheless, to demonstrate the environmental performance for conventional production, ICRAF (1996) compared the nutrients of selected top soils for secondary forests and cocoa-dominated tree based home gardens in southern Cameroon. The results show that soil pH, organic matter, calcium and magnesium are higher in cocoa dominated home gardens than in secondary forests (Table 19.2).

Table 19.1 Identified environmental impacts related with cocoa production

Conventional production	Production under voluntary standard system
Use of agrochemicals alters ecosystems and contaminates both surface and ground water[a]	Use of biological methods e.g. Integrated Pest Management (IPM), organic manure reduces environmental degradation[a,b]
Soil productivity is not optimised[b,c]	Soil productivity is optimised[b,c]
Little or no expertise, thus increases the vulnerability of the natural environment[d]	New expertise in farming techniques that preserve the natural environment[d]
Pod wastes are a nuisance and provide breeding sites for black pod disease, viruses and pests[e]	Pod wastes are recycled as organic fertiliser and applied on farms[e]

Sources: [a]de Battisti et al. (2009), [b]Jaffee (2008), [c]Krain et al. (2011), [d]Wanji (2009) and [e]Potts and Giovannucci (2012)

Table 19.2 Selected soil properties at 0–20 cm depth in cocoa dominated home gardens and secondary forest in southern Cameroon (ICRAF 1996)

	Soil properties									
	PH (1:1 water: soil)		Organic matter (%)		Ca ($Cmol_c\ kg^{-1}$)		Mg ($Cmol_c\ kg^{-1}$)		K ($Cmol_c\ kg^{-1}$)	
Study site	Hg	Sf	Hg	Sf	Hg	Sf	Hg	Sf	Hg	Sf
Yaounde	6.9	5.2	4.4	2.5	10.8	2.6	2.1	1.0	0.4	0.11
Mbalmayo	6.8	6.5	4.1	4.8	11.4	5.2	2.0	1.8	0.6	0.15
Ebolowa	6.5	4.8	4.7	3.2	11.8	3.0	2.5	0.9	1.4	0.15

Note: *Hg* Cocoa dominated home garden, *Sf* secondary forest

IRAD (1997) compared the total biomass for different land use systems in the Mekoe region of southern Cameroon and ranked cocoa agro-forestry third in comparison to long-term fallow and primary forest respectively (Table 19.3). According to Kotto-Same et al. (1997), cocoa agro-forest contained 62 % of the carbon stock found in primary forest (Duguma et al. 2001).

Furthermore, IRAD (1997) concluded that above-ground plant biodiversity and below-ground micro-fauna for cocoa agro-forestry are higher than in food crop fields, and are comparable to those in short- and mid-term fallows as shown in Table 19.3. According to UNEP (2009), the cocoa agro-forestry is relatively sustainable and provides habitats for animal and bird life. They also act as ecological corridors that reduce habitat fragmentation and offer opportunities for carbon sequestration (UNEP 2009). Furthermore, they provided greater benefits to erosion control than most other agricultural ecosystems. Increased levels of erosion were observed immediately after the cocoa plantations were established but once established and the canopy closed, erosion drastically reduced with time in the shade-grown cocoa producing areas (UNEP 2009). Even though not all components (such as deforestation, species diversity, etc.) related to sustainability and biodiversity were measured and analysed, cocoa agro-forestry produced some positive results when compared with other agricultural systems.

Table 19.3 Total vegetation biomass (t ha^{-1}) in various land use systems in Mekoe, southern Cameroon (IRAD 1997)

	Biomass in t ha^{-1}				
Land use	Tree	Understorey	Litter	Root	Total
Primary forest	485	2.9	8.7	44.6	541.2
Food crop field	45	7.6	–	32.7	85.3
Young fallow (less than 5 years)	–	2.6	11.8	27.7	42.1
Mid fallow (5–10 years)	54	4.5	14.2	34.6	107.3
Old fallow (10–20 years)	400	4.3	12.3	44.2	460.7
Cocoa agroforest (26 years)	250	1.6	11.7	41.2	304.4

19.4.2 Social Impacts

Cocoa production plays an important role in the livelihood and the stability of the local communities in terms of job creation and bringing income to the local population. The social benefits and impacts resulting from cocoa production are to some extent inter-related and can be mutually reinforced. For example, a market development that leads to a decrease in price would affect the income of the farmer and also increase the rate of migration (especially of youths) from rural to urban areas. Moreover, as migration increases, fewer people are left to work on the relatively labour intensive cocoa plantations which could lead to an increase in the level of child labour. This has the reverse effect when the market impact is positive. Table 19.4 summarises some identified social impacts of cocoa production in Cameroon.

In Cameroon, it was also identified that voluntary standards often come with additional costs, for example an audit cost for certification which is often relatively high for farmer groups not yet benefiting from the standard market. Many farmers are not willing to accept some standards for fear that reduced inputs will be reflected in a loss of production and reduced productivity. According to Ebong (2008), a recurrent complaint by producers groups are the difficulties of complying with voluntary standard system's generic and product standards that appears to be dictated by the consumer preferences of developed nations, particularly those relating to environmental protection, which in some cases run contrary to what is allowed or encouraged by national producer-support policies and services e.g. the list of chemicals forbidden under standards.

On the other hand, the cocoa agro-forestry system has also brought additional benefit through wood and non-wood products which serve a variety of purposes, such as providing food, medicine, construction materials and industrial applications. And these are additional sources of income for farmers.

Table 19.4 Identified social impacts related with cocoa production in Cameroon

Production under conventional methods	Production under voluntary sustainability standards
Community projects are not initiated because of lack of premium[a]	New social or community development projects (e.g. infrastructure projects, schools, health centers, etc.) due to gains through premium[a]
Gender differences in land ownership, access to extension services, marketing and control proceeds[b]	Positive attitude changes and more transparency in resource management[b]
Little or no awareness about environmental issues[a]	An increased awareness of environmental issues[a]
Fluctuating price and a low income for farmers[a]	Better price and increased farmer incomes[a]
Social circumstances are not secure[a]	Secure social circumstances[a]
Lack of quality control thus leading to poor quality of the cocoa bean exported[a]	Better quality of cocoa beans exported[a]
Traceability is not easy[c,d]	Traceability is much easier[c,d]
No guarantee to consumer[c,d]	Guarantee to consumer[c,d]

Sources: [a]KPMG (2012) [b]Lyon et al. (2010), [c]Liu et al. (2004) and [d]Bolwig et al. (2007)

19.5 Conclusion and Recommendations

The Cameroonian government has an objective to increase the production of cocoa to about 320,000 tonnes per year by 2015, and achieving this goal would necessitate the creation of over 100,000 ha of additional cocoa plantations if yields are not improved (UNEP 2009). Therefore, allowing the creation an additional 100,000 ha of cocoa plantation means deforestation and the lost of biodiversity would continue to be encouraged. Nonetheless, the cocoa agro-forestry practices in Cameroon have shown an important potential to conserve agro-biodiversity and biodiversity. It plays an important role regulating the environment media such as the soil and the air. However, the use of synthetic fertiliser, insecticides, fungicides and child labour render the system unfriendly to the environment and unsustainable. Theoretically, voluntary standard systems sound promising but the practical experiences with respect to their environmental performance are not known in Cameroon. Therefore, it is recommended that the environmental and sustainability impacts for these standards are measured and quantified for easy comparison with conventional methods of cocoa production.

Additionally, many of the requirements in voluntary standard systems concerning social aspects in communities are basically similar to those governing cooperative societies in Cameroon, as stipulated in Law No. 92/006 of 14 August 1992 on Cooperative Societies and Common Initiative Groups, and Executive Directive No. 92/455/PM on the application of the same law (Nyambo 2008; Ebong 2008). Therefore, similarities in local policies, international policies and economic factors should be given equal attention with technical or management feasibilities to enable practical ways for producers to build up their development capacity so that they can qualify for certification and a consumer market for producers interested in developing international trade.

References

Bakala J, Kone S (1998) Lutte chimique la pourriture brune des Cabosses du Cacao: le forum R., un nouveau fungicide a 28 jours de fréquence de traitements, une grande première au Cameroun. Communication présentee lors du Séminaire international sur les maladies et les insects nuisibles du Cacaoyer a Yamoussoukrou, Côte d'Ivoire, 19–24 Janvier, 1998

Bisseleua DHB (2007) Ecological, social and economic determinants in cocoa production systems in southern Cameroon, Dissertation to obtain the Ph.D. degree in the International PhD Program for Agricultural Sciences in Göttingen (IPAG) at the Faculty of Agricultural Sciences, Georg-August-University Göttingen, Germany. http://ediss.uni-goettingen.de/bitstream/handle/11858/00-1735-0000-000D-F12C-1/bisseleua_daghela.pdf?sequence=1. Last accessed 31 Aug 2013

Bobo SK, Waltert M, Sainge MN, Njokagbor J, Fermon H, Mühlenberg M (2006) From forest to farmland: species richness patterns of trees and understorey plants along a gradient of forest conversion in Southwestern Cameroon. Biodivers Conserv 15(13):4097–4117

Bolwig S, Gibbon P, Odeke M (2007) Certified organic exports production – implications for economic welfare and gender equity amongst smallholder farmers in tropical Africa. Trade & Development Research Unit, Danish Institute for International Studies, Copenhagen

Carey C, Guttenstein E (2008) Governmental Use of Voluntary Standards: Innovation in Sustainability Governance, ISEAL Alliance, London, United Kingdom. http://www.isealalliance.org/sites/default/files/R079_GUVS_Innovation_in_Sustainability_Governance_0.pdf. Last accessed 31 Aug 2013

de Battisti B, MacGregor AJ, Graffham A (eds) (2009) Standard bearers: horticultural exports and private standards in Africa. International Institute for Environment and Development, London

Drum commodities (2012) An Introduction to the Cameroon cocoa Industry. Somerset, England. http://www.drumcommodities.com/assets/39/Cameroon_Cocoa_Industry_Introduction_May_2012.pdf. Last accessed 31 Aug 2013

Duguma B, Gockowski J, Bakala J (2001) Smallholder Cacao (*Theobroma cacao* Linn.) cultivation in agroforestry systems of West and Central Africa: challenges and opportunities. Agrofor Syst 51(3):177–188

Ebong MN (2008) The International Fairtrade Movement and Prospects for Cameroonian Producer Organisations. Prepared for Cameroon government and Agence Française de Développement (DFD) through its Cameroon PARI project (Projet d'Appui au Renforcement Institutionnel des Organisations Paysannes). http://chede.org/chede/wp-content/uploads/2010/06/03-10-08-Final-FT-report-English.pdf. Last accessed 31 Aug 2013

FAO – Food and Agricultural Organisation (2001) World Markets for Organic Fruit and Vegetables – Opportunities for Developing Countries in the Production and Export of Organic Horticultural Products. International Trade Centre, Technical Centre For Agricultural and Rural Cooperation, Food and Agricultural Organisation of the United Nations, Rome

ICCO – International Cocoa Organization (2006) International Cocoa Organization annual report 2005/2006. London, United Kingdom, 43 pp. http://www.icco.org/home/icco-documentation.html. Last accessed 31 Aug 2013

ICRAF – International Centre for Research in Agroforestry (1996) Annual report 1996. International Centre for Research in Agroforestry, Nairobi

IRAD – Institut de la Recherche Agricole pour le Développement (1997) Progress report on slash-and burn agricultural research in Cameroon., Institut de la Recherche Agricole pour le Développement, P.O. Box 2067, Yaounde, Cameroon, 124 pp; cited in Kotto-Same JPL,

Woomer Moukam A, Zapfack L (1997) Carbon dynamics in slash- and burn agriculture and land use alternatives in the humid forest zone of Cameroon. Agriculture, Ecosystems and Environment 65(3):245–256

Jaffee D (2008) Better, but not great: the social and environmental benefits and limitations of Fairtrade for indigenous coffee producers in Oaxaca, Mexico'. In: Ruben R (ed) The impact of Fairtrade. Academic Publishers, Wageningen, pp 195–222, Chapter 9

Kanmegne J (2004) Slash and burn agriculture in the humid forest zone of southern Cameroon: soil quality dynamics, improved fallow management and farmers' perception. PhD Dissertation. Wageningen: Wageningen University. 184 p. http://library.wur.nl/WebQuery/wurpubs/336542. Last accessed 31 Aug 2013

KPMG – KPMG Advisory N.V. (2012) Cocoa Certification Study on the costs, advantages and disadvantages of cocoa certification commissioned by The International Cocoa Organization (ICCO). Amsterdam

Krain E, Milijard E, Konan E, Servat E (2011) Trade and pro-poor growth: introducing rainforest alliance certification to cocoa production in Ivory Coast. Deutsche Gesellschaft für Internationale Zusammenarbeit

Leplaideur A (1985) Les systemes agricoles en zone forestiere: les paysans du Centre et du Sud Cameroun. CIRAD-IRAT, Paris

Liu P, Andersen M, Pazderka C (2004) Voluntary standards and certification for environmentally and socially responsible agricultural production and trade. Food and Agriculture Organization, Chapter 1

Lyon S, Bezaury JA, Mutersbaugh T (2010) Gender equity in fairtrade–organic coffee producer organizations: cases from Mesoamerica. Geoforum 41(1):93–103

NCCB – National Cocoa and Coffee Board (2008) ONCC Infos, Bulletin d'Informations de l'ONCC, N 000, August 2008. Douala, Cameroon

Nyambo TJ (2008) The legal framework of civil society and social movement in Cameroon. Chapter 2 in. Emmanuel Yenshu Vubo (ed) Civil society and the search for development alternatives in Cameroon, Dakar

Potts J, Giovannucci D (2012) COSA: global findings v. 1. Committee on Sustainability Assessment/SECO (unpublished report)

Righarts AMT (2009) Partnerships for sustainable agriculture and poverty reduction in Africa. A case study into the opportunities and challenges in creating sustainable partner-ships within the cocoa sector in Cameroon to upgrade the livelihoods of small-scale farmers, Master Thesis, Open University, the Netherlands. http://dspace.ou.nl/bitstream/1820/3014/1/mwAMTrighartsmei09.pdf. Last accessed 31 Aug 2013

Sonwa DJ (2004) Biomass management and diversification within cocoa agroforest in the humid forest zone of Southern Cameroon. PhD thesis, Faculty of Agriculture, University of Bonn, Germany, Cuvillier Verlag, Goettingen

Sonwa DJ, Nkongmeneck BA, Weise SF, Tchatat M, Adesina AA, Janssens MJJ (2007) Diversity of plants in cocoa agroforests in the humid forest area of Southern Cameroon. Biodivers Conserv 16(8):2385–2400

UNEP – United Nation Environmental Programme (2009) Integrated Assessment of Trade-Related Policies and Biological Diversity in the Agricultural Sector in Cameroon: Impacts of liberalization in the cocoa sector. http://www.unep.ch/etb/initiatives/Executive%20Summaries%20and%20briefs/Cameroon%20Ten%20Pager.pdf. Last accessed 31 Aug 2013

Wanji T (2009) Report of Feasibility Study on Opportunities in Organic Cocoa Production in Cameroon, March 2009 (unpublished report)

Zapfack L, Engwald S, Sonke B, Achoundong G, Birang M (2002) The impact of land conversion on plant biodiversity in the forest zone of Cameroon. Biodivers Conserv 11(11):2047–2061

Chapter 20
Transnational Initiatives to Promote Sustainable Cocoa Production and Trade: The Case of the German Initiative on Sustainable Cocoa (GISCO)

Elena Rueda, Ulrich Helberg, Vera Morisse, and Eberhard Krain

20.1 Introduction

As a result of liberalisation, the global market place has evolved towards a complex system of border-crossing structures. Production, manufacturing and marketing processes within the same value chain are subject to different regulatory frameworks that vary in stringency depending on the location where these processes take place, often to the detriment of social welfare and the environment.

Furthermore, this complexity leads to information asymmetries which particularly affect both ends of the chain: at the product's origin, in many cases, producers and manufacturers lack skills and information on export markets and depend on middlemen to access them instead. At the other end of the chain, it is difficult to retrace social and environmental accountability in order to satisfy the end consumers' growing transparency demands. Beyond managing reputational risks, securing access to the necessary qualities and quantities of raw material is a growing concern for industry players in general. The cocoa sector in particular is already facing complex challenges at the farming level that put positive long-term market development at risk, especially concerning mainstream segments.

Voluntary standards can play a crucial role in bridging the regulatory and informational gaps in transnational value chains; contributing to transparency, better governance and increased manageability of processes to achieve sustainability goals. Businesses can improve their market positions by supporting improvements in the producing countries that lead to economic, environmental and social sustainability and at the same time, managing risks more effectively. Meanwhile, cocoa producers benefit from higher incomes and improved living conditions, arising from amongst other things, a long-term improved process quality.

E. Rueda (✉) • U. Helberg • V. Morisse • E. Krain
Programme for Social and Environmental Standards, Deutsche Gesellschaft für Internationale Zusammenarbeit (GIZ) GmbH, P.O. Box 5180, 65726 Eschborn, Germany
e-mail: elena.rueda@giz.de; ulrich.helberg@helberg-consult.com; vera.morisse@giz.de; eberhard.krain@giz.de

C. Schmitz-Hoffmann et al. (eds.), *Voluntary Standard Systems*, Natural Resource Management in Transition 1, DOI 10.1007/978-3-642-35716-9_20,

The purpose of this chapter is to present hands-on experiences, lessons learned and current developments regarding multi-stakeholder based value chain governance structures that can contribute to broadening the application of sustainable production, trading and consumption practices in a consistent way.

Section 20.2 describes the challenges and sustainability constraints that both cocoa producing and consuming countries are currently facing. Section 20.3 highlights the role of voluntary standards as a means of fostering sustainable practices in a manageable way. Section 20.4 provides an insight into the recently established German Initiative on Sustainable Cocoa and embeds voluntary standards in the broader context of an intended multi-stakeholder value chain governance approach that brings in small, medium-sized and large players from private, public and civil society groups.

20.2 Background of the Cocoa Sector

Cocoa is one of the major agricultural commodities worldwide and provides a revenue base for 40–50 million people. Cocoa is mainly produced in West and Central Africa with approximately 73 % of global production coming from African countries, followed by Asia and Oceania (14 %) and Latin America (13 %) (WCF 2012).

The economies of some West and Central African countries are highly dependent on cocoa production. According to statistics provided by the International Trade Centre (ITC) cocoa exports accounted for more than 40 % in Ghana and 37 % in Côte d'Ivoire of the total export revenue in 2011 (ITC 2012).

Cocoa is grown mainly by smallholders on 1–3 ha of land and is often the only source of income for family farms (WCF 2012). The vast majority of cocoa farmers live in remote areas. Poor infrastructure and a low level of farm organisation make it difficult and costly to access technical advice, agricultural inputs and financial services. Limited knowledge of good agricultural practices results in low yields and often, poor quality cocoa beans. Unresolved property rights and share-cropping systems discourage producers from investing in the renewal of cocoa plantations and sustainable farm management practices. The lack of incentives for investment results in decreasing soil fertility and unsustainable farm practices. As a result, the quality of cocoa beans frequently produced fails to meet the specifications set by global markets. Over-aged cocoa plantations and decreased soil fertility led to an expansion of plantations into other food crop areas and forests, contributing to a high degree of rainforest disappearance in West Africa and a higher dependency on the commodity. The negative impacts of climate change on cocoa production are also contributing to the vulnerability of the cocoa sector. The growing demand of the world markets in terms of quality and quantity can hardly be met under these circumstances.

Table 20.1 Production of cocoa beans, in thousand tonnes

	2010/ 2011	Percentage of total global production	Estimates 2011/ 2012	Percentage of total global production	Forecast 2012/ 2013	Percentage of total global production
Africa	3,226	74.8	2,905	71,3	2,796	69.8
Côte d'Ivoire	1,511		1,486		1,470	
Ghana	1,025		879		820	
Nigeria	240		230		210	
Cameroon	229		207		210	
Others	221		104		86	
America	562	13.0	639	15.7	664	16.1
Brazil	200		220		230	
Ecuador	161		190		190	
Others	201		229		224	
Asia and Oceania	527	12.2	531	13.0	563	14.1
Indonesia	440		450		475	
Papua New Guinea	48		45		45	
Others	39		36		43	
World total	4,314	100	4,075	100	4,003	100

Source: ICCO (2012)

20.2.1 Main Producing Countries

World cocoa production rose from nearly 3.2 million tonnes in the 2002/2003 cocoa season to an estimated 4 million tonnes during the 2011/2012 season (ICCO 2013). This represents an average annual growth rate of 3.3 %. An all-time record output of over 4.3 million tonnes was achieved during the 2010/2011 season (ICCO 2012). The production of cocoa beans by producing regions and countries is summarised in Table 20.1.

20.2.2 The Livelihoods of Smallholder Cocoa Producers

Small-scale farmers account for between 90 and 95 % of global cocoa production—cocoa often being the only source of cash income for the family farms. The share of proceeds received by producers varies widely from country to country, ranging between 48 % to over 70 % of the shipping price of raw cocoa (ICCO 2010). The reasons for the major national and regional differences include the degree of liberalisation or regulation of the cocoa sector, the lack of market transparency and infrastructure.

The majority of cocoa farmers live below the poverty line equating to two US dollars a day per person. Aggravated by limited market access, lack of efficiency and transparency in supply chains and the underlying low productivity and professionalism, this poverty often results in the use of child labour. An estimated 1.7 million children work on cocoa farms in Ghana and Côte d'Ivoire, the two major cocoa producing countries, of which, 50 % of workers are exposed to hazardous work (Tulane University 2011).

Poor living conditions deter the younger generation from taking over their parents' farms (World Bank 2013). Furthermore, as many studies suggest, cocoa competes with other, more lucrative agricultural crops such as rubber and oil palms, which often replace it as the major cash crop in some areas (Kuklinski and Oghenerobo 2013; Varlet and Kouame 2013).

20.2.3 Challenges and Opportunities for the Global Cocoa and Chocolate Industry

In 2011, trading volume of cocoa futures on the Intercontinental Exchange (ICE) was 4.95 million metric tonnes, outpacing production by 750,000 metric tonnes (WCF 2012). In general terms, with an average yield of 300–500 kg of cocoa beans per hectare of land cultivated, only about 20–50 % of the potential yield is being achieved (World Bank 2013). Production is forecast to grow by 5.6 % by 2012/2013, but demand is already outstripping supply: according to statistics produced by the International Cocoa Organisation (ICCO), there is an annual shortfall of around 55,000 tonnes of high-quality cocoa (GIZ 2011). Although ICCO expects an almost parallel increase in supply and demand, production will nevertheless remain at about 5–7 % below demand.

In recent years, public attention has focused increasingly on issues of child labour and ecologically questionable production processes. Meanwhile, stagnating supply threatens the availability of the raw material, increasing the risk to the reputation and secure production base of the international cocoa sector. Given these challenges, more and more cocoa processors and chocolate manufacturers have made commitments to purchase sustainably produced and certified cocoa. Consumers, processors, manufacturers and governments in both producing and consuming countries have realised that rendering the cocoa sector more sustainable will safeguard the future supply chain of affordable, high quality cocoa beans, improve the livelihoods of small holder cocoa producers and help alleviate consumers' concerns about child labour and environmental degradation.

Despite these challenges, there are substantial opportunities: yield potentials are significantly under-exploited and consumers are increasingly willing to pay the premium for cocoa produced in an environmentally and socially sustainable manner.

20.3 The Significance of Sustainability Standards for Cocoa Supply Chains

In recent decades, trade has been an important driver for globalisation, with all its benefits and challenges. International and national regulatory systems should offer a framework for making globalisation happen in a socially and ecologically acceptable way. However, they are often limited in their scope of influence or do not adequately address social and ecological factors.

The context in which globalisation occurs must be reshaped for trade to promote growth and reduce poverty in the long term. To this end, companies, civil society groups, policy makers and service providers have gathered together to contribute to the development of a framework for sustainable trade and production. This joint effort has been taking place over the past two decades and has resulted in the creation of voluntary sustainability standards whose aim is to enable production and trade to become more transparent, economically viable, socially sound and environmentally friendly.

Experience in recent years has shown that the development and implementation of sustainability standards, as well as the promotion of corresponding initiatives, help prevent economic growth from happening at the expense of the environment and social justice. Furthermore, they contribute towards the saving of resources for the future (see e.g. Krain et al. 2011).

Sustainable production of cocoa has become an increasingly important driver for the development of the cocoa sector. The threats on supply shortages together with the growing social and environmental consciousness of final consumers has brought a growing market of sustainably produced and certified cocoa in consumer countries. New initiatives have emerged, including commitments by cocoa importers and chocolate manufacturers to purchase sustainably certified cocoa and to provide services to producers (e.g. Mars Inc. and Hershey to procure 100 % of their cocoa supply from sustainably certified sources by 2020). This trend provides an opportunity for cocoa producers to improve their livelihoods and for the cocoa sector to make its sustainability efforts more visible. To realise this opportunity, producers need support to meet requirements in order to access these emerging markets. Increasing awareness of the environmental consequences of food production (deforestation, biodiversity loss, soil erosion, contamination with pesticides) results in a growing demand for environmentally-friendly cocoa products. Similarly, the cocoa sector's labour practices are receiving public attention. Particularly since the international media began reporting on child labour in West Africa, resulting in increased consumer demand for cocoa products originating from socially sustainable producers who respect international social standards.

Recognising that only sustainable cocoa production can guarantee a thriving cocoa and chocolate industry worldwide, multinational companies have engaged in initiating projects to improve sustainability. In some, but not all cases, this is made visible by certification labels on the final products.

Some of the most significant tools towards a more sustainable cocoa sector are the implementation of sustainability standards (such as Fairtrade, Sustainable Agriculture Network/Rainforest Alliance and UTZ Certified) and capacity building measures along the cocoa value chain. Furthermore, training farmers to become business-minded entrepreneurs (e.g. through the Farmer Business School) helps them improve farm management, manage family income and motivates them to make necessary investments (GIZ 2011). Meanwhile, almost all trading companies, processors and chocolate manufacturers are engaged in large scale certification programmes to source sustainably produced, certified cocoa beans.

20.3.1 The Main Sustainability Standards in the Cocoa Sector

The main voluntary sustainability standards systems (VSS) in the cocoa sector are Fairtrade International, Rainforest Alliance and UTZ Certified. All three promote good agricultural, environmental and social practices. The environmental aspects in their standards protect producers' health and safety and ban the use of dangerous chemicals. The standards systems share the goal of transforming the world's production systems and value chains to make them more sustainable and believe that certification can help with that transformation. They are all members of the ISEAL Alliance, the global association for sustainability standards which works with companies, non-profit organisations and governments to support their use of voluntary standards. ISEAL's role is to represent the common interests of the organisations that set the standards and to ensure the standard's quality in terms of impact, transparency, credibility, governance etc. by defining good practices. In 2004, ISEAL launched the 'Code for Good Practice for Setting Social and Environmental Standards'.

Although the standards setting organisations share the same goal, a variety of complimentary approaches gives producers and buyers alternatives and the opportunity to select the system or combination of systems that best suits their interests and needs.

In order to be part of the Fairtrade system[1], traders and producers have to meet certain criteria which are defined in the Fairtrade Standards. Fairtrade's independent certification company, FLO-CERT, manages the process of auditing and certification to ensure compliance with their standards. The 'Fairtrade Minimum Price' acts as a safety net should market prices fall below a sustainable level. 'The Fairtrade Premium' is paid in addition to the sales price to farmer organisations but not to the individual farmer. By the end of 2011, 71 smallholder cocoa organisations representing 142,000 certified farmers received more than €7.6 million in 'Fairtrade Premiums' (approximately US$9.9 million). Fairtrade certifiable cocoa

[1] www.fairtrade.net.

volume rose to more than 124,000 metric tonnes (MT) worldwide at the end of 2011. However, only 35 % of the total certifiable volume was sold as certified cocoa (Fairtrade Labelling Organisations International e.V. 2011). By the end of 2012, the volume of certifiable cocoa had grown to approximately 150,000 MT, of which about 50,000 MT entered the market as certified produce (Meyer 2013).

Rainforest Alliance certification (RA) is based upon the 'Sustainable Agriculture Network (SAN) Standard'.[2] The SAN is a coalition of conservation groups that links farmers with consumers by means of the 'Rainforest Alliance Certified Seal of Approval'. The SAN standard does not incorporate the concept of guaranteed and fixed premiums. However, RA facilitates market-based premiums agreed upon in a transparent manner between producers and traders. Once a premium is offered for RA certified beans, RA typically encourages the group to invest in infrastructure to comply with the continuous improvement concept of the standard, run the 'Internal Management System' (IMS) and pay additional prices to the producer to motivate them to continue implementing the best practices on their farm. In 2010, 60 producer groups with 22,750 farmers produced 55,000 tonnes of certified cocoa. The volume of certified cocoa for 2011 is reported to be 98,400 MT (KPMG 2012).

UTZ Certified is the youngest of the three standards systems and labels for sustainable farming worldwide.[3] Its mission is to create a world where sustainable farming is the norm. The UTZ Certified Code of Conduct for Cocoa is the principle document used for the certification of cocoa producers. The UTZ Code of Conduct does not incorporate guaranteed and fixed premiums. However, in practice, premiums for certified cocoa are usually paid to producers.

UTZ increased the volume of certified cocoa from 70,000 MT in 2010/2011 to over 534,000 MT in 2012/2013, of which 118,000 MT was marketed as being certified. The number of certified smallholders increased during the same period from 40,000 to 256,000. About €13.2 million (US$17.2 million) were paid as premium in 2012 (UTZ Certified 2012).

When discussing the benefits of implementing sustainability standards systems, it is important to understand that we refer to improvements in the process quality that are not necessarily visible to consumers, but nevertheless, may lead to better product quality. Sustainability standards provide farmers with valuable tools to steer, monitor and manage continuous improvement processes on their farms (i.e. Internal Management System). The growing market recognition and consumer willingness to acknowledge sustainably produced cocoa is further triggering the transformation of the highly fragmented and complex cocoa supply chains into more transparent ones. Curious as to the social and environmental conditions under which the production process has taken place, consumers demand that the chocolate industry be able to retrace the sources of raw cocoa contained in the final products. Satisfying this demand requires the availability of information all the way back to the point of origin. This information can be made accessible through the traceability

[2] RA: www.rainforest-alliance.org, SAN: www.sanstandards.org.

[3] www.utzcertified.org.

systems provided by some certification systems. Furthermore, single companies are increasingly interested in establishing more stable and direct business links with the producer organisations and smallholders responsible for primary production. These developments provide an opportunity for cocoa producers to benefit from market recognition, access better business conditions and thus, improve their livelihoods.

While it is crucial, especially for the poor and vulnerable farmers, to make full use of certification potentials many producers do not have access to certification systems as information about requirements and markets, as well as capacity building for certification is hardly accessible. There are also enormous constraints concerning accessibility to certification for the majority of small scale farmers who are not organised, who have no land titles and no access to finance.

There is a clear opportunity for different standards initiatives to collaborate to increase the efficiency of standards implementation, thus, decreasing burdens and costs for producers, especially in cases where producers may wish to pursue multiple certifications. All this suggests a clear need for stronger collaboration between the standards systems in the cocoa sector.

In February 2011, Fairtrade, the Sustainable Agriculture Network (SAN)/ Rainforest Alliance and UTZ Certified issued a joint statement, in which they promised to increase collaboration, "…to reduce the level of complexity and costs for farmers. We are developing tools and materials to enable adherence to multiple standards. We share a commitment to independent third party certification and to increasing efficiencies in auditing. We are committed to seeking further cooperation in the field to benefit farmers" (Fairtrade et al. 2011).

All three standards systems are partners in the 'Certification Capacity Enhancement project' (CCE).

20.3.2 Certification Capacity Enhancement: Fostering Cooperation for a Common Capacity Building Approach

The CCE project is a multi-stakeholder initiative within the West African cocoa sector that has been developed jointly by the voluntary standards systems Rainforest Alliance, UTZ Certified and Fairtrade International, the development organisations Solidaridad, the Sustainable Trade Initiative (IDH) and Deutsche Gesellschaft für Internationale Zusammenarbeit (GIZ) GmbH, in collaboration with industry partners Armajaro, Archer Daniel Midlands, Continaf, Kraft Foods (now Mondelez) and Mars Inc. CCE aims to contribute towards a sustainable cocoa sector by enhancing farmer capacity to meet the agronomic, environmental and social requirements of the three voluntary standards systems and to improve productivity. First training activities (training of trainers) started in early 2011 and training of farmers is meanwhile carried out in Ghana, Côte d'Ivoire and Nigeria by the project partners. By the end of 2012, about 30,000 producers had been trained by the cocoa industry partners with the CCE materials to prepare them for certification (GIZ 2013).

A common certification training curriculum has been developed that enhances the efforts of standard initiatives, governmental extension services, and both public and private training programmes to facilitate access to certification by providing training materials for certification trainers and cocoa producers. Whether the producer groups aim for multiple (several standards) or single (one standard) certification, the same curriculum serves all prerequisites as it covers the requirements of all three sustainability standards. In addition to preparing cocoa farmers for certification, the curriculum also aims to enhance productivity, quality and economic viability of cocoa production to support sustainable development of the West African cocoa sector. Due to the training alignment, producers have enhanced opportunities to access certification and to increase their productivity. The common training approach not only reduces the effort and time taken to prepare cocoa farmers for certification but also reduces the transaction costs as compared to separate trainings for each certification. The CCE project has built certification capacity by training government trainers and private extension services, NGOs and industry partners in the three countries using the CCE curriculum. Farmers are now being instructed on a large scale by these trainers.

In March 2013 the CCE Initiative was officially integrated into the German Initiative on Sustainable Cocoa as one core element of GISCO's activities.

20.4 The German Initiative on Sustainable Cocoa (GISCO)

The German Initiative on Sustainable Cocoa (GISCO) is an initiative of the Federal Ministry of Food and Agriculture (BMEL), the Federal Ministry for Economic Cooperation and Development (BMZ), the Association of the German Confectionery Industry (BDSI) and its member companies, the Federal Association of the German Retail Grocery Trade (BVLH) and its member companies and civil society (see Fig. 20.1).[4] GISCO aims at promoting sustainable cocoa production, addressing economic and ecological aspects and contributing to improvement of the livelihoods of people involved in cocoa production. It brings together the German public and private-sector actors interested in the production, trading and processing of cocoa, civil society groups, producer organisations and producer countries in order to connect and further develop the manifold activities already under way. The following chapter highlights the importance of the German market in the global context of the cocoa sector and shows its crucial significance in

[4] According to BVLH, a large proportion of the confectionery remains in the German market and is distributed via wholesalers and retailers, with around 28 % of the final products carrying generic brands. It is therefore highly relevant that this stakeholder group (wholesale and retail) also form part of the Initiative. Christian Mieles, Federal Association of the German Retail Grocery Trade (BVLH).

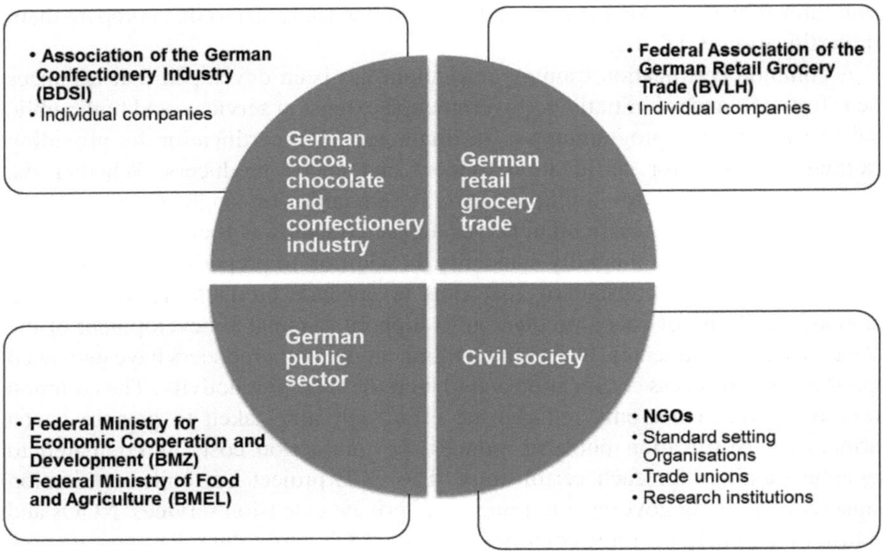

Fig. 20.1 Membership profile of the German Initiative on Sustainable Cocoa (illustration based on GISCO's 'Rules and Regulations')

fostering sustainable cocoa production at the global level following a broad sector-wide approach.

20.4.1 GISCO's Background, Initiators and Partners

Europe imports about 58 % of the approximately 4.3 million tonnes of cocoa produced worldwide. More than 10 % of the global cocoa harvest is processed in Germany. The confectionery industry is the third largest sector of the German food industry and generates around 10 % of its turnover (Association of the German Confectionery Industry 2012). This demonstrates the importance of cocoa for the German and global economies.

Chocolate and confectionery manufacturers (including brand names and generic brands) have a vital interest in securing sufficient quantities of high-quality raw cocoa or semi-finished cocoa products. Poor yields resulting from inappropriate production methods and an absence of young farmers create alarm in the international chocolate industry as this poses a threat of raw material shortages in the long term. Consumers all over the world are becoming increasingly restrictive, not only with regards to quality but also in terms of social and environmental acceptability of products. Civil society and media accusations in relation to the use of child labour, to various degrees, on West African cocoa farms and the deforestation in favour of new cocoa plantations, tarnishes the reputation of all those involved in the cocoa value chain.

Single companies put programmes in place to support growers in applying sustainable methods, in making use of voluntary standards to optimise processes and in bettering the traceability of their cocoa supply chains. Nevertheless, there are certain hurdles that affect the whole sector that cannot be tackled by individual players alone.

Standard systems also have limitations, not only in terms of their capacity to scale-up but also in terms of properly addressed long-term prevention and remediation measures for certain issues such as child labour.

Against the backdrop of the sectoral challenges explained in Sect. 20.2 and given the importance of the cocoa sector for both local and global economies, the promotion of sustainable cocoa farming has become clear. There is a need for a wide-scale initiative that brings together the cocoa industry and trade, civil society groups, standard setting organisations and government representatives in an effort to engage in dialogue with representatives of producing countries.

Conscious of this fact, a core group of German actors from industry, retail, civil society and governmental bodies called for a global consultative meeting on Sustainable Cocoa Production with proposals for the establishment of a Forum on Sustainable Cocoa (now GISCO). The meeting was held on 16 March 2012 in Berlin, bringing together all actors involved in the value chain. The event was convened by the BMZ and the BMEL. Around 50 representatives of various stakeholder groups from industry and retail, civil society organisations, producing countries, standard setting organisations and the public sector were in attendance.

As a result of their deliberations the following aspects were concluded as crucial for a consistent way forward:

- It is important that ongoing sustainability initiatives in the cocoa sector (according to ICCO, numbering well over 60) are well networked and coordinated. To this end, the forum should collaborate closely with international and European stakeholders and also with cocoa producing countries, providing leadership at international level.
- Efforts towards greater sustainability are most urgent. However, the forums' activities should be aligned with the sovereign activities and action plans of each producing country.
- Promoting sustainable production is not enough. There is a need to define targets and assess initiatives in relation to these. For this purpose, widening and optimising the base for data collection and impact analysis, setting criteria for evaluating projects and disseminating best practice knowledge are areas that deserve closer attention.
- Concerns on the availability of sufficient quantities of good quality raw cocoa in the long term need to be addressed. Much still remains to be done in terms of production potential, i.e. to increase the quality and quantity of cocoa produced per unit of land. Most small-scale farmers who are currently producing only

about 500 kg/ha could easily increase the yield at least threefold by using good planting material and applying sustainable farming practices.
- There is a need to establish a discussion platform for all market players. The public sector should also be engaged as a crucial player for entering into systematic dialogue with producing countries on greater sustainability.

Consequently, supported by the feedback, expectations and positive response of the broadly represented actors involved at the global level in the March meeting, the 'German Initiative on Sustainable Cocoa' was launched on 13 June 2012 in Berlin. By June 2013, GISCO counted 76 members.

20.4.2 Concept Approach, Expected Results and Activities

The aim of this sector-wide and broad-based initiative is to contribute to increased systematic support for cocoa farmers who wish to increase sustainability, in close cooperation with cocoa producing countries as they implement the sector's sustainability strategies and development plans.

The first goal to be achieved is to reach a consensus on methods, means and paths to sustainability in cocoa growing. This will involve all actors and stakeholders agreeing on an effective, best-practice-based model for gradually encouraging cocoa producers to adopt sustainable production practices.

As this can only succeed as a joint effort, constructive exchange with producing countries on challenges, experiences and opportunities for development have taken place since the beginning of the initiative. Consequently, extensive dialogue processes in the producing countries started with Côte d'Ivoire and, to a lesser extent, with Ghana and form a major element of GISCO's activities. These activities are intended to involve cocoa producing countries in developing and designing the model and integrating this approach into their strategies and action plans.

Large-scale implementation should bring about a significant increase in the proportion of cocoa in confectionery products that is certified as being sustainable according to current standards. Rather than certification as such, enabling and encouraging cocoa farmers to adopt sustainable production methods will be the main goal. Measures implemented on the ground will therefore not only deal with the gradual adaptation of cocoa production to sustainable practices but also with the establishment of broad-based structures and capacities that can support the spread of knowledge and enable cocoa growers to implement these practices. The model will include specific provisions for monitoring as well as instruments for its implementation across the entire value chain.

In order to avoid a duplication of efforts, GISCO links existing projects without establishing parallel structures and processes. Its measures are to be embedded in existing national and international initiatives, particularly in producing countries. Moreover, GISCO can develop its own ideas for projects for implementation, for example, as part of the bilateral programme of cooperation undertaken by the BMZ and the BMEL or in collaboration with other stakeholders such as companies, foundations, international organisations and other donors. Such projects will support GISCO's work, complement existing initiatives and implement the approaches devised in a practical manner, along with good practices in sustainable cocoa production. To this extent, networking and cooperating with existing initiatives to promote sustainable cocoa production is crucial.

Experiences and good practice from previous projects will be collected and reviewed, and will benefit initiatives by producers and businesses that seek to make their cocoa supply chain more sustainable. In order to also enhance uptake on the demand side, GISCO shall provide businesses, especially small- and medium-sized enterprises, with guidance on how to integrate sustainability issues into cocoa procurement.

The setting of the German Initiative on Sustainable Cocoa allows for the integration of different approaches in an overall sector strategy. With the integration of the CCE Initiative as a core element (see Sect. 20.3.2), GISCO acknowledges CCE as a best practice example to be widespread at a sectoral level.

Given the different profiles of actors involved, intrinsic 'push-and-pull' effects can be attained to leverage each other, driving innovation (push) and attracting additional actors/participants to engage (pull). Striving for horizontal cooperation ensures the channelling of coordinated efforts into high risk areas affecting all involved actors, leveraging the efficiency and effectiveness of single efforts. The horizontal approach generates a 'pull-effect' when important players join and the sectoral need for a common approach becomes obvious. This approach, therefore, requires the existence of a 'critical mass' which must be powerful enough to attract those rather reluctant 'followers' to the system. This context is ideal for integrating and anchoring proven and acknowledged best practices like CCE in the overall value and supply chain systems of the sector, following a *common approach*.

Within a horizontal pre-competitive cooperation approach, negotiations on common goals take place at every single stage of the value chain on macro and meso levels among government authorities and associations of producers, traders, processors and retailers. Single measures are then put in place within a vertical integrative approach all along the value chain in line with consolidated goals in a common, sector-wide approach while integrating all stakeholders, from local producers to end markets.

Furthermore, striving for a vertical integration of sustainability issues in *individual* supply chains can bring a 'push-effect' into the system. While parting from competitive motivation, individual interests may contribute to the whole when addressing less explored sustainability areas that, whilst still being crucial, do not necessarily belong to the common set of priorities, thus, playing a pioneering role when testing new approaches that trigger change. The leverage effect is then achieved at the horizontal level of cooperation, once the results of the test phase prove that there are risks, opportunities and needs arising from the addressed topic that affect or benefit the whole sector. The development and testing of the common CCE curriculum can serve as an example to explain the above described effect. Implementing the project with pioneering companies has built the basis for developing and testing solutions that can now be applied in a broader sectoral context, once the consciousness of the needs and potentials of such an approach has grown among a larger number of crucial players. Current discussions may result in an uptake of necessary updates and further development of CCE by the German Initiative on Sustainable Cocoa.

Crucial to the whole system is the provision of plausible tailor-made solutions with a clear return on investment for players to get involved in the long-run. While the *common approach* requires a high level of negotiation when identifying common goals, as well as risk and opportunity areas where all involved actors will be willing to invest; the *individual approach* requires an in-depth analysis at corporate level to optimise the strategic position with a clear competitive advantage for engaging in 'new land'.

Using a close cooperation with current sustainability standard setting organisations to enable a broad implementation of the newly developed CCE Curriculum (see Sect. 20.3.2), the on-going process of developing a 'European Standard for Sustainable Cocoa and its Traceability' by the European Committee for Standardisation (CEN) is also to be taken into account into GISCO's work, ensuring that it simplifies matters for farmers from the start of the value chain. In doing so, GISCO intends to address the farmer needs and limitations to engage in sustainability efforts that have been mentioned in Sect. 20.3.1.

20.4.3 Governance Structure

GISCO is under the aegis of the German Federal Government. Any individual or organisation can become a member if they have a profound, measurable and proven interest in sustainability in the cocoa sector. They must contribute to GISCO financially or by providing other inputs (e.g. specialist knowledge and expertise), in addition to taking part in the activities on a continuous basis and be willing to become involved in a stakeholder group. GISCO lays down clear rules and builds on a clear declaration from its members on their willingness to provide measurable support for sustainability in the cocoa sector.

The German Initiative on Sustainable Cocoa comes together at least once a year for an 'Annual General Meeting' where they discuss and decide on activities and budgets, set the rules of procedure and decides on representatives for the 'Steering Committee'.

The Steering Committee sets out the framework for GISCO's work and makes strategic decisions if GISCO itself does not take these into account. The Steering Committee is comprised of eight members. Two members from each of the following four stakeholder groups are represented: the cocoa, chocolate and confectionery industry, the food industry, civil society organisations and the German Federal Government. Decisions are made by consensus.

If the need arises, GISCO can appoint a consultative group to advise on particular technical issues (e.g. project proposals, design of instruments for promoting

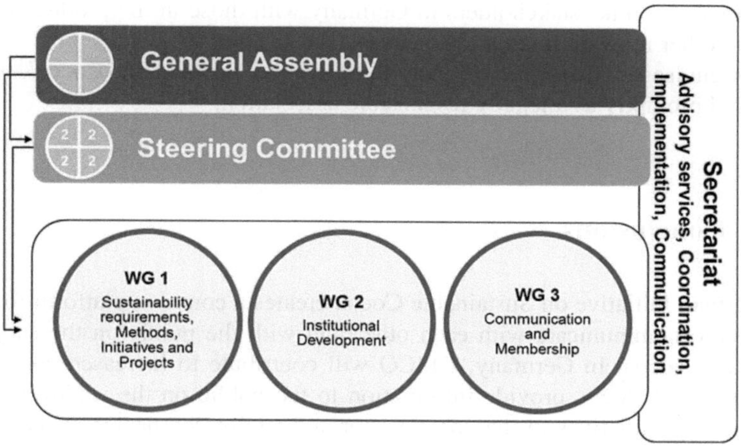

Fig. 20.2 Governance Structure of the German Initiative on Sustainable Cocoa (illustration based on GISCO's 'Rules and Regulations')

sustainability, training materials) as well as the preparatory strategic decisions of the Steering Committee. In particular, representatives of cocoa producing countries should be involved in this group.

The Steering Committee also convenes 'Working Groups' in connection with specific areas of activity, especially to obtain scientific advice and support for projects, and to involve specific actors. In the first phase of the project, three working groups are being established to tackle different issues, including the institutional development of the Initiative, the communication and work on sustainability requirements, methods, and projects. To this end, GISCO is counting on members with many years of hands-on experience in promoting sustainable practices to make a key contribution, both strategically and content wise. Input from international initiatives and representatives of cocoa farmers and producer countries on how to shape GISCO's activities is seen as fundamental.

The working groups, see Fig. 20.2, support the initiative by establishing a consensus on how to achieve sustainable cocoa production by agreeing on a model to gradually encourage cocoa producers to adopt sustainable production practices. Partnerships with governments of producer countries alongside cooperation and networking with existing German, international and local sustainability initiatives in cocoa-producing countries are crucial to GISCO's work in order to promote existing best-practice approaches and to prevent overlaps and duplication of effort.

The GIZ GmbH acts as the Secretariat for the period from June 2012 to May 2014 initially. The Secretariat implements the resolutions adopted by the Steering Committee and the General Assembly, reports to these bodies and supports them in their tasks in accordance with the present 'Rules and Regulations'. GISCO's

Secretariat also links stakeholders in Germany with those in the producing countries and other relevant international actors.

It is envisaged that the 'German Initiative on Sustainable Cocoa' will be registered promptly as a legally responsible association.

20.5 Conclusions

The German Initiative on Sustainable Cocoa creates a common platform for those involved to communicate with each other and with the public on the subject of sustainable cocoa. In Germany, GISCO will contribute to increased cooperation between stakeholders, provide information to the public on the achievements of sustainability initiatives in the cocoa sector and advise companies in the confectionery industry and food trade on how to address sustainability issues in their own value and supply chains.

The Initiative strives for close cooperation in order to exchange information and knowledge, develop coherent approaches, link up activities in a purposeful manner and capture the value of best practice experience in all realms, taking particular regional features into account.

A project with a public-private partnership funding of about €400,000 p.a. (approximately US$522,332) supports the start-up phase of the Initiative for the first 2 years. A permanent, privately run organisational structure with set membership fees is currently being developed and will ensure long-term continuation of GISCO's activities.

References

BDSI – Association of the German Confectionery Industry (2012) About the Association. http://www.bdsi.de/en/wir_ueber_uns_en/der_verband_en.html. Last accessed 20 Sept 2013

Fairtrade Labelling Organisations International e.V. (2012) Monitoring the scope and benefits of Fairtrade, 4th edn

Fairtrade, SAN/Rainforest Alliance and UTZ Certified (2011) Joint Statement, published at ISEAL website: http://www.isealalliance.org/online-community/news/historic-joint-statement-fairtrade-sanrainforest-alliance-utz-certified. Last accessed 20 Sept 2013

GIZ – Deutsche Gesellschaft für Internationale Zusammenarbeit GmbH (2011) Cocoa with a future, GIZ expertise and experience

GIZ – Deutsche Gesellschaft für Internationale Zusammenarbeit GmbH (2013) CCE Project Fact Sheet 2013

ICCO – International Cocoa Organisation (2010) The world cocoa economy: past and Present. September 2010

ICCO – International Cocoa Organisation (2012) Quarterly Bulletin of Cocoa Statistics XXXIX (1) [Cocoa year 2012/2013, published: 28-02-2013]

ICCO – International Cocoa Organisation (2013) Quarterly Bulletin of Cocoa Statistics XXXIX (2) [Cocoa year 2012/2013, published: 29-05-2013]

ITC – International Trade Centre (2012) Trade map – trade statistics for international business development. http://www.trademap.org. Consulted 07 Oct 2012

KPMG Advisory N.V (2012) Study on the costs, advantages and disadvantages of cocoa certification: report prepared for the International Cocoa Organization. http://www.icco.org/about-us/international-cocoa-agreements/cat_view/30-related-documents/37-fair-trade-organic-cocoa.html. Last accessed 20 Sept 2013

Krain E, Millard E, Konan E, Servat E (2011) Trade and pro-poor growth: introducing Rainforest Alliance Certification to Cocoa Production in Côte d'Ivoire, OECD and World Trade Organization, Deutsche Gesellschaft für Internationale Zusammenarbeit (GIZ) GmbH, Eschborn

Kuklinski F, Oghenerobo A (2013) CCE Baseline study report on Cocoa producers' livelihood perceptions in five pilot areas in Ghana, Côte d'Ivoire and Nigeria (Internal project document)

Meyer R (2013) Report by the Chief Executive Officer, FLO-CERT GmbH at the ICCO International Workshop on Cocoa Certification Douala, Cameroon

Tulane University (2011) Oversight of Public and Private Initiatives to Eliminate the Worst Forms of Child Labour in the Cocoa Sector in Côte d'Ivoire and Ghana. March 31, 2011

UTZ Certified (2012) 10 years in Coffee, Cocoa and Tea. From Good to Better: UTZ Certified Annual Report 2012. https://www.utzcertified.org/en/newsroom/utz-in-the-news/26582784-out-now-utz-certified-annual-report-2012. Last accessed 20 Sept 2013

Varlet F, Kouame G (2013) Etude de la Production de Cacao en Zone Riveraine du Parc National de Tai, Rapport Final, Programme de Développement Economique en Milieu Rural et Biodiversité (PRODEMIR), Deutsche Gesellschaft für Internationale Zusammenarbeit (GIZ) GmbH, Abidjan

World Bank (2013) Growing Africa: unlocking the potential of agribusiness. http://siteresources.worldbank.org/INTAFRICA/Resources/africa-agribusiness-report-2013.pdf. Last accessed 20 Sept 2013

WCF – World Cocoa Foundation (2012) Cocoa Market Update; March 2012. http://worldcocoafoundation.org/wp-content/uploads/Cocoa-Market-Update-as-of-3.20.2012.pdf. Last accessed 20 Sept 2013

Chapter 21
VSS and Climate Change in the Coffee Sector: The 4C Climate Module

Kerstin Linne

21.1 Introduction

Climate change is putting at risk the livelihoods of many farmers. Coffee production depends on stable climatic and environmental conditions. Smallholder coffee farmers always had to manage changes in yields and therefore income and they have developed their own coping strategies as a result. Now they are facing a new challenge: climate change and its impacts on their production systems. Rising temperatures are impacting negatively on coffee quality and are triggering newer and more numerous incidences of disease and pest infestation. Changes in rainfall patterns are disrupting flowering cycles and erratic rains are impeding maturation of coffee berries leading to a reduction in coffee quality and quantity.

This was the rationale behind the Development Partnership on "Climate Change Adaptation and Mitigation in the Kenyan Coffee Sector" (Sangana PPP) between Sangana Commodities Ltd., the Kenyan subsidiary of the ECOM Group, the Deutsche Gesellschaft für Internationale Zusammenarbeit (GIZ) GmbH on behalf of the German Federal Ministry for Economic Cooperation and Development, the 4C Association, Tchibo GmbH and the World Bank. The project was implemented from 10/2008 to 09/2011 together with the Kenyan coffee smallholder cooperative Baragwi Farmers' Cooperative Society.

This chapter presents the project approach as well as its results. Section 21.2 presents the project approach and partners, including their interests and roles in the Sangana PPP. Emphasis is given to the 4C Association and its Code of Conduct as a vehicle to facilitate climate change adaptation in coffee production. Section 21.3 looks into the project results including identified impacts of climate change in Kenya, training development for climate change adaptation and mitigation for coffee smallholder farmers as well as the 4C Climate Add-on Module. The chapter

K. Linne (✉)
Green Line Consulting, Bremen, Germany
e-mail: info@greenlinne.com

C. Schmitz-Hoffmann et al. (eds.), *Voluntary Standard Systems*, Natural Resource Management in Transition 1, DOI 10.1007/978-3-642-35716-9_21,

ends with some achievements of the project in Sect. 21.4, outlining additional information on further work related to this topic.

21.2 Project Approach and Partners

Coffee is the world's second largest traded commodity and around 125 million people worldwide depend on the income from coffee for their livelihoods (Fairtrade Foundation 2012). At the same time the coffee plant is very sensitive to minor changes in climatic conditions. Impacts of climate change are felt by all coffee actors along the supply chain. To tackle this challenge, cooperation and joint efforts are necessary to identify and implement effective response strategies. Thus several partners joined together in the Sangana PPP to consolidate their resources in meeting these challenges.

21.2.1 Project Approach

The aim of the Sangana PPP was to support coffee producers to adapt their production to the changing climate and to create and use synergies between adaptation and mitigation. To tackle this goal, an additional component to the existing 4C Code of Conduct was developed enabling coffee producers to respond to climate change.

The 4C Code of Conduct counts three dimensions: social, environmental and economic. The project developed an additional voluntary dimension covering climate relevant aspects (see Fig. 21.1). *Voluntary*, in this context, means that any coffee producing group opting for verification under the 4C Code of Conduct will have to gradually comply with the existing three dimensions whereas they can opt to comply with the climate dimension. Therefore, the *Climate Code* does not affect existing *4C Units*, i.e. producers or producer groups verified under the 4C Code of Conduct, and producers interested in 4C verification do not have to comply with the Climate Code in order to achieve verification. The Climate Code includes agricultural practices for adaptation and mitigation, training for producers and verifiers, verification instruments and a climate data base. It was tested together with the Baragwi Farmers' Cooperative Society Ltd. as a pilot group.

21.2.2 Project Partners

Sangana Commodities Ltd. is the Kenyan subsidiary of the Swiss ECOM Agroindustrial Corporation Ltd. and serves as a major exporter of Kenyan coffee. Sangana ensured the active participation of the producers, sharing specific

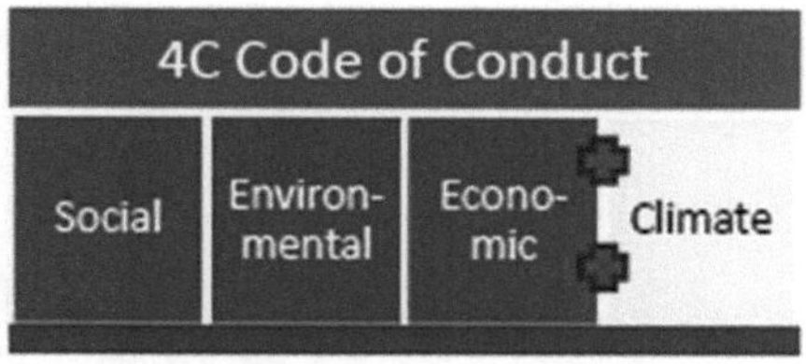

Fig. 21.1 The 4C Code of Conduct plus the climate dimension (*Source*: Linne et al. 2011)

knowledge on Kenyan coffee production and trained producers in sustainable production techniques. Sangana's interest in the project was to learn more about the impacts of climate change on coffee production and to help find potential solutions.

The Deutsche Gesellschaft für Internationale Zusammenarbeit (GIZ) GmbH is a federally owned enterprise supporting the German government in achieving its objectives in the field of international cooperation for sustainable development. It assists people and societies in developing, transition and industrialised countries in shaping their own futures and improving living conditions. In this project GIZ coordinated overall implementation, shared expertise on climate change adaptation and supported the development of the 4C Climate Module. Their interest was to develop internationally acknowledged and accessible tools for climate change adaptation.

The 4C Association is an independent membership organisation that provides standards for sustainable economic, social and environmental practices in coffee production. Within the 4C Association, producers, trade, industry and civil society from around the world work together for increased sustainability in the global coffee sector. In the Sangana PPP, the 4C Association actively engaged in the development of the climate add-on module, verified the pilot group under the 4C standards and trained 4C acknowledged auditors in the implementation of the add-on module on climate change. Their interest was to broaden their existing approaches so as to cater for climate change challenges.

Tchibo GmbH is an international company and operates in many more business sectors than the traditional selling of coffee. Tchibo supported the pilot group in achieving certification under the Sustainable Agriculture Standard and ensured the dissemination of findings and lessons learnt throughout the coffee sector. Their interest was to explore adequate ways and means to secure coffee production against the negative impacts of climate change.

The World Bank is a vital source of financial and technical assistance to developing countries around the world. The World Bank's role within the project was to supply expertise on climate change mitigation in the coffee sector and to identify agricultural practices which reduce or remove greenhouse gas emissions. Their interest was to see if and how smallholder producer structures may be connected to the carbon markets.

The pilot group, Baragwi Farmers' Cooperative Society Ltd. is located in the Kirinyaga District in central Kenya and has 13,000 members delivering coffee cherry to 12 cooperative owned wet mills. The cooperative actively participated in

Fig. 21.2 Vision and mission statements of Baragwi Farmers' Cooperative Society Ltd (personal photo Kerstin Linne, 2010)

project activities and included the climate change work in their long-term planning. Their interest was to reduce negative impacts on their production brought about by climate change (Fig. 21.2).

21.2.3 The 4C Code of Conduct

The 4C Code of Conduct is a baseline standard for mainstream sustainability throughout the coffee sector. It is based on 28 social, environmental and economic principles for coffee production, processing and trading. The Code works within a traffic light system, which means that for each principle there are red, yellow and green criteria and indicators stated. Green practices present the ultimate status to be attained; yellow practices indicate that there is room for improvement and red practices need to be abandoned. For verification purposes, an average status of yellow needs to be achieved where red practices can be balanced by green practices. Additionally the 4C Code of Conduct includes ten 'unacceptable practices' that must be eliminated before any coffee actor can become a member of the 4C Association and before coffee can be sold as 4C Compliant. An illustrated guide to the 4C Code of Conduct is available at the 4C website.[1]

[1] http://www.4c-coffeeassociation.org/uploads/media/4C_Code-of-Conduct_IllustratedGuide_en.pdf.

The 4C voluntary sustainability standard is based on a business-to-business model. This means that 4C Compliant Coffee is passed along the coffee supply chain together with a copy of the producers' 4C Certificate to show that the 4C Code of Conduct has been complied with. There is no label that communicates this compliance to consumers. However any member of the 4C Association may opt to include a 4C membership statement on a pack of their coffee.

21.3 The Project: Sangana PPP

Within the agricultural sector there is considerably more need for climate change adaptation than for mitigation. Though 31 % of global emissions are coming from agricultural activities and deforestation, coffee farmers, especially smallholders, are highly vulnerable to changing climatic conditions (IPCC 2007). Therefore the Climate Module developed within the framework of the Sangana PPP aims to support coffee producers to adapt to these changes in climate.

However, due to the World Bank's focus on climate change mitigation, the module also aims to explore mitigation effects achieved by the implementation of adaptation measures. An easy example of this double effect is the shade tree. From an adaption perspective a shade tree helps to protect the microclimate by regulating temperatures in the area, by enhancing water infiltration in the soil and by generating organic matter for e.g. composting. From a mitigation perspective, a shade tree generates biomass and therefore sequesters greenhouse gas emissions. There are other examples where adaptation measures can generate mitigation effects and the Climate Module deals directly with this link.

21.3.1 The 4C Add-On Climate Module

The 4C Add-On Climate Module is based upon four pillars (see Fig. 21.3):

1. The Climate Code
2. Training for producers and auditors
3. Instruments for auditing compliance under the Climate Code
4. A climate database

The core of the module is the Climate Code stating principles and criteria broken down into indicators and structured using a traffic light system just as the other dimensions of the 4C Code of Conduct (see Fig. 21.4).

The Climate Code has four categories:

1. Enabling Environment
2. Natural Resource Management
3. Soil and Crop Management
4. GHG Emissions and Stocks

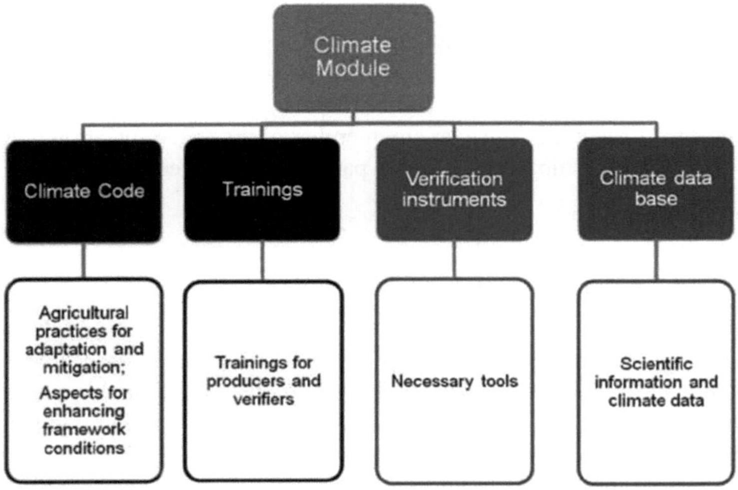

Fig. 21.3 The structure of the Add-On Climate Module (*Source*: Linne et al. 2011)

Category	Principle	Criteria / Indicator			Comments
		Green	Yellow	Red	
Enabling Environ-ment	Capacity building on climate change adaptation and mitigation is accessible	Action plan to address climate vulnerabilities and risks is being implemented	Action plan to address climate vulnerabilities and risks has been elaborated, implementation has not taken place	Action plan to address climate risks and vulnerabilities has not yet been developed	See manual for participatory workshop on climate risks + vulnerabilities; New
	How to verify	Participatory WS on climate risks + vulnerabilities carried out (participants list) Action Plan developed (Action Plan) Action Plan being implemented (Field)	Participatory WS on climate risks + vulnerabilities carried out (participants list) Action Plan developed (Action Plan)	Participatory WS on climate risks + vulnerabilities not carried out	

Fig. 21.4 An example from the Climate Code (*Source*: Linne et al. 2011)

The first category "Enabling Environment" looks into options on how to strengthen the producer organisation and enhance their framework conditions e.g. through capacity building, enhanced access to information or setting up early warning systems. The second category, "Natural Resource Management", covers topics such as biodiversity, the extraction of timber and water management. Issues regarding soil conservation, different coffee varieties or agrochemicals are dealt with in the "Soil and Crop Management" category. Category four focuses on data collection rather than the implementation of pure mitigation measures. It requires the monitoring of biomass on the farm as well as identifying emission hot spots and potential reduction measures.

21.3.2 Verification Under the Add-On Climate Module

In order to start working on climate change issues, a producer organisation first has to identify the need to act, i.e. they have to be aware of climate change impacts on their production. To support them in this task, the Climate Module offers introductory training, looking into climate change adaptation as well as mitigation.

Once a producer organisation decides to become verified under the 4C Climate Code, a two day participatory analysis is carried out in order to identify which present challenges at the organisation are climate related. The outcome of this analysis is a short action plan in which the producer organisation prioritises activities to be implemented to address climate change challenges. The activities stated in this action plan are then further enriched according to the Climate Code. This ensures implementation of the Climate Code due to ownership of the producer organisation.

As climate change is a continuous process, this needs to be mirrored in the analysis of climate change impacts to reflect the dynamism of the problem. Therefore the Climate Code recommends a participatory analysis, every time there is an upcoming audit for the 4C Code of Conduct. Such an audit takes place every 3 years, while internal inspection is done on a yearly basis. By re-checking how climate change impacts the production every 3 years, new climate challenges can be determined and progress on adaptation activities can be monitored.

21.3.3 Baragwi Farmers' Cooperative Society Ltd. and Climate Change

Optimum climatic conditions for *Coffea arabica*, as produced by Baragwi, include temperatures between 15 and 24 °C and between 1,500 and 2,000 mm of rainfall per year. The Sangana PPP together with the International Centre for Tropical Agriculture (CIAT) developed future scenarios to project future suitability for coffee production in Kenya (Eitzinger 2010). According to these models, by 2050, Kenya will have less seasonality in its climate indicated by predicted values for maximum mean temperature increasing to 31.2 °C (currently 28.6 °C) and minimum mean temperature increasing to 12 °C (currently 9.8 °C). Rainfall is also predicted to increase from current measurements of 1,405–1,575 mm by 2050. However, expected distribution of this rainfall is not necessarily favourable for coffee. These changes will lead to a shift of optimal coffee producing zones from currently 1,600 masl (metres above sea level) to higher altitudes at 1,700 masl. General suitability of the existing coffee regions will decrease. Current climatic suitability for coffee production in Kenya is between 50 and 70 %, by 2050 suitability is predicted to be between 30 and 60 %. For Baragwi the biggest impacts were predicted at the wet mills of Githiururi, Rwambiti, Kianyaga, Kianjiru and Gichugu, which are the wet mills located in the lower altitudes. This is mainly caused by excessively high temperatures, especially in the lower regions, leading to less climatic suitability for coffee production.

After the introductory training session on climate change adaptation and mitigation where producer representatives learn about these two different concepts, Baragwi indentified deforestation, pests and diseases, poor farming practices and erosion as leading to strong climate vulnerabilities in their catchment area and changing weather patterns as their biggest climate risk. The 4C Climate Code, through its principles, criteria and indicators offered guidance to the cooperative on how to tackle these challenges. The audit of the Baragwi Cooperative Society Ltd. under the 4C Climate Code was carried out by the certification body *AfriCert Ltd.* in July 2011. Baragwi already held the 4C Certificate, i.e. was an official 4C Unit, and was then the very first cooperative to have achieved verification under the 4C Climate Add-On.

21.3.4 Emissions in Coffee Production at Baragwi

Together with the Sustainable Food Lab as one of the developers of the Cool Farm Tool,[2] the Sangana PPP looked into estimating the greenhouse gas emissions of one kilogram of coffee cherry at Baragwi. The results showed that there are significant differences between individual producers based on the applied production techniques (see Fig. 21.5). On average, one kilogram of coffee cherry at Baragwi emits −0.5 kg of carbon dioxide equivalents, while most farmers fall between 0.90 and −2.72 kg CO_2e emissions per kg coffee cherry. Fertiliser production and induced emissions from fertiliser application, together with crop residue management, are the primary emission sources. However, carbon sequestration from above ground biomass and management practices such as incorporation of residues, compost and manure largely offset these emissions.

Based on these results the efficient use of chemical fertilisers, optimised use of organic fertilisers, composting as well as shade management seem to be the most promising strategies to further reduce emissions in Baragwi's coffee production.

---

[2] www.coolfarmtool.org.

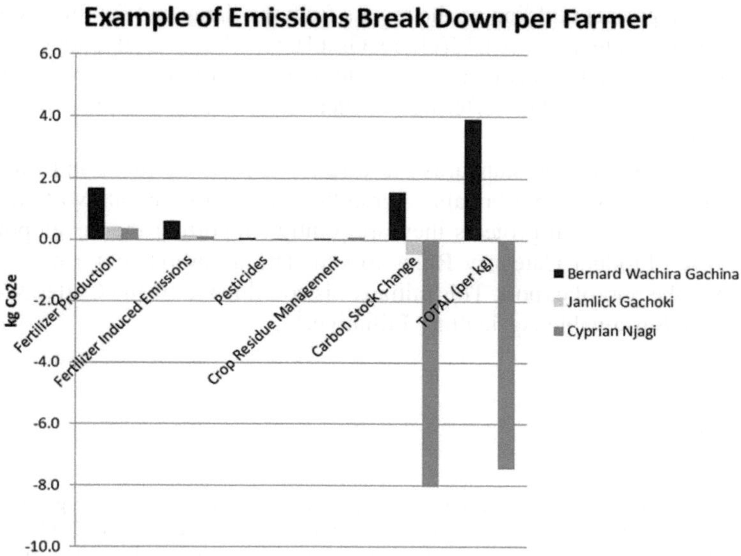

Fig. 21.5 An example of emissions by Baragwi Farmers (*Source*: GIZ and Sangana Commodities 2011)

21.4 Achievements

When the project started, the farmers already knew that changes in their local precipitation patterns and temperature were occurring. The Sangana PPP simply supported the producers in analysing and structuring their knowledge and gave guidance for finding effective responses via the Climate Code. Due to the sensitisation and training carried out with the Baragwi Farmers' Cooperative Society Ltd. and the implementation of the 4C Climate Code, farmers have changed some of their practices and some changes have occurred at the cooperative's organisational level.

Some tangible achievements of the project include the set up of an indigenous tree nursery to increase the shade cover within the region, the proper disposal of solid waste, e.g. coffee pulp, now being used for making compost and most of the farmers having adopted farming skills that conserve water and soil such as bench terracing, strip grassing, mulching, reduced tillage and integrated pest management.

Generally the project has led to an increase in the awareness of climate change and environmental issues. Water levels of local river streams are beginning to rise due to improved conservation of riparian areas and the yields (coffee quantity and quality) have started to increase, as reflected in the rising incomes received by farmers.

Since the Sangana PPP, considerable progress has been made regarding climate change in the coffee sector. GIZ, in addition to Tchibo GmbH and Ecom, are active in a follow-up project building on the results from the Sangana PPP. In this project, the Coffee&Climate initiative,[3] Tchibo GmbH and Ecom collaborate with the Neumann Group and the members of the International Coffee Partners. The project aims to build up a toolbox offering practical advice and guidance on specific adaptation measures.

Furthermore the coffee standards Fairtrade, 4C, UTZ Certified and Rainforest Alliance on behalf of the Sustainable Agriculture Standards, are all working on the topic. Besides their own projects they are jointly supporting the development of Green Coffee Product Category Rules to calculate a carbon footprint for green coffee up to the port of export. This initiative is coordinated by the Coffee Working Group of the Sustainable Agriculture Initiative.[4]

References

4C Association (2009) The 4C code of conduct. http://www.4c-coffeeassociation.org/our-services/4c-code-of-conduct/standards-setting-procedure.html. Last accessed 20 Sept 2013

Eitzinger A (2010) Climate change adaptation and mitigation in the Kenyan coffee sector. International Center for Tropical Agriculture (CIAT), Decision and Policy Analysis group. http://dapa.ciat.cgiar.org/climate-change-adaptation-and-mitigation-in-the-kenyan-coffee-sector/. Last accessed 20 Sept 2013

Fairtrade Foundation (2012) Fairtrade and Coffee, Commodity Briefing, Fairtrade Foundation UK. http://www.fairtrade.org.uk/includes/documents/cm_docs/2012/F/FT_Coffee_Report_May2012.pdf. Last accessed 20 Sept 2013

GIZ and Sangana Commodities (2011) Carbon and coffee, Kenya: cool farming options pilot with Sangana Commodities Ltd. and GIZ. CFT Summary Report. http://www.coolfarmtool.org/reports/GIZ_Sangana_CFO_Report_carbon_and_coffee.pdf. Last accessed 20 Sept 2013

IPCC – Intergovernmental Panel on Climate Change (2007) Fourth assessment report. http://www.ipcc.ch/publications_and_data/publications_and_data_reports.shtml#.UbBZ85Wj9DQ. Last accessed 20 Sept 2013

Linne K, Nzioka C, Mullan J, Daniels S (2011) Climate change adaptation and mitigation in the Kenyan coffee sector, guide book – Sangana PPP – 4C climate module. GIZ and Sangana Commodities Ltd. http://www.4c-coffeeassociation.org/uploads/media/SanganaPPP_GuideBook_4CClimateCode.pdf. Last accessed 20 Sept 2013

[3] www.coffeeandclimate.org.

[4] http://www.saiplatform.org/activities/alias/climate-change/coffee-pcr-project.

Chapter 22
The Role of VSS in Enhancing the Contribution of Fisheries and Aquaculture to Sustainable Development

Mark Prein and Uwe Scholz

22.1 Introduction

This chapter presents an overview of initiatives by GIZ[1] and its partners in respect to the introduction of voluntary sustainability standards (VSS) in the fisheries and aquaculture sectors in developing countries since 1999. In the capture fisheries sector these range from the implementation of the FAO Code of Conduct for Responsible Fisheries, to standards of the Marine Stewardship Council (MSC), Naturland Wildfish and the Marine Aquarium Council for ornamental fish. Countries concerned were Kenya, the Philippines, Tanzania, Senegal, Sri Lanka and Vietnam. In the aquaculture sector VSS comprised of the Naturland Organic Aquaculture standard for shrimp (*Penaeus monodon*) and pangasius (*Pangasianodon hypophthalmus*) and the GlobalG.A.P. and Aquaculture Stewardship Council (ASC) standards for pangasius. Projects supported were mainly in Ecuador, Thailand and Vietnam. In the Philippines, a local organic standard, based on Naturland criteria, was developed and applied for tilapia (*Oreochromis niloticus*) and milkfish (*Chanos chanos*).

[1] In 2012 GIZ was formed following a merger between three institutions tasked with implementing development policy for the German Federal Government on behalf of the Ministry of Economic Cooperation and Development: DED (Deutscher Entwicklungsdienst), GTZ (Deutsche Gesellschaft für Technische Zusammenarbeit) and InWEnt (Gesellschaft für Internationale Weiterbildung und Entwicklung).

M. Prein (✉)
Rural Development and Agriculture Division, Programme on Sustainable Fisheries and Aquaculture, Deutsche Gesellschaft für Internationale Zusammenarbeit (GIZ) GmbH, Eschborn, Germany
e-mail: mark.prein@giz.de

U. Scholz
Rural Development and Agriculture Division, Deutsche Gesellschaft für Internationale Zusammenarbeit (GIZ) GmbH, Eschborn, Germany
e-mail: uwe.scholz@giz.de

C. Schmitz-Hoffmann et al. (eds.), *Voluntary Standard Systems*, Natural Resource Management in Transition 1, DOI 10.1007/978-3-642-35716-9_22,

The main focus of this chapter is to discuss approaches taken versus their situational context, the achievements attained and lessons learned. Objectives are as follows:

- to outline the general situational background that necessitate initiatives such as VSS in the fisheries and aquaculture sectors of developing countries;
- to present the rationale for a government-owned institution that is implementing German development policy, such as GIZ, for engaging in the introduction of VSS in the fisheries and aquaculture sectors in developing countries;
- to reflect on different standards, the context in which the introductions were attempted and the approaches for their introduction; and
- to present and discuss the lessons learned from these initiatives.

The chapter starts by describing the role of VSS in fisheries and aquaculture in the context of developing countries, the rationale for engaging with this, as well as a description of the perspective of development policy as maintained by GIZ. Section 22.2 describes the six standards and approaches implemented in the area of capture fisheries: one on co-management of marine protected areas, another on the marine ornamental (aquarium) fish trade. Section 22.3 presents reflections on the four standards for aquaculture that were applied, of which two are conventional standards and two are organic standards.

The chapter ends with conclusions and recommendations in Sect. 22.4 highlighting the factors leading to success in those situational contexts and standards that were trialled.

22.1.1 Rationale for Introducing VSS in Fisheries and Aquaculture

Globalisation has brought considerable changes to global markets for fish and fish products. These are characterised by increased competition as well as the need for policy decisions and interventions by governments to ensure an increased contribution to sustainable development in the fisheries and aquaculture sectors (OECD 2008).

The total global seafood trade more than doubled from 69 million tonnes in 1976 to 154 million tonnes in annual landings in 2011 (FAO 2012). There are two main modes of production: capture fisheries (harvest) and aquaculture (fish farming). Capture fisheries production is the largest contributor and has fluctuated between 80 and 90 million tonnes since the mid-1980s.

Aquaculture production was rather unimportant in terms of production figures up to the 1970s with 3.5 million tonnes per annum which represented about 5 % of global supply. However, since then the sector experienced a virtual revolution, in 2011, aquaculture made up 41.3 % of total seafood supply with production of 63.6 million tonnes (excluding aquatic plants and non-food products) (FAO 2012). The sector has thus grown more rapidly than other animal food-producing sectors, namely at an average rate of 8.8 % per year, compared with only 1.2 % for capture fisheries and 2.8 % for terrestrial farmed meat production systems over the same period.

Asian countries dominate production with the top eight nations being China, India, Vietnam, Indonesia, Bangladesh, Thailand, Myanmar, and the Philippines, with the other two being Norway and Egypt. In Asia, more than 65 % of aquaculture production comes from freshwater environments (FAO 2012).

Due to increased aquaculture production, global seafood supply has continued to increase since 1990 and therefore has been sufficient to increase global per capita consumption of seafood from 14.4 kg in the 1990s to 18.6 kg in 2010 (FAO 2012; Asche 2011). Globally, fish constitutes the main animal source food for over one billion people of which most are in developing countries (Tacon and Metian 2013).

Capture Fisheries

Worldwide, 90 % of all artisanal fishers are living in developing countries. They catch about 50 % of global seafood landings which meet two-thirds of direct human consumption (FAO 2013). Also, 39 % of the artisanal catches do not remain in their country of origin but are traded globally. The trend is increasing with particular effects on the fishing communities in developing countries.

It is today seen that a major proportion of global fisheries are not managed according to modern scientific and sustainability principles. In respect to developing countries, reasons for this include lack of information and capacity to draft management and investment plans, let alone budget and capacity to monitor or enforce them. In countries in which fisheries are better managed, a wide variety of management systems exist. These range from community-based management systems at the local level all the way to directive systems assigning fishing rights to users, where in some cases these are tradable. Fishers however, while trying to make a living, behave individually within their particular management system. Their actions may be irrational from a broader perspective, causing considerable losses to the fishery, and consequently to the fishers' income as a whole (Asche 2011).

Aquaculture

Aquaculture production includes a variety of species like algae, seaweed, molluscs, crustaceans, low-value finfish (e.g. carp), medium-value finfish (e.g. tilapia), and high-value species (e.g. shrimps, sea bass, sea bream, groupers and salmon). High-value species tend to play a more significant role in the international trade of aquaculture products (Asche 2011).

The on-going intensification of aquaculture results in higher stocking densities and feeding rates, often causing nutrient overload in public water bodies and the environment when ponds are emptied for harvesting. Non-indigenous species sometimes escape from the farming facilities, infiltrate natural populations, interbreeding or competing with local native species, putting biodiversity at risk. Last but not least, the construction of large aquaculture facilities in coastal zones has in some cases resulted in the clearance of mangrove forests of high ecological value, depriving local communities of their traditional access to these areas (FAO 2011).

For developing countries, aquatic production is important from a trade, employment and nutritional point of view. In 2004, the number of fish farmers accounted for one-quarter of the total number of fish workers worldwide. The majority of these fish farmers are located in developing countries, principally in Asia. Significant increases in aquaculture production over recent decades reflect the strong expansion of aquaculture activities which includes a diversification of cultured species and systems. A common trait has been the gradual spatial expansion and intensification of production systems, creating local income opportunities with spill-over benefits. However, in the absence of local regulatory frameworks and enforcement capacity, the expansion of aquaculture is also often characterised by negative attributes such as land conversions, environmental and social impacts as well as increased use of agrochemicals which raise concerns about food safety.

Rationale for VSS

The increasing international trade and awareness of consumers about social and environmental consequences, or food safety/health risks connected to imported foods, has shown that there is a justified role for an introduction of VSS into the seafood sector (Roheim et al. 2012). Furthermore, such standards promote the efficiency of international trade (OECD 2010, 2012). In recent years, the number of standards for food production and processing has increased substantially. Due to the variety of actors and interests involved in standard formulation and implementation, there are large differences among standards in respect to their scope and objectives (Nadvi and Wältring 2002). The benefits of fisheries and aquaculture certification (such as ecolabelling) include: potential for premium market prices, access to new markets, safeguarding existing market channels, preferred supplier status and the potential to attract ethical investment of local community social and economic infrastructure (Scholz 2006; Blueyou and ENDA/REPAO 2007; MRAG 2009; WWF 2009; OECD 2010, 2011, 2012; FAO 2011; Prein et al. 2012).

Ecolabelling of seafood started in the capture fisheries sector as an initiative of the private sector in response to the widely perceived failure of fisheries management (Gardiner and Viswanathan 2004). The concept of sustainability gained increasing importance in the entire seafood sector and developed into a major driving force. Subsequently, the concept of ecolabelling was extended to aquaculture production. However, there are two different foci: ecolabelling of capture fisheries mainly target fish stock conservation, while ecolabelling of aquaculture focuses on the elimination of potential negative externalities of aquaculture production (environmental, economic, social, and animal welfare). Therefore, aquaculture certification schemes promote an alternative 'scope' of sustainability than certification schemes for capture fisheries, thereby potentially adding confusion in the market (Roheim et al. 2012; OECD 2010; MRAG 2009).

22.1.2 VSS in Fisheries and Aquaculture in Developing Countries

From a development perspective, consideration of environmental and social issues in aquaculture is a particular challenge for non-industrialised countries and should be linked to gaining market access into developed countries (WWF 2009; OECD 2010, 2012; Jonell et al. 2013). The Deutsche Gesellschaft für Internationale Zusammenarbeit/German International Cooperation (GIZ) has been assisting several developing countries to move into seafood certification programs in the fisheries and aquaculture sectors, e.g. through support of small and medium enterprises (SMEs) development programs. Promoting seafood certification programs can be seen as an effective tool to implement and support the articles and provisions that are made with the FAO Code of Conduct for Responsible Fisheries with regards to fisheries and aquaculture (FAO 1995). Under the framework of several projects and programmes GIZ supports the stakeholders' efforts with a broad array of services in fisheries and aquaculture, in the latter for organic and non-organic standards (Nolting and Schirm 2003a; Finkel 2004, 2005; Scholz 2006; Nolting and Prein 2008; Jarchau et al. 2009; Deichert and Linh 2013) as follows:

- advice to Governments and projects on the effective application of standards;
- information and training for personnel and cooperation partnerships with other German development organisations;
- documentation and assessment of good practices in defining and implementing standards;
- facilitating and setting up strategic alliances among government, business and industry, non-governmental organisations, trade unions and development organisations;
- assisting disadvantaged groups in developing countries seeking to take part in defining standards;
- promotion and build-up of local advisory and auditing capacities;
- assistance in gaining economic benefits from standards, e.g. through contact management among actors in the supply chain and through marketing advice; and
- advice to government agencies, business enterprises and standard initiatives on the participatory development and implementation of standards.

22.2 Capture Fisheries Standards

Aside from recommendations to consumers on which seafood species to avoid and which can be consumed due to their basis on sustainable exploitation or production (e.g. Greenpeace, WWF, Monterrey Bay Aquarium, various consumer fora, etc.) a range of seafood sustainability standards for capture fisheries have been developing since the mid-1990s (Gardiner and Viswanathan 2004; MRAG 2009; WWF 2009; OECD 2010, 2011, 2012). These are usually based on the FAO Code of Conduct for Sustainable Fisheries (FAO 1995).

22.2.1 Marine Stewardship Council

The Marine Stewardship Council[2] (MSC), was established in 1997 by the World Wide Fund for Nature (WWF) and Unilever as an independent non-profit organisation setting a standard for sustainable fishing. MSC certified products are fully supply-chain traceable and display the well-known blue MSC label to enable customers to make informed choices through their purchasing behaviour. Globally, to date almost 18,000 seafood products carry the MSC logo with Germany being in the lead in respect to volume of products labelled. However, during the first decade since its foundation, no finfish fishery in a developing country was certified (Bush et al. 2013).

The MSC has a Developing World Working Group (DWWG) in which GIZ staff has been members since its inception in 2007.

MSC Pre-Assessment of the Lake Victoria Nile Perch Fishery

Lake Victoria is the most productive freshwater fishery in Africa and is shared between Kenya, Tanzania and Uganda who have entrusted the management of the Lake's Fisheries to the Lake Victoria Fisheries Organisation[3] (LVFO). The lake provides an important source of income, employment, food and foreign exchange to East Africa. Lake Victoria fisheries support almost two million people with household incomes, and meet the annual fish consumption needs of almost 22 million people in the region (Geheb et al. 2008; Kolding et al. 2008, 2014). At its peak, the fishery yield was over 900,000 tonnes per year, valued at US$340 million generated at the shore. A further US$250 million a year was earned with exports from the Nile perch fishery.

In recent years, fish stocks in Lake Victoria are being increasingly exploited mainly through overfishing and illegal, unreported and unregulated (IUU) fishing as well as anthropogenic effects on the environment and insufficient governance by stakeholders of the lake's fishery (Marshall 2010; Gitonga 2012; Kolding et al. 2008, 2014; Witte et al. 2013; Nunan 2013). Nile Perch biomass has experienced a particular decline from 1.3 million tonnes to an alarming 300,000 tonnes.

In response to very critical media reports (e.g. the movie *Darwin's Nightmare* which largely exaggerates the situation at the lake and draws unproven connections between fish and arms trading in the region) and public debates about the status of

[2] www.msc.org.

[3] www.lvfo.org.

the fishery in Lake Victoria and its relevance towards livelihoods of the population around the Lake, GTZ decided in 2006 to pilot the certification of the Nile perch fishery on Lake Victoria to contribute toward improved governance, production and benefits of the dependent populations (Scholz 2006).

An MSC Pre-Assessment (Moody Marine Ltd. 2008) was commissioned by GTZ with co-funding from the German Federation of Fish Importers and Fish Processors[4] ('Bundesverband Fisch') in collaboration with the LVFO. The resulting roadmap towards improved governance was adopted by the LVFO Scientific Committee. Additionally, the Nile perch was an introduced species (i.e. not indigenous to the lake prior to 1960) and the statutes of the MSC did not permit the fishery to be certified. This rule was modified in 2011 as a consequence of which the Nile perch fishery could be certified today if its governance and management regime were compliant to MSCs standard.

A lake-wide planning workshop with the active participation of fisheries ministers of all three lake-side countries was held in November 2009 in Nairobi to agree on the implementation of the roadmap of steps and measures towards lake-wide certification and thereby restoration of the Nile perch stocks on Lake Victoria. Representatives of the three countries pledged funding for and implementation of the roadmap. However, this was not followed through.

22.2.2 Naturland Wildfish

Given the lack of initiatives for certification of small scale capture fisheries on finfish in developing countries (Gardiner and Viswanathan 2004; Scholz 2006), Naturland and GTZ in 2006 decided to formulate a standard on capture fisheries[5] using the Lake Victoria Nile perch (*Lates niloticus*) fishery as a case study. Naturland e.V., a proven cooperation partner in various projects, came into play due to the refusal of MSC to get involved with the non-indigenous species and probably due to the on-going negative PR, e.g. by Greenpeace and WWF.

In 2007 the ministers of the three riparian countries assigned the Lake Victoria Fisheries Organization (LVFO) to evaluate the possibility of introducing an ecolabel for the small-scale fisheries in the region. A Public Private Partnership (PPP) project between the Deutsche Gesellschaft für Technische Zusammenarbeit (GTZ) GmbH, and the European importer ANOVA Seafood and the Tanzanian processor Vicfish Ltd. was designed and implemented to prove the feasibility of ecolabelling the Nile perch small scale fisheries in Lake Victoria. Two years later, in 2009, the Nile perch fishery in Bukoba (Tanzania) was announced the first ecolabelled small-scale fishery in a developing country.

---

[4] www.fischverband.de.

[5] www.naturland.de.

From the beginning, the tool of the Standard Round Table (SRT) was employed to ensure stakeholder involvement. In the pilot project, all important actors, including representatives of the fishing community, government officials and the private sector participated in meetings to plan and formulate the Naturland standards (which did not exist prior to this pilot activity and therefore had to be defined as a shared and consensus-seeking exercise, Jarchau et al. 2009):

- The *ecological criteria* ensure that the fishery is operated as such that it minimises its impact on the ecosystem, e.g. the application of selective fishing gear and sanctuaries;
- The *social criteria* include acceptable working conditions, adequate lodging and housing, establishment of gender-separated sanitary facilities; access to education for the children, health services and affordable transport to services for the fisher community;
- The *economic criteria* include access to financial services, information about fish prices paid by factories, the implementation of a "Community Development Trust Fund" for community development projects and the payment of a price premium.

By 2009, all criteria were fulfilled and the Nile Perch Fishery in Bukoba (Tanzania) was announced the first ecolabelled small-scale fishery in a developing country involving more than 800 certified fishermen and improving the working and living conditions of 5,000 people. Additionally, the Naturland certification was regarded as complimentary to the MSC pre-assessment that was supported by GTZ as a long-term objective for lake-wide certification.

In 2009 and 2012, GIZ commissioned impact assessments which revealed that the certification had positive effects on the value chain actors, notably the fishing communities and the processors (Yang 2009; De Beule and Ayoku 2011; De Beule 2012). Furthermore, up-scaling has started at further sites in Tanzania (De Beule 2012).

22.2.3 Code of Conduct for Responsible Fisheries and Lake Victoria

The global and voluntary Code of Conduct for Responsible Fisheries (CCRF)[6] was adopted in 1995 by the FAO Conference and sets out principles and international standards of behaviour for responsible practices with a view to ensuring the effective conservation, management and development of living aquatic resources, with due respect for the ecosystem and biodiversity. This 'normative' code recognises the nutritional, economic, social, environmental and cultural importance of fisheries and the interests of all those concerned with the fishery sector. The code

[6] www.fao.org.

takes into account the biological characteristics of the resources and their environment and the interests of consumers and other users, including special requirements for developing countries (FAO 1995).

Through its program in support of responsible fisheries, GIZ activities supported the implementation of the CCRF, mainly through bilateral development cooperation projects between Germany and partner countries in which the fisheries sector plays an important role, e.g. in Cape Verde, the Philippines, Mauretania, Senegal and Sri Lanka. These mainly comprised of awareness-raising and training in the elements of the CCRF and formulations (where none existed) or revisions (where existing) of national fisheries legislations and fisheries management plans setting sustainable fisheries management at their core, as well as striving for coherence with other legislations and sectors.

Parallel to the pilot project on Naturland Wildfish certification and MSC Pre Assessment, GTZ strengthened LVFO's capacity in stock management by supporting in policy development and planning, especially for the successful implementation of the Nile Perch fishery management plan. This included setting up the conditions for a regional Fish Levy Trust Fund to ensure donor independent fish stock surveys, preparation of information sheets on species and fishery specific licensing aimed at increasing the understanding of local governments and Beach Management Units (BMUs), and most important enabling frameworks for fisheries management on Lake Victoria using a co-management and ecosystem-based approach. Moreover, first steps were prepared to harmonise national fishery specific licensing documentation for legal adoption in each Partner State. One considerable success was to include the private sector (fish factory owners) into the management structure of LVFO as it is in their own interest to govern the resource sustainably.

Around Lake Victoria, the formation of the Regional Beach Management Unit Network (RBN) which took place in March 2007, was the culmination of considerable efforts made at all levels over the prior 3 years. This network had a membership of at least 215,000 individuals who had registered as members of Beach Management Units (BMUs). Of this total, 23 % were women and 49 % were crew members. Both groups being the most marginalised in the fisheries sector are invariably targeted as key beneficiaries by international and national poverty reduction initiatives (Scullion 2009, 2010).

Technical and financial support provided by GIZ enabled the RBN to be formed at a time when public–private dialogue was (and continues to be) critically important for Nile perch stock recovery. The inauguration of the RBN now enables it to formally engage as a member of the regional technical advisory committees of the Lake Victoria Fisheries Organization (LVFO), the Scientific Committee and the Fisheries Management Committee. The East African Industrial Fishing and Fish Processing Association (EAIFFPA) representative of the fish processors and export industry also joined the RBN as a new member of the Scientific Committee (SC) and of the Fisheries Management Committee (FMC) of the LVFO. Up to 2011, GIZ provided technical mentoring support for a public–private dialogue process during the first FMC and SC meetings with integrated membership of the private sector (Scullion 2009, 2010). The dialogue was based on the Nile Perch

Fisheries Management Plan (NPFMP) as the framework guiding Nile perch stock recovery and management (Scullion 2010; Schwab 2010).

This was a landmark achievement for the LVFO in consolidating its co-management approach through institutional restructuring to create a strong partnership with the private sector at regional level. The LVFO is one of only two Regional Fisheries Management Organizations in the world to accommodate the private sector in a joint formal regional advisory capacity. GIZ support on behalf of the BMZ also guided the filling of remaining gaps in the institutional co-management framework by the formation of Fisheries Co-management Committees and the networking of Nile perch suppliers (Scullion 2009, 2010).

GIZ has focused the attention of LVFO on the need to streamline objectives and proposed actions defined in the different regional policy documents and plans of action to specifically focus on the NPFMP as the principal guiding framework for Nile perch research and management. The restructuring of the SC and FMC reports to align them more closely to the NPFMP and to the Lake Victoria Fisheries Management Plan 2009–2014 was an indication that the process had begun. A further sign of NPFMP implementation is the acceptance of a draft Nile Perch Research Plan for 2010. This was the first lake- and species-specific research plan to be developed in the region (Scullion 2010).

GIZ technical support also convinced the LVFO of the need to refine objectives and specific national actions included in the "Operation Save Nile Perch" to be supported by emergency government funding of US$1.8 million that was launched in 2009 by the ministers of all three countries on Lake Victoria. While short-term emergency funding of this kind should be welcomed, GIZ has repeatedly advocated for the LVFO to initiate sustainable financing mechanisms rather than continue to rely on donor support. In 2010, GIZ raised this issue again and the SC and FMC, including its members from the private sector, finally recognised the urgency of this issue and recommended immediate actions to be taken to set up the Fish Levy Trust Fund and to ensure sustainable financing of BMUs.

Based on feedback from numerous inquiries, it is recognised that the pilot projects at Lake Victoria serve as orientation for other initiatives of small scale fisheries in the developing world moving towards certification, e.g. pole-and-line tuna fishery in the Maldives.

GIZ sponsored a think tank event on 'Ecolabelling of African Fisheries' in Nairobi in November 2012. The purpose was to define the key policy pathways on ecolabelling in African fisheries and utilise ecolabelling as a mechanism to improve fisheries governance and trade with key inputs through studies commissioned by GIZ (De Beule 2012; Gitonga 2012). The outputs were also targeted as an input to the second Conference of African Ministers of Fisheries and Aquaculture (CAMFA) to be held in 2014.

22.2.4 Marine Aquarium Fish Trade

Together with Indonesia, the Philippines supply an estimated 85 % of the world's saltwater ornamental aquarium fish. Over the past decades, cyanide fishing, despite being illegal, has become a common method used to catch ornamental aquarium fish on the reefs. Around 4,000 aquarium fish collectors operated across the country in the early 2000s (Nolting and Schirm 2003a) providing a source of livelihood. In 1998 ornamental aquarium fish exports to developed countries from the Philippines were worth more than US$6.4 million per year and steadily growing.

The collection and trade of marine ornamental fish provides livelihood for thousands of fishermen. The persistent demand for decorative species in world markets has created a multi-million dollar business.

In the 1980s, cyanide fishing was the main fishing method in the Philippines. Sodium cyanide was used to stun the fish making them easier to catch. However, sodium cyanide is a powerful poison that affects the respiratory system of organisms and can cause death up to several weeks later. During the transportation and export of cyanide-stunned fish, the mortality rate can reach up to 50 % or higher. Next to this, the substance is also lethal to the coral polyps and thus causes wide spread destruction of reefs.

The Marine Aquarium Council (MAC) had established a label for the ornamental aquarium fish trade with essential quality criteria:

- Zero use of cyanide
- Environmentally friendly collecting methods
- Minimised mortality through good husbandry and handling
- Best water quality
- Appropriate and safe working conditions
- Coral reef stewardship and management

In cooperation with selected Philippine/German private enterprises GTZ initiated a Private Public Partnership (PPP) project where the MAC criteria and standards were reviewed and adopted with the aim of raising the quality and sustainability of the Philippine ornamental fish trade. The German company FLORA 2000, together with its Philippine partner company AquaEx, set up a MAC-certified pilot export facility in the Philippines which complied with the MAC standards for marine ornamental trade. The company MarineFauna Inc. was located on Cebu Island. The PPP project was implemented from 2000 to 2002.

The objectives of the Public–Private Partnership were to create incentives for all stakeholders, improving the working conditions for ornamental fish collectors, conduct training and education of the collectors, introduce quality criteria for a sustainable collection and trade. The results and impact of the PPP were:

- A training program on sustainable, environmentally sound collection techniques was established;
- 900 collectors were trained and qualified in applying the standards;

- Higher survival rate for aquarium fish through water quality management (Most non-certified Philippine export companies do not observe proper water quality management);
- MAC-certified companies comply with new standards for best water quality;
- Monitoring of physico-chemical water parameters reduces the mortality rate.

The situation in 2010 (8 years after the end of the PPP project) was that since August 2005 MarineFauna Inc. had discontinued certification according to MAC. This decision was made voluntarily and based on different reasons:

- MAC certified fish are limited to certain number of species and only 2–4 % of these species are traded by MarineFauna Inc.
- MarineFauna Inc. cannot comply with the minimum traceability requirement of MAC to identify any supplied MAC certified fish by an individual collector;
- MarineFauna Inc. criticise that the MAC label does not guarantee that the fish was caught sustainably.

In the PPP, GIZ also supported trials for coral reef rehabilitation, eco-tourism, and the drafting of guidelines (Heeger and Sotto 2000).

22.2.5 Best Practices: Marine Protected Areas

GIZ (including DED) has been engaged in the Philippines Integrated Coastal Zone Management (ICZM) and fisheries management since the early 1980s, and in particular in the establishment and governance of Marine Protected Areas (MPAs) since the 1990s (Scholz 2008).

The Philippines are characterised by overpopulated coastal zones resulting in massive pressure on coastal resources, which has led to a decrease in fish stocks with a severe negative impact on local fishing communities. In general, it can be considered that the fish stocks in the 1990s were only 10 % of the level of the 1970s (Aliño 2002; Dalzell 1988).

In Negros Oriental, the establishment of more than 30 community-managed marine protected areas (MPAs) constituted one part of a provincial Integrated Coastal Zone Management (ICZM) program designed to address this situation. These comprise coral reef and mangrove habitats where fishing and mangrove logging activities are prohibited (Wiedemeyer et al. 2003).

MPAs Within Integrated Coastal Zone Management

The establishment of MPAs was only one part of the Philippine ICZM program. Efficient ICZM interventions require the integration of additional components:

- Watershed management to avoid soil erosion (to prevent sediment intake into MPAs, which negatively affects coral survival).

- The consideration of different ecosystems in ICZM planning (natural coral reefs, mangroves and artificial reef habitats).
- Solid waste management.
- Application of an interdisciplinary approach (sector-oriented community mobilisation that efficiently links environmental and social aspects).

Since 1993, the coastal communities, with support from governmental (GO), non-governmental (NGO) as well as international development organizations have achieved improvements in Integrated Coastal Zone Management e.g. in the province of Negros Oriental. After 3 years of protection, it was possible to measure a 240 % increase in catch rates in the surroundings of Marine Protected Areas (MPA). By 2003, members of the fishing communities were successfully protecting and managing 30 designated coral reef and mangrove habitats (Wiedemeyer et al. 2003).

Significant positive changes were also recorded under the Environment and Rural Development Programme and its Coastal Fisheries Resources Management Component, which mainly operated in the Provinces of Leyte, Southern Leyte and Negros Occidental (Scholz 2008); see also Fig. 22.1. After 6 years of implementation, the concept of Coastal Fisheries Resources Management has meanwhile raised awareness and received a widespread acceptance by key stakeholders. In the Province of Negros Occidental (Region 6), 18 LGU partners have formed three coastal alliances in order to apply uniform conservation regulations and techniques. A Provincial Technical Working Group regularly meets to monitor progress, to clarify roles and to harmonise services. In the Province of Southern Leyte (Region 8), GTZ assisted the establishment of a Provincial Coastal Fisheries Resources Management Unit, which is in the meantime able to provide support services throughout the province, i.e. the seven partner LGUs which are currently supported. In both regions, members of the fisherfolk participate through LGU and Provincial Fisheries and Aquatic Resources Management Councils (FARMCs). At present, the established and supported Marine Protected Areas (MPAs) cover an area of 40,000 ha in both regions. The management is governed by a legal framework and by financial allocations from the LGU Annual Investment Plans. Prior to the establishment of an MPA, intensive consultation with peoples' organisations and fisherfolk took place. As a result of the protection efforts fish catch has increased by a factor of between two to five in surrounding areas.

The fishing community is also involved in resource protection efforts through participation in Bantay Dagat (community fish wardens) patrols. Their members are mandated by law to carry out the dangerous task of enforcement of fisheries regulations, also often assisted by the Philippine National Police. As a result, e.g. Province of Southern Leyte was considered as a 'dynamite free' zone in 2009.

Recently the MPA approach has been expanded to encompass responses and adaptations to climate change (Scholz 2008; Heine 2009; Salzer 2012).

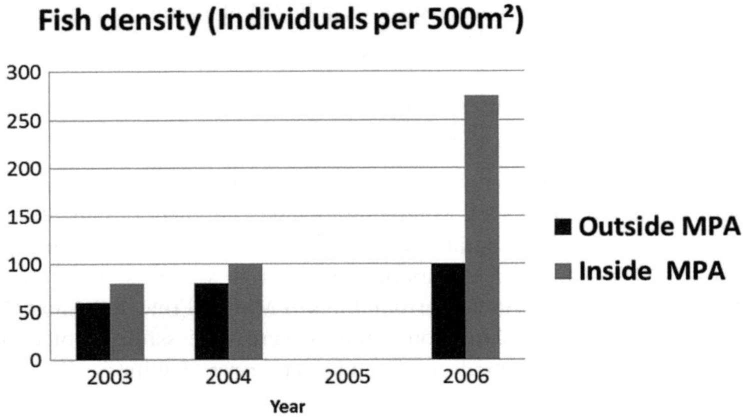

Fig. 22.1 Impact of the establishment of a marine protected area (MPA) on a coral reef in terms of abundance of fish species targeted by artisanal fishers. Southern Leyte, Philippines (*Source*: GIZ Philippines 2011)

22.3 Aquaculture Standards

In the industrialised countries, various aquaculture standards have been developed for specific market segments. Examples are GlobalG.A.P. as a business-to-business standard, the Aquaculture Stewardship Council (ASC) and Best Aquaculture Practices (BAP) standards, which enable consumers to identify sustainably farmed products, and organic labels such as Naturland. However, there is a noticeable trend at present for the major marketing chains to develop their own labels (FAO 2011).

22.3.1 GlobalG.A.P. and ASC

The two most widespread standards in aquaculture in Europe are the GlobalG.A.P. Aquaculture Standard and the Aquaculture Stewardship Council (ASC) standards. In North America, the Best Aquaculture Practice (BAP) standards set by the Global Aquaculture Alliance (GAA)[7] are the most widespread.

The GlobalG.A.P. Aquaculture Standard

The GlobalG.A.P.[8] Aquaculture Standard applies to a diversity of fish, crustaceans and molluscs and extends to all hatchery-based farmed species, as well as the

[7] www.gaalliance.org/bap.

[8] www.globalgap.org.

passive collection of seedlings in the planktonic phase. The standard covers the entire production chain, from broodstock, seedlings and feed suppliers to farming, harvesting, processing and post-harvest handling operations. It serves as a practical manual for any aquaculture producer, ensuring food safety, minimal environmental impact and compliance with animal welfare and worker health and safety requirements (GlobalG.A.P. 2012). The GlobalG.A.P. Aquaculture Standard sets specific criteria for:

- Site Management
- Reproduction
- Chemicals
- Occupational Health and Safety
- Fish Welfare, Management and Husbandry
- Harvesting
- Sampling and Testing
- Feed Management
- Pest Control
- Environmental and Biodiversity Management
- Water Usage and Disposal
- Post-Harvest—Mass Balance and Traceability
- Post-Harvest—Operations
- Social Criteria

Aquaculture producers are also required to source the compound feed used at the aquatic farming and hatchery levels from reliable suppliers.

The Aquaculture Stewardship Council Standard

The Aquaculture Stewardship Council[9] (ASC) is an independent non-profit organisation promoting best practice within aquaculture. It was established in 2010 by the World Wide Fund for Nature[10] (WWF) and the Sustainable Trade Initiative[11] (IDH). The outcome of the so called "Aquaculture Dialogues" was the creation of eight global standards that define how responsible farming for 12 different types of fish and shellfish (salmon, shrimp, tilapia, trout, pangasius, abalone, mussels, clams, oysters, scallops, cobia and seriola) should be conducted. The cultivation of these fish can have a huge impact on natural surroundings and the environment. Their market value and (potential) international commercial value are also high. It was precisely for such reasons that these 12 species were chosen. Promoting farming

[9] http://www.asc-aqua.org/.

[10] http://wwf.panda.org/.

[11] http://www.idhsustainabletrade.com/.

practices that minimise their impact on the environment and communities is the ultimate goal. The ASC standard is guided by seven principles (ASC 2012):

1. Comprehensive legal compliance;
2. Conservation of natural habitat and biodiversity;
3. Conservation of water resources;
4. Conservation of species diversity and wild population through prevention of escapes;
5. Use of feed and other inputs that are sourced responsibly;
6. Good animal health (no unnecessary use of antibiotics and chemicals);
7. Social responsibility for workers and communities impacted by farming (e.g. no child labour, health and safety of workers, freedom of association, community relations).

GlobalG.A.P. and ASC Certification of Pangasius in Vietnam

In Vietnam, the export of aquaculture products (notably the striped catfish, *P. hypophthalmus*, locally known as "tra", internationally known as "Pangasius") has exceeded US$3 billion and the revenues from the sector, both from local sales as well as from export, are an important part of the economy. Particularly in rural areas, the income generation and value adding potential is considerable. However, concerns on the sustainability of the sector have been raised for several years. Local stakeholders, international buyers and consumer protection organisations are increasingly concerned with food safety issues and the impact on the environment given the intensive nature of production methods (Anh et al. 2010; Bosma et al. 2009; Da Silva et al. 2010).

In 2006 GIZ facilitated trials on the EurepG.A.P./GlobalG.A.P. standard for Pangasius, assessing the potential for certification of existing farms in Vietnam (Van 2006; Becker et al. 2009; Schütz 2006).

In 2012, Vietnamese pangasius production reached 1,255,500 tonnes and exports valued US$1,744 million. Vietnam accounts for more than 90 % of world's pangasius exports and the sector generates about 30 % of the national revenues from seafood products. Vietnamese pangasius is exported to 136 countries. Largest importers are the EU and the US and account for almost half of all Vietnamese Pangasius exports.

Pangasius consumers are increasingly demanding higher food quality and safety. Moreover, the demand for a more sustainable production system that takes effects on the environment into account is increasing (Anh et al. 2010; Bosma et al. 2009; Da Silva et al. 2010). Therefore sustainably produced seafood is more and more important for producers to access markets. In Europe certification and ecolabelling of pangasius is key to maintain and increase market share.

GlobalG.A.P. was established as a 'Business to Business' standard. Therefore it does not have a label or logo but only a product code. It is now considered to be a food safety consumer standard. GlobalG.A.P. has a pangasius standard for individual farm certification as well as for group certification. Certification is based on the farm as the legal entity. With the GlobalG.A.P. version 4 standard, GRASP is a voluntary addition for social criteria. For the group certification, the Internal Control System (ICS) therefore needs to be managed by a legal entity as well.

The ASC (Aquaculture Stewardship Council) standard for pangasius was created by the Pangasius Aquaculture Dialogues (PADs) in 2007 and was initiated and managed by the WWF and the Dutch Sustainable Trade Initiative (IDH). The ASC standard for pangasius was published in August 2010. Similar to the MSC (Maritime Stewardship Council), ASC is a 'Business to Consumer' standard, recognised by the ASC logo. ASC certification is farm-based, which means that separate farms under one company can be individually certified.

The Sustainable Pangasius Supply Chain Programme (SPSP) is a Private Public Partnership (PPP) initiated in July 2010 in the Mekong Delta (Deichert and Linh 2013) of Vietnam. Public participants include:

- GIZ (Deutsche Gesellschaft für Internationale Zusammenarbeit);
- IDH (Dutch Sustainable Trade Initiative);
- The World Wildlife Fund for Nature (WWF US);
- GlobalG.A.P. Germany;
- Tra Vinh People's Provincial Committee (PPC).

Private participants include:

- ANOVA Seafood BV (Netherlands);
- Two large-scale pangasius farms and processing companies:
- Docifish Joint Stock Corporation in Dong Thap Province;
- Vinh Quang Seafood Corporation in Tien Giang Province;
- Up to 30 small-scale pangasius farmers in Tra Vinh province, which are represented by the Tra Vinh Fishery Association (FA);
- Two processing companies in Tra Vinh province, who committed to buy the certified fish from the small scale fish farmers.

The overall objective of SPSP was that "*farmers, processors and feed producers operating along the Pangasius supply chain in selected provinces of the Vietnamese Mekong Delta comply with internationally accepted and accredited certification schemes for sustainable production and sell certified Pangasius to European buyers*".

To achieve this objective, two targets were identified for the SPSP:

1. Efficient supply chain management to achieve an accredited certification scheme for the medium- and large-scale farmers.
2. Testing the viability of small-scale farmers adopting an accredited certification scheme and participating in efficient supply chain management.

Achievements

By 2013 medium sized cooperating farms had attained GlobalG.A.P. and ASC certification. These farms produced and marketed around 10,000 t/y of certified pangasius. For the smallholder farmers, an internal control system (ICS) was established, with training for the farmers and pre-assessment being conducted.

Lessons Learned

- When establishing a pangasius supply chain with small scale producers the value chain actors up to the retail actors should be identified in advance. Ideally all the actors should make clear commitments in a written agreement. This implies that all actors decide which certification they would like to achieve. In any case the producer should make a well informed decision, which certification standard he/she is aiming for.
- Therefore understanding the difference between the standard's criteria and current gaps on the farm is important. In our case it was the difference between GlobalG.A.P. and ASC standard criteria for pangasius, as the farms were initially certified according to GlobalG.A.P. and only a year later started the process towards ASC certification.
- A premium price is probably the main motivating factor for the producer and the other actors of the supply chain to decide on certification. Experiences show that actual premium price information is not available, especially with a continuously fluctuating price in the market and a fairly high price differentiation according to quality of whole freshly harvested fish. Therefore, it is very challenging to keep small-scale farmers motivated to seek certification.
- Especially in a project situation, the small scale farmers seem to focus primarily on closing the investment (i.e. 'hardware') gaps, and often underestimate challenges that are included to close the 'software' gaps, i.e. enhancing their aquaculture knowledge and farm management skills including record keeping.
- Record keeping and management were found to be the most challenging gaps in implementation not only for large-/medium-sized farms but also for small-scale farms.
- A group certification implies a number of additional challenges, first and foremost with regard to the establishment of an Internal Control System (ICS). As the group certification approach was new to all involved parties, information regarding steps and tools needed to be identified prior to starting the process and explained to all identified actors.
- When pursuing group certification, one should set clear criteria for farmers to join from the beginning. One could start with a fairly large group, as several farmers would drop out along the process, often because farmers might stop rearing pangasius completely for some time, when the price is too low. Therefore it is advisable to start with closing the 'software' gaps first (i.e. enhancing aquaculture knowledge and farm management skills) otherwise some of the 'hardware' investments might have been done in vain. The farmers have to be ready to invest time and labour as their own contributions to achieve the changes to meet the necessary criteria.

- Small farmers are highly flexible in their operation. They will readily switch from pangasius farming to other fish species such as tilapia or gourami for which the market prices are higher, or at least not as volatile as they are for pangasius. However, when the prices for pangasius increase again, many small farmers will eagerly switch back to growing that species.
- Many criteria of the GlobalG.A.P. and ASC standards are very difficult to comply with for the small scale farmers, as they were often developed for larger industrialised operations in the first place. If certain criteria appear quite unfeasible for the actual small holder situation, it increases the task of keeping the farmers motivated. One way to do so is by explaining repeatedly the benefits from improved farm management, if they close the software gaps, and which ultimately may also result in tangible economic benefits (Deichert and Linh 2013).

22.3.2 Organic Aquaculture Standards

Organic aquaculture refers to the production processes and practices of ecological production management systems that promote and enhance biodiversity, biological cycles and biological activity (Tveterås 2000; Bergleiter 2003; Bergleiter et al. 2009). A range of private, national and regional standards exist (Tacon and Brister 2002; Subasinghe et al. 2004; Bergleiter et al. 2009; CEC 2009; IFOAM 2010; Prein et al. 2012; Jonell et al. 2013). Similar to the promotion of organic agriculture in the context of addressing the situation of resource-poor farmers in developing countries, organic aquaculture has been introduced mainly in developing and transitional economies in Latin America and Asia and implemented to a considerable degree by small scale farmers (Subasinghe et al. 2004; Funge-Smith and Halwarth 2004; Bergleiter et al. 2009; Prein et al. 2012; Jonell et al. 2013; Xie et al. 2013).

22.3.3 Naturland Organic Aquaculture

All initiatives on organic aquaculture by GIZ together with partner farms and institutions were established in collaboration with Naturland e.V. and implemented their guidelines (Naturland 2007). Naturland is a German non-profit organisation established in 1982 to promote certified organic food production and has developed standards for organic aquaculture. An independent third party organisation conducts the inspections and submits the reports to Naturland. Following a schedule of successful inspections and meeting administrative requirements, the actual certificate is then finally issued by Naturland and is valid for 1 year (Bergleiter 2003; Nolting and Prein 2008; Bergleiter et al. 2009; Stamer 2009; Bergleiter 2011).

Organic Shrimp in Ecuador

The first ever pilot activity with GIZ involvement on certified organic aquaculture was conducted from 1999 to 2001 in Ecuador in cooperation with Naturland and stakeholders from the private and public sectors of the farmed shrimp value chain in Ecuador to initiate eco-friendly and profitable shrimp production (Nolting and Schirm 2003b). It comprised the formulation of the detailed criteria of a standard of organic shrimp aquaculture and the first certification of a shrimp farm. The 'Round Table approach' was an important mechanism to ensure that a stakeholder-inclusive process towards consensus-building for a workable standard was ensured.

New certification standards for organic shrimp (white leg shrimp *Litopenaeus vannamei*) aquaculture were developed and tested in cooperation with Naturland, selected shrimp farmers, and importers from Ecuador, Europe and GTZ. As a result, the first certified, eco-labelled shrimp from Ecuador were imported to Europe in 2001. Today, several certified shrimp farming companies in Ecuador and a growing number in other Latin American countries produce white leg shrimp from aquaculture that is in compliance with the Naturland certification standards for organic farming. These shrimp from organic aquaculture are now being sold successfully in the European market in a total volume of over 25,000 t/y (Censkowsky 2012).

Lessons Learned

With the application of standards and criteria for sustainable shrimp production, positive impacts on the production and management methods of these related businesses and industries can be achieved (Nolting and Prein 2008).

Shrimp hatcheries have given up the collection of wild shrimp larvae in order to meet Naturland standards for organic farming. Feed producers felt obliged to responsibly select and use appropriate raw feed materials and feed additives. In order to fulfil the certification standards they have abandoned the use of artificial feed ingredients and mixture of prophylactic antibiotics and chemotherapeutics.

Social criteria ensure that conditions for farm employees follow local laws and regulations and in some aspects are outright prescriptive (e.g. accommodation, sanitary facilities, and health provisions). For shrimp farmers and the associated industries, the production of eco-labelled shrimp offers an alternative to the normal progression of intensive farming systems. The production and export of high-value eco-labelled shrimp products from developing countries serves as an important source of secured employment and income.

Supporting small scale producers in developing countries has been identified as an appropriate strategy to introduce socio-economic and environmental standards in other farming sectors (e.g. for coffee production). However, in general, small scale farmers have fewer difficulties in adopting organic principles (due to previous extensive management).

With the introduction of eco-labelled shrimp, increased competitiveness among shrimp producers can be observed. This has influenced the production techniques of related industries and businesses. Globally, this can contribute significantly to a product range diversification, which will drive competition among traders and markets for certified and non-certified shrimp products.

Organic Shrimp in Thailand

The well-known expansion of the Thai shrimp farming industry into becoming the globally dominant exporter of farmed shrimp and shrimp products is well documented. Until 2004, the indigenous black tiger shrimp (*P. monodon*), was the major species cultured. However, since the mid 1990s, as in other shrimp farming countries around the world, the Thai shrimp farmers were faced with numerous emerging diseases which affected survival as well as retarding growth (e.g. monodon stunting syndrome), and also caused survival problems for larvae in hatcheries. Within this scenario, farmers introduced an alien species, the white-leg shrimp (*L. vannamei*) to the country from its native Latin America. This species was rapidly adopted by farmers and hatchery operators due to a number of favourable attributes, which increased the annual share of production volume dramatically. Over just 4 years, until in 2006 it made up 98 % of the entire volume of cultured shrimp in Thailand.

A small number of farmers decided to continue farming black tiger shrimp, but faced difficulties in marketing them. One farmer, Mr. Prayoon Hongrat, owner of Sureerath farm with 30 years of experience in shrimp farming, decided to design and build a new farm which went against the common practice of ever increasing intensity. His new farm of 2.2 km^2 had 143 ponds (each of approximately 0.6–0.7 ha size, stocked at low densities of 13–15 post larvae per m^2) with settling ponds, canals and a large water storage pond replanted with mangroves. The entire water management system is such that it is completely recycled and designed in such a way that gravity flow is used as much as possible to reduce pumping costs. Additionally, and as an innovation, Mr. Hongrat grew filamentous algae (*Enteromorpha intestinalis*) in the ponds before stocking shrimp (similar to the function of a pasture) to feed shrimp in the first two months after stocking, thereafter followed by low levels of additional pellet feed.

Upon request from the Thai Department of Fisheries, GTZ's country program on "Enhancing the Competitiveness and Eco-Efficiency of Thai SMEs" saw the opportunity for Sureerath Farm to explore new market avenues through organic certification because the farm had design and management features that already complied in large parts with organic standards. GTZ facilitated the certification by Naturland, which commenced in early 2006 and was completed in June 2007 (Klinkhardt 2007). For several years, Sureerath farm served as an example, and a program was launched by the Thai Department of Fisheries to assist other farms to remodel their operations and achieve organic certification. Several organic shrimp farmer groups and associations were subsequently formed and attained certification against a new national organic standard (Ruangpan 2007; 2008).

Lessons Learned

In the first years the farm operators spent considerable efforts in scaling up their organic production and experienced many setbacks in establishing market linkages. The growth period until harvest took much longer than in non-organic intensive systems (8–10 vs. 4–6 months) so that the farm could only produce 1.2–1.3 crops per year instead of two. In 2007 the total production of the farm was around 200 t/y (compared to the pre-certification level of 800 t/y). Overall farm profitability was reduced, despite the 'organic premium' of 20–30 % envisaged by the farmers but not provided by the market.

Problems in sourcing organic feed: Despite the well-developed nature of the Thai shrimp feed production industry, as well as the Thai seafood processing industry, but due to relatively small amounts of organic feed required and relatively small amounts of organic shrimp harvested at the moment (by the first and only organic shrimp farm), there were initial problems with sourcing a supplier of certified organic shrimp feed, and with sourcing processors to deliver the required services under certified conditions. This caused the farm owners to start to establish their own feed mill and processing facility on the farm premises, which is designed to be able to handle feed requirements and processing volume of other members of the organic shrimp farmer association in their area.

Organic Pangasius Catfish in Vietnam

In 2004, GTZ initiated a Public Private Partnership (PPP) on organic catfish production. In this, GTZ and private sector companies engaged in jointly financed projects that had sound economic principles and at the same time high development potential. The overall objective was to establish organic catfish production with a small pilot group of Vietnamese producers and processors as a proof of concept (Finkel 2005, 2006; GIZ 2012).

The two German partners were an importer of seafood, Binca Seafood GmbH, a relatively small company which saw the potential for developing organic aquaculture fish as a high-end niche market, and Naturland.

The local partners in An Giang province in the Mekong Delta, are small-scale catfish farmers and processors, organised within the An Giang Fisheries Association. In times when the traditional catfish export markets suffered from protectionism as well as food safety scandals, the organic catfish production represented a unique opportunity to open up new export markets.

Lessons Learned

As a result, 70 tonnes of organic pangasius were exported to Germany in 2005, which increased to 400 tonnes in 2006, and is expected to have doubled since then.

Farmers were, and still are, faced with insufficient amounts of certified feed. These were partly addressed by involving local producers of raw materials and feed, which is an ongoing process (Finkel 2006). The inavailability of sufficient amounts of certified organic aquafeed is a strong limiting factor for further expansion of organic aquaculture in general, not just that of pangasius in Vietnam. Since 2012 the two farms have their own feed production capacity operating according to the Naturland organic standard.

Organic Aquaculture: Philippines

Aquaculture is one of the priority programs for development in the Philippines. Aside from the local supply of aquaculture products, the exported volume has been one of the top products in foreign exchange earnings for the country. Aquaculture development in the Philippines has seen unsustainable practices develop, notably in the coastal shrimp farming sector. Diseases spread and caused the near-collapse of the shrimp industry in the mid and late 1990s. From 2007 to 2010, GTZ through its Environment and Rural Development Program (EnRD) in the Philippines, and the German Development Service (DED), supported the Office of the Provincial Agriculturist (OPA) on the Island of Negros in the central Philippines in the introduction of organic aquaculture to fish and shrimp farmers on the island.

A first step was a scoping event involving all interested stakeholders, titled the "First Philippine Organic Aquaculture Symposium", held in Bacolod City in October 2007 (GIZ/EnRD 2007). This involved a range of local stakeholders, including farmers, processors and other service providers, as well as speakers from Naturland, and from the successful organic aquaculture introductions in Vietnam and Thailand.

A second step was a survey on the suitability of existing farms for certification according to organic aquaculture standards (Kühlmann and dela Fuente 2008). The aim of the process was to develop a gradual strategy towards successful implementation of organic aquaculture, learning from existing successful examples elsewhere. In 2008, supported by GIZ, staff of Philippines-based conformity assessment bodies were trained at Naturland in Germany to be farm inspectors according to the Naturland organic aquaculture standard.

Although considerable interest existed among Philippine shrimp and fish farmers (brackish water milkfish in ponds, marine black tiger shrimp in ponds, tilapia in net cages and ponds, and groupers in marine net cages), the process towards certification did not continue. Local markets did not demand organic products while export markets were well-served with organic products at prices with which the Philippine producers could not compete.

22.4 Conclusions and Recommendations

The promotion of seafood certification programs is an effective market-based tool to implement and support the FAO Code of Conduct for Responsible Fisheries. This experience gained by GIZ stems from its initiatives with seafood products destined for export, notably to Europe. In these cases, market demand in importing countries for various types of standards has led to their introduction and subsequent spread. This chapter has presented experiences in developing countries with a number of new introductions and in some cases first-time formulations of such standards. The initiatives were motivated by the policy objective of development cooperation, aiming at equitable benefits for all stakeholders involved.

In the case of implementing approaches for the management of communal resources (i.e. in capture fisheries) such as MPAs, or fisheries sustainability standards such as the MSC or Naturland Wildfish, much wider aspects need to be considered and ensured such as strong buy-in by all stakeholders involved, e.g. fishers, beach management units, national fisheries departments, as well as processors) in order to generate strong governance of the resource at the local level (Schlager and Ostrom 1992; Berkes et al. 2001; Basurto and Ostrom 2009; Fox et al. 2012).

The implementation of VSS for private production enterprises (i.e. in aquaculture) such as GlobalG.A.P. (as a Business-to-Business standard), ASC (as a Business-to-Consumer on-pack label), Naturland Organic Aquaculture (as a consumer-facing label) requires strong commitment by aquaculture operators (aquapreneurs) and strong support services (i.e. advisory and training services, industry associations, fisheries departments).

Overall, assurance of integrity through enforcement programs by standard-setting organisations, through independent, unannounced and rigorous third-party inspections and measures towards attaining compliance prior to accreditation of conformity assessment bodies (certifying bodies) such as through witness audits.

22.4.1 Summary

An overview is presented of initiatives by GIZ and its partners in respect to the introduction of voluntary sustainability standards (VSS) in the fisheries and aquaculture sectors in developing countries since 1999. In the capture fisheries sector these were (1) the implementation of the FAO Code of Conduct for Responsible Fisheries, (2) the use of Marine Protected Areas, (3) the Co-management approach to coastal fisheries, (4) the Marine Stewardship Council (MSC) standard, (5) the Naturland Wildfish standard, and (6) the Marine Aquarium Council standard for ornamental fish. Countries in which these initiatives were conducted were Kenya, the Philippines, Tanzania, Senegal, Sri Lanka and Vietnam. In the aquaculture sector the VSS that were implemented by GIZ and its partners comprised of (a) the

Naturland Organic Aquaculture standard for shrimp (*P. monodon*) and (b) for pangasius (*P. hypophthalmus*), (c) the GlobalG.A.P. and (d) Aquaculture Stewardship Council (ASC) standards for pangasius. Projects in which these certifications were piloted were mainly located in Ecuador, Thailand and Vietnam. In the Philippines, a local organic standard, based on Naturland criteria, was developed and applied for tilapia (*O. niloticus*) and milkfish (*C. chanos*).

In the chapter, the rationale is outlined for a government-owned institution that is implementing German development policy, such as GIZ, for engaging in the introduction of VSS in the fisheries and aquaculture sectors in developing countries. The initiatives towards the formulation and/or implementation of VSS in the fisheries and aquaculture sectors of selected developing countries are presented in view of their situational context, the achievements attained and lessons learned. Each of the cases presents the general situational background that necessitated the VSS initiatives. The processes and experiences with the different standards are briefly discussed together with the lessons learned and recommendations from these pilot activities.

References

Aliño PM (2002) An overview of Philippine fisheries. Marine Science Institute, University of the Philippines, Quezon City

Anh PT, Kroeze C, Busch SR, Mol APJ (2010) Water pollution by Pangasius production in the Mekong Delta, Vietnam: causes and options for control. Aquac Res 42(1):108–128

ASC – Aquaculture Stewardship Council (2012) ASC fact sheet. Aquaculture Stewardship Council, Utrecht, the Netherlands. http://www.asc-aqua.org/. Last accessed 22 Sept 2013

Asche F (2011) Green growth in fisheries and aquaculture. Organisation for Economic Co-operation and Development (OECD), Trade and Agriculture Directorate, Fisheries Committee, Paris, TAD/FI(2011)6

Basurto X, Ostrom E (2009) Beyond the tragedy of the commons. Economia delle fonti di energia e dell'ambiente 52(1):35–60

Becker D, Pham NT, Hoang DT (2009) Value chain promotion as a tool for adding value to agricultural production: case study Pangasius. GTZ, Hanoi

Bergleiter S (2003) Organic aquaculture: completing the first decade. Org Stand 30:14–16

Bergleiter S (2011) Organic aquaculture – from a "niche" to the "whole cake"? Ecol Farm 30 (2):14–17

Bergleiter S, Berner N, Censkowsky U, Julià-Camprodon G (2009) Organic aquaculture 2009 – production and markets. Organic Services GmbH, Munich, and Naturland e.V., Graefelfing

Berkes F, Mahon R, McConney P, Pollnac R, Pomeroy R (2001) Managing small-scale fisheries: alternative directions and methods. International Development Research Centre, Ottawa

Blueyou and ENDA/REPAO (2007) Feasibility study: eco-labelling in artisanal coastal fisheries in Senegal. Research commissioned by GTZ (unpublished report)

Bosma RH, Hanh CTT, Potting J (2009) Cradle-to-gate: environmental impacts of pangasius farming in the Mekong Delta. Aquac Eur 34(4):13–17

Bush SR, Toonen HM, Oosterveer PJM, Mol APJ (2013) The 'devils triangle' of MSC certification: balancing credibility, accessibility and continuous improvement. Mar Policy 37:288–293

CEC – Commission of the European Communities (2009) Commission Regulation (EC) No. 710/2009 of 5 August 2009 amending Regulation (EC) No. 889/2008 laying down detailed rules for the implementation of Council Regulation (EC) No. 834/2007, as regards laying down detailed rules on organic aquaculture animal and seaweed production. In: Official Journal of the European Union L 204: 15–34

Censkowsky U (2012) Organic aquaculture perspectives in Asia. Presentation at the Organic Aquaculture Session, Sustainable Seafood Summit, Hong Kong, 6–8 September 2013, Organic Services GmbH, Munich

Dalzell P (1988) Small pelagic fisheries investigations in the Philippines. Part II: the current status. FishByte 6(4):2–4 [December 1988: International Center for Living Aquatic Resources Management (ICLARM), Metro Manila, Philippines]

Da Silva SS, Ingram BA, Nguyen PT, Bui TM, Gooley GJ, Turchini GM (2010) Estimation of nitrogen and phosphorus in effluent from the striped catfish farming sector in the Mekong delta, Vietnam. Ambio – J Hum Environ 39(7):504–514

De Beule H (2012) Follow up of the impact of eco-labelling of Nile Perch in Tanzania: ecological and socio-economical impact in the Bukoba, Mwanza and Musoma region, between 2010 and 2012. Unpublished Survey Report, GIZ, Eschborn

De Beule H, Ayoku D (2011) Impact study of the eco-labelling of Nile Perch Project in Bukoba, Tanzania ecological & socio-economic impact between 2009 and December 2010. Unpublished Survey Report, GIZ, Eschborn

Deichert G, Linh NT (2013) Moving Pangasius smallholders towards certification. Sustainable Pangasius Supply Chain Program (SPSP), Unpublished manuscript, GIZ, Eschborn

FAO – Food and Agriculture Organisation of the United Nations (1995) Code of conduct for responsible fisheries. FAO, Rome

FAO – Food and Agriculture Organisation of the United Nations (2011) Technical guidelines on aquaculture certification. Version adopted by the 29th Session of Committee on Fisheries (COFI) held in Rome, 31 January to 04 February 2011, FAO, Rome

FAO – Food and Agriculture Organisation of the United Nations (2012) The state of world fisheries and aquaculture 2010. FAO Fisheries and Aquaculture Department, FAO, Rome

FAO – Food and Agriculture Organisation of the United Nations (2103) International guidelines for securing sustainable small-scale fisheries. Zero draft, FAO, Rome

Finkel T (2004) The value chain approach for organic catfish in Vietnam within a public private partnership. GTZ, Hanoi (unpublished)

Finkel T (2005) GTZ initiatives in aquaculture in Vietnam. German Technical Cooperation (GTZ), Vietnam Country Office, Small and Medium Enterprises Development Programme (SMED), Value Chain Component, Hanoi

Finkel T (2006) Lasting impacts of GTZ's public private partnership project on organic pangasius production: from organic niche markets to conventional mass markets with farmers' improved awareness of fish safety – a series of small scale projects initiated by GTZ in Vietnam. German Technical Cooperation (GTZ), Vietnam Country Office, Small and Medium Enterprises Development Programme (SMED), Value Chain Component, Hanoi

Fox HE, Mascia MB, Basurto X, Costa A, Glew L, Heinemann D, Karrer LB, Lester SE, Lombana AV, Pomeroy RS, Recchia CA, Roberts CM, Sanchirico JN, Pet-Soede L, White AT (2012) Reexamining the science of marine protected areas: linking knowledge to action. Conserv Lett 5(1):1–10

Funge-Smith S, Halwarth M (2004) The role of organic aquaculture in food security and poverty alleviation. In: Subasinghe S, Singh T, Lem A (eds) The production and marketing of organic aquaculture products: proceedings of the global technical and trade conference, 15–17 June 2004, Ho Chi Minh City, INFOFISH, Kuala Lumpur, pp 11–17

Gardiner PR, Viswanathan KK (2004) Ecolabelling and fisheries management. Penang, Malaysia, WorldFish Center Studies and Reviews 76

Geheb K, Kalloch S, Medard M, Nyapendi A-T, Lwenya C, Kyangwa M (2008) Nile perch and the hungry of Lake Victoria: gender, status and food in an East African fishery. Food Policy 33 (1):85–98

Gitonga NK (2012) Ecolabelling as a tool to improve fisheries trade and governance of Lake Victoria: final consultancy report. GIZ and FishAfrica, Fisheries and Aquaculture Consultants
GIZ/EnRD (2007) Proceedings of the first Philippine organic aquaculture symposium 2–3 October 2007, Bacolod City, Negros Occidental, GIZ and EnRD, Manila
GIZ Philippines (2011) Marine protected areas: CFRM interventions in Southern Leyte. Map of locations and graph of fish densities within and outside MPAs. GIZ, Philippines
GIZ – Deutsche Gesellschaft für Internationale Zusammenarbeit GmbH (2012) Growing business with smallholders: a guide to inclusive agribusiness. GIZ, Bonn and Eschborn, BMZ, Bonn, and ENDEVA UG, Berlin
GlobalG.A.P. (2012) The global partnership for safe and sustainable aquaculture. Foodplus GmbH, Cologne, Germany
Heeger T, Sotto F (eds) (2000) Coral farming: a tool for reef rehabilitation and community ecotourism. BMU and GTZ Tropical Ecology Program, Manila
Heine B (2009) Adaptation to climate change in coastal areas of the Philippines and Viet Nam. GIZ, Eschborn
IFOAM EU Group (2010) Organic aquaculture EU Regulations (EC) 834/2007, (EC) 889/2008, (EC) 710/2009: background, assessment, interpretation. In: Szeremeta A, Winkler L, Blake F, Lembo P (eds) International federation of organic agriculture movements EU group, Brussels and CIHEAM/IAMB, Valenzno, Bari
Jarchau P, Nolting M, Wiegler K (2009) Eco-labelling of small-scale fisheries in developing countries: the example of the Nile perch fisheries in Bukoba, Tanzania. Lessons Learned No. 10, Sector Project Promotion of responsible Fisheries, GTZ, Eschborn
Jonell M, Phillips M, Patrik Rönnbäck P, Troell M (2013) Eco-certification of farmed seafood: will it make a difference? Ambio – J Hum Environ 42(6):659–674
Klinkhardt M (2007) Sureerath prawns: big black tiger prawns producer goes "organic". Eurofish Magazine No. 2/2007: 32–33 (April)
Kolding J, van Zwieten P, Mkumbo O, Silsbe G, Hecky R (2008) Are the Lake Victoria fisheries threatened by exploitation or eutrophication? Towards an ecosystem-based approach to management. In: Bianchi G, Skjoldal HR (eds) The ecosystem approach to fisheries. FAO, Rome
Kolding J, Medard M, Mkumbo O, van Zwieten P (2014) Status, trends and management of the Lake Victoria Fisheries FAO Fisheries Technical Paper 579 (in press)
Kühlmann K-J, dela Fuente L (2008) Organic aquaculture development in the Philippines – the healthy way forward. Aquac Eur 33(2):28–29
Marshall B (2010) Is the Nile perch fishery on Lake Victoria sustainable? LVFO, Jinja. Presentation at the Seafood Summit 2010, Paris
Moody Marine Limited (2008) Pre assessment report for the Nile Perch fishery on Lake Victoria. Ref 80269. Moody Marine Limited, Derby
MRAG (2009) Review of fish sustainability information schemes – final report. Marine Resources Assessment Group Ltd., London
Nadvi K, Wältring F (2002) Making sense of global standards. INEF Report, No. 58/2002. Institut für Entwicklung und Frieden, INEF, Gerhard-Mercator-Universität, Duisburg
Naturland (2007) Naturland standards for organic aquaculture. Naturland – Verband für ökologischen Landbau e.V. (Naturland – Association for Organic Agriculture, Registered Association), Gräfelfing
Nolting M, Schirm B (2003a) Marine ornamental fish trade in the Philippines – new ecological and quality standards. Sector Project: Policy Advice for Sustainable Fisheries, Lessons Learned No. 2, Deutsche Gesellschaft für Technische Zusammenarbeit (GTZ), Eschborn
Nolting M, Schirm B (2003b) Shrimp production in Ecuador: new ecological and quality standards. Sector Project: Policy Advice for Sustainable Fisheries, Lessons Learned No. 4, Deutsche Gesellschaft für Technische Zusammenarbeit (GTZ), Eschborn
Nolting M, Prein M (2008) Organic certification of aquaculture products – a chance for sustainable aquaculture development. Aqua Cult Asia Pacific Mag 4(5):9–13 (September/October 2008)

Nunan F (2013) Wealth and welfare? Can fisheries management succeed in achieving multiple objectives? A case study of Lake Victoria, East Africa. Fish and Fisheries. doi:10.1111/faf.12012

OECD (2008) Globalisation and the fishing industry. Paper presented at the 100th session of the Committee for Fisheries as TAD/FI(2007)15

OECD (2010) Economics of fisheries and aquaculture certification: revision. Fisheries Committee, Trade and Agriculture Directorate, TAD/FI(2010)14, OECD, Paris

OECD (2011) Fisheries and aquaculture certification. OECD Publishing, Paris. http://dx.doi.org/10.1787/9789264119680-en

OECD (2012) Eco-labelling and certification in the fisheries sector: summary report of the OECD/FAO round table. In: OECD review of fisheries 2011: policies and summary statistics, OECD Publishing, Paris. http://dx.doi.org/10.1787/rev_fish-2011-6-en

Prein M, Bergleiter S, Ballauf M, Brister D, Halwart M, Hongrat K, Kahle J, Lasner T, Lem A, Lev O, Morrison C, Shehadeh Z, Stamer A, Wainberg AA (2012) Organic aquaculture: the future of expanding niche markets. In: Subasinghe RP, Arthur JR, Bartley DM, De Silva SS, Halwart M, Hishamunda N, Mohan CV, Sorgeloos P (eds) Farming the waters for people and food: proceedings of the global conference on aquaculture 2010, Phuket, 22–25 September 2010, FAO, Rome and NACA, Bangkok, pp 549–567

Roheim CA, Sudhakaran PO, Durham CA (2012) Certification of shrimp and salmon for best aquaculture practices: assessing consumer preferences in Rhode Island. Aquac Econ Manag 16 (3):266–286

Ruangpan L (2007) Thailand's road map for organic aquaculture. Aqua Cult Asia Pacific Mag 3 (3):8–10 (May/June 2007)

Ruangpan L (2008) Paper presented at the satellite conference on organic aquaculture of 16th IFOAM organic world congress, Cattolica/Rimini

Salzer W (2012) Adaptation to climate change in coastal areas. Fact Sheet on International Climate Initiative Projects in the Philippines, GIZ, Manila

Schlager E, Ostrom E (1992) Property-rights regimes and natural resources: a conceptual analysis. Land Econ 68(3):249–262

Scholz U (2006) Ecolabelling as a way out of the Nile perch dilemma? Eurofish Magazine 6/2006, pp 52–56

Scholz U (2008) There is more in the sea than fish: the growing importance of the "blue" sector in Philippine environmental programmes. EUnews 6(2):5, June 2008, Delegation of the European Commission to the Philippines

Schütz K (2006) Facilitation of the market for business development services to overcome difficulties of the Pangasius fish farmers in An Giang province, Vietnam. Sectoral Project Agricultural Trade, Trade Programme, Deutsche Gesellschaft für Technische Zusammenarbeit (GTZ), Eschborn

Schwab P (2010) First ecolabelled artisanal fishery in developing country – Nile Perch at Lake Victoria, Unpublished manuscript, GIZ, Eschborn

Scullion J (2009) Promotion of responsible fisheries on Lake Victoria, Final Consultancy Report, GIZ, Eschborn

Scullion J (2010) Promotion of responsible fisheries on Lake Victoria, Final Consultancy Report, GIZ, Eschborn

Stamer A (2009) Aspekte nachhaltiger Fischzucht: Ökologische Aquakultur als Alternative. Ökologie & Landabau 151(37):18–20

Subasinghe S, Singh T, Lem A (eds) (2004) The production and marketing of organic aquaculture products: proceedings of the global technical and trade conference, 15–17 June 2004, Ho Chi Minh City, INFOFISH, Kuala Lumpur

Tacon AGJ, Brister DJ (2002) Organic aquaculture: current status and future prospects. In: Scialabba NEH, Hattam C (eds) Organic agriculture, environment and food security. FAO, Rome, pp 163–175

Tacon AGJ, Metian M (2013) Fish matters: importance of aquatic foods in human nutrition and global food supply. Rev Fish Sci 21(1):22–38

Tveterås S (2000) Assessment of the sustainability of organic salmon farming. Working paper 2000:18/Discussion paper 2000:4, Centre for fisheries economics, Institute for Research in Economics and Business Administration (SNF), Bergen

Van CT (2006) Pangasius value chain in An Giang. In: GTZ-VN-SMED-VCC (eds) GTZ SME Development Programme, Value Chain Component. German Technical Cooperation (GTZ), Vietnam Country Office, Small and Medium Enterprises Development Programme (SMED), Value Chain Component, Hanoi, pp 13–14

Wiedemeyer W, Waltemath M, Wendelken O, Schirm B (2003) Management of marine protected areas (MPA) in Negros Oriental, Philippines. Sector Project: Policy Advice for Sustainable Fisheries, Lessons Learned No. 3, Deutsche Gesellschaft für Technische Zusammenarbeit (GTZ), Eschborn

Witte F, Seehausen O, Wanink JH, Kishe-Machumu MA, Rensing M, Goldschmidt T (2013) Cichlid species diversity in naturally and anthropogenically turbid habitats of Lake Victoria, East Africa. Aquatic Sci 75(2):169–183

WWF – World Wide Fund for Nature (2009) Assessment study of on-pack, wild-capture seafood sustainability certification programmes and seafood ecolabels. Accenture, WWF International

Xie B, Qin J, Yang H, Wang X, Wang Y-H, Li T-Y (2013) Organic aquaculture in China: a review from a global perspective. Aquaculture 414–415:243–253

Yang P (2009) Impacts of certification on the fisheries sector: a case study from Bukoba, Tanzania. MSC thesis (Diplomarbeit), Faculty of Economics, Leibniz University Hannover

Chapter 23
Measuring Sustainability in the Construction and Real Estate Sector: A Case Study of the DGNB Certification System

Christine Lemaitre

23.1 Introduction

This chapter focusses on the system developed and run by the German Sustainable Building Council, DGNB, for the certification of sustainable buildings and urban districts and sets the DGNB approach within the context of the wider agenda of sustainable development.

The chapter is an introduction to the DGNB certification system, which is aimed at aiding designers and construction companies to create sustainable buildings and urban districts, as well as maintaining and improving existing sustainable buildings. A key issue, which will be elaborated in greater detail, relates to the way in which the DGNB system goes beyond conventionally 'green' considerations, e.g. the environmental pollutants and the responsible procurement of building materials, and facilitates an evaluation which reflects societal objectives for social, economic and environmental sustainability and captures life-cycle performance and environmental impacts.

The background section of this chapter, seeks to set the creation of the DGNB certification system within the historical context of perceptions and attitudes towards sustainability at that time of its inception. This will be followed in Sect. 23.3 by a brief overview of the DGNB System and its structure in terms of the six topics addressed, with Sect. 23.4 elucidating the individual criteria within those topics in greater depth.

Section 23.5 addresses the role of certification systems in promoting best practice, followed by a discussion of certification systems contributions towards greater efficiencies in achieving excellent performance levels against the sustainability objectives. This is followed by an outline of the international development and implementation of the DGNB system in Sect. 23.7 and finally the conclusions

---

C. Lemaitre (✉)
German Sustainable Building Council (DGNB), Kronprinzstraße 11, 70173 Stuttgart, Germany
e-mail: c.lemaitre@dgnb.de

C. Schmitz-Hoffmann et al. (eds.), *Voluntary Standard Systems*, Natural Resource Management in Transition 1, DOI 10.1007/978-3-642-35716-9_23,

and recommendations in Sect. 23.8, which will seek to summarise the key points of this chapter.

23.2 Background

The construction and real estate sector is witnessing a paradigm shift, whereby the focus is on using energy efficiently, conserving resources, safeguarding health at home and at work, supporting long-term value retention and mitigating risk.

In late 2009, Roland Berger Strategy Consultants conducted a survey of more than 40 key players in the real estate sector in Germany, Austria and Switzerland (Henzelmann et al. 2010). Published in 2010, this research revealed that clients and investors viewed sustainability in real estate management as a long-term trend, with 73 % of respondents expressing a willingness to accept higher investment costs for the use of sustainable real estate properties, and a further 86 % willing to accept higher rents (Henzelmann et al. 2010, pp. 8–9).

This research chimed with the findings of an earlier survey by Jones Lang LaSalle in 2007, which found that respondents' willingness to accept higher rents for office space was closely correlated to running costs, such as for energy use, rather than merely for sustainability as a good within itself (Barthauer 2007, p. 6).

The study reflected this rising interest in sustainable building design at a time when a variety of building rating tools had already been developed. Already in 1990, the British *Building Research Establishment's Environmental Assessment Method* (BREEAM) was developed in the United Kingdom, followed by the *Haute Qualité Environmentale* (HQE) standard in France in 1996, and *Leadership in Energy and Environmental Design* (LEED) which was established in the United States in 1998. The LEED system focussed mainly on the ecological quality of buildings and used a points system to score aspects such as the use of environmentally sound materials, indoor air quality or the optimisation of energy performance. Similarly, the BREEAM system took buildings' ecological quality into account and reflected aspects such as health and comfort, environmental pollution and the design process in the overall evaluation.

The respondents to the Roland Berger study expected the proportion of certified properties to rise significantly in the following 5 years (Henzelmann et al. 2010, p. 24). Significantly, the Roland Berger research also identified a concern that certificates on the market at the time placed too much emphasis on environmental concerns, and not enough on their economic sustainability (Henzelmann et al. 2010, p. 16).

Launched later than the other systems, in 2009, the DGNB's second-generation system responded to this perception with a holistic approach to building performance, including—and lending significant weight to—economic criteria, such as a life-cycle costing over a 50 year cycle in the evaluation.

In the years which have passed since the publication of the Roland Berger research, the prevailing conditions and interests have further evolved and as a result, the notion that buildings need to be planned, built and operated more sustainably has gained acceptance throughout the sector. A broad consensus has emerged, which views sustainable building as an intelligent response to the environmental and socio-economic context, whereby the design preserves environmental resources and provides users with a safe, healthy and comfortable environment.

In 2013, Jones Lang LaSalle highlighted the need for investors to "act now to future-proof their assets" with "83 % of clients positioning sustainability as a top priority for office real estate", going on to state that "From almost nowhere a decade ago, sustainability is now a key consideration for office real estate" (Jones Lang LaSalle 2013).

The DGNB criteria can be used to identify efficient, inexpensive steps during the planning phase—with the added benefit of pre-certification. Moreover, the DGNB pre-certificate lends to investor confidence that the building's performance targets will be reached upon completion, even during the early design stages and that the system supports the sustainable maintenance and operation of the buildings.

Meanwhile, the DGNB system has gone from strength to strength and, with more than 400 pre-certificates and certificates for new buildings and urban quarters, currently leads the market for building certification in Germany (according to DGNB data).

23.3 The DGNB System

The DGNB System was first developed for new office and administrative buildings. Subsequently, other schemes for completely different types of buildings were developed from this starting point. The evaluation is based on a set of criteria covering six topics related to sustainable construction, comprising:

- Environmental Quality
- Economic Quality
- Sociocultural and Functional Quality
- Technical Quality
- Process Quality
- Site Quality

The six topics are weighted according to their importance in the overall evaluation of the building. Economic quality, environmental quality, sociocultural and functional quality, and technical quality each make up 22.5 % respectively of the building's total performance index, with process quality contributing 10 %. In the building-based schemes, the quality of the location is evaluated separately and is not included in the score relevant to certification (see Fig. 23.1). Within the schemes for urban districts, the quality of the location is integrated into to assessment of the other five topics, rather than as a stand-alone category.

Each of the six evaluation topics is covered by a range of individual criteria, such as total primary energy demand, noise protection, and land use. For each criterion,

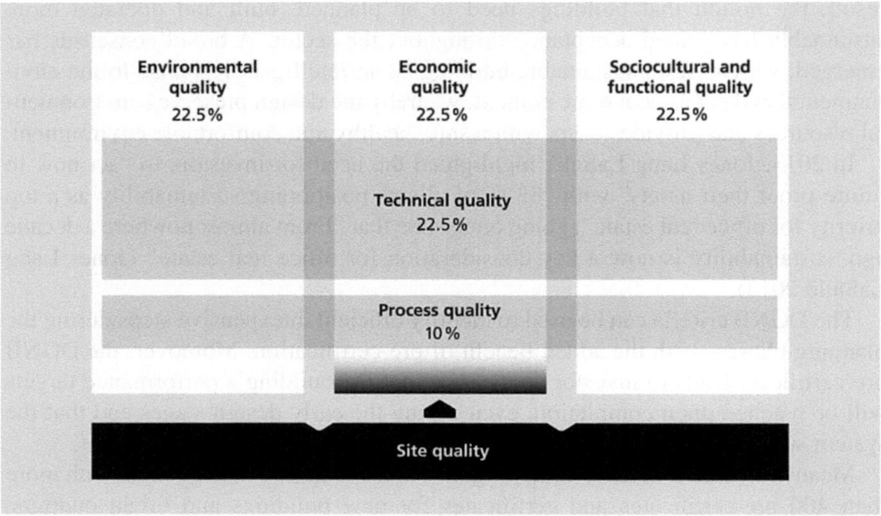

Fig. 23.1 Weighting of the DGNB topics (*Source*: DGNB 2012)

measurable target values are defined; measurement methods and documentation required for verification are clearly outlined. A maximum of ten points is given for each criterion. All criteria are weighted for the evaluation in two steps. Regardless of the specific scheme, each criterion has a weighting factor and can be multiplied by a factor of up to three within its broader category. This weighting factor reflects the criterion's social and political relevance. A building's primary energy demand is thus viewed as more important than its performance in terms of noise protection. At the scheme level, the system's methodology allows for further fine-tuning. Here, the weighting is determined according to a use-specific adaption factor that can increase a criterion's value by as much as threefold. This weighting factor can also be set to zero in order to remove certain criteria where appropriate—for example, indoor air quality is not considered relevant to highway bridges.

Whereas the weighting of the topic areas is fixed throughout all the DGNB certification schemes, the criteria within these topic areas can be weighted differently according to use-specific factors relevant to buildings of different uses. Each type of building thus has its own criteria set and weighting factor which is optimally adapted to the specific requirements, challenges and opportunities of its dedicated use.

Depending on the extent to which the overall performance benchmarks were fulfilled, a bronze, silver, or gold certificate is awarded. With an overall degree of fulfilment of at least:

- 50 %, a bronze certificate is awarded,
- 65 %, a silver certificate is awarded, and
- 80 %, a gold certificate is awarded.

As stated above, striking the best balance between often conflicting objectives for sustainability across the wide range of topics is central to the DGNB approach. For this reason, the award of a bronze, silver, or gold certificate is also subject to the building achieving a certain minimum score across all of the topic areas. This minimum score is set at:

- 35 % across all topics for a bronze certificate,
- 50 %, across all topics for a silver certificate, and
- 65 %, across all topics for a gold certificate.

In this way, the DGNB offers the appropriate certificate for each type of building—and yet all buildings are evaluated on the same basis, facilitating the consistent use of the system and thereby also providing comparability across different building types and uses.

23.4 The DGNB Criteria

In the 2012 version of "New office and administration buildings", DGBN certification within the six topic areas is based on 41 criteria, 37 of which describe building quality and the remaining four describe the site quality.

23.4.1 Environmental Quality

The evaluation of environmental quality contributes 22.5 % to the overall score. Within this share, a major proportion is accounted for by the Life Cycle Assessment (LCA), which captures the global environmental impacts of production, construction, use, and end-of life associated with each of the building materials used. Another criterion within this topic evaluates the local environmental impact associated with the materials and construction methods chosen. The responsible procurement of materials such as timber and stone is a further criterion within this section. In evaluation of the efficient use of natural resources, further criteria evaluate the primary energy use throughout the building's life cycle, as well as drinking water demand and waste water volume and land use.

23.4.2 Economic Quality

The assessment of economic quality contributes 22.5 % to the total score and includes the calculation of overall life cycle cost, with this criterion being allocated the single greatest weighting of any criterion within the entire DGNB system. This aspect demonstrates the DGNB system's emphasis on gaining a holistic understanding of building performance, including the aspect of economic sustainability. Further aspects included in the evaluation of economic sustainability include the design's flexibility and adaptability—for example, in terms of the potential to respond to changing use requirements by reconfiguring office space layouts at minimum cost.

Finally, another aspect included in the evaluation of the building's economic sustainability is its commercial viability in terms of responding to changing economic conditions and accommodating third-party use.

23.4.3 *Socio-Cultural and Functional Quality*

The topic of socio-cultural and functional quality accounts for 22.5 % of the overall score and includes the aspect of user comfort, which is addressed by criteria addressing thermal comfort throughout the seasons. This topic also includes indoor air quality which is subject to air quality measurement and which forms one of only two knock-out criteria, i.e. criteria without which the certification a building is not possible. The other criterion which must be met if a building is to gain a certificate is also part of the socio-cultural and functional topic area and relates to barrier free access.

Further criteria focussing on user comfort evaluate acoustic comfort, visual comfort, and occupant control—in other words, the facility for building users to individually influence the conditions within their immediate environment. Other aspects included in the assessment include the quality of outdoor spaces associated with the building, e.g. roof terraces, balconies, terraces and courtyards. Safety and security for potentially vulnerable user groups is addressed, as is the possibility for certain areas of the building to be accessible to the general public, as well as facilities for cyclists.

Finally, three criteria within this topic specifically address the quality of the design, e.g. by focussing on the evaluation of design quality throughout the design procurement process, by addressing the integration of public art and the quality of the building layout.

23.4.4 *Technical Quality*

The assessment of technical quality accounts for 22.5 % of the overall score and includes aspects such as fire prevention, noise protection and the quality of the building envelope. Focussing on the building's performance in use, criteria address the adaptability of technical systems to meet changing requirements, the ease of cleaning and maintaining the building, as well as the ease of deconstruction and disassembly of building components at the end of their lifetimes. Finally, sound emissions from the building are also taken into account in evaluating its technical performance.

23.4.5 *Process Quality*

Criteria addressing process quality account for 10 % of the total score and address aspects such the drafting of a comprehensive project brief and integrated design, i.e. the formation of interdisciplinary teams in developing the initial design and contributing to its further development. Other criteria address the quality of the design concept and the inclusion of sustainability considerations in tendering processes. The documentation of building systems is addressed; with an eye to facilitating better and more sustainable practice in facility management. Finally, a further three criteria address the environmental impact of construction, quality assurance in construction and systematic commissioning processes.

23.4.6 *Site Quality*

The criteria for site quality do not contribute to the score which is relevant for certification. In this separate evaluation, criteria address the quality of the local environment, the public image of the area and local social conditions, access to public transport as well as access to local facilities amenities. Finally, the evaluation also includes an assessment of environmental risks within the local environment which could jeopardise the building's safe construction and operation, such as landslides, flooding, and unexploded ordnance and so on.

23.5 Promoting Best Practice

Any criteria-based system has to strike a delicate balance between breaking complex and interlinked characteristics of a building's design and construction down into manageable topics for assessment, whilst simultaneously facilitating an integrated and holistic overall evaluation.

Another key issue to be resolved is whether to focus on inputs or outcomes. Whilst an input-driven system can be easier to communicate and apply without undue reflection, this can also lead to formulaic—and thereby ultimately inappropriate solutions to highly differentiated contextual situations. The DGNB system reconciles these issues by creating a framework of reference and benchmarks for the building's overall performance, rather than prescribing specific solutions to individual issues.

This approach allows developers and designers the greatest possible freedom in finding solutions which meet the overall target of excellence whilst appropriately reflecting contextual factors and specific priorities. As a result, the DGNB system fosters and encourages the processes whereby innovation and technical progress in the construction sector gives birth to new approaches and solutions. The system is updated and adapted continuously; step-by-step with expanding horizons in terms of building performance.

With the years following its launch in 2008, the DGNB system has been regularly updated and adapted to reflect the changing context in terms of best practice and legislation. In doing so, it has remained true to one of the core concepts in the field of certification: creating an incentive and reward for outperforming legal regulation and minimum standards. Instead of regulation initiating a 'race-to-the-bottom' to merely fulfil statutory minima, the incentive of certification helps lend competitive advantage to market leaders and generates an upward drift through best practice to excellence (Fig. 23.2).

The DGNB approach reflects the construction sector's evolving concept of sustainability and future-proofing. Coupled with the option for pre-certification during the planning stage, the DGNB system can be used to identify efficient, inexpensive steps during the design development—with the added benefit of lending a marketing boost to a quality-assured process. From the early design stages, the DGNB pre-certificate can thereby lend investors' confidence that the building's performance targets will be reached upon completion, and that the system supports the sustainable maintenance and operation of the buildings.

Given the environmental, economic, and social aspects of sustainability in mind, it is hardly surprising that the pursuit of sustainability in all areas often reveals a number of conflicts between individual decision-making processes. These conflicts require balanced and interdependent decision-making which can be facilitated by integrating all of the aspects under consideration for long-term sustainability and allocating a specific weighting to the complex and interlocking formulae which contribute to the calculation of the overall score. Throughout a project's development, the process of conducting the audit and examining the individual design decisions in terms of their effect upon the achievement of the DGNB benchmarks requires and rewards an integrated approach to building design and thereby maximises the potential to reconcile individually conflicting targets to arrive at the best overall balance in terms of the building's performance during construction, operation, and in its end-of-life phase.

The process of certification also offers a range of other benefits, which can be summarised under the headlines of reduced risk and value stability.

As a tool geared to active engagement during the design development, the DGNB criteria-set can make a significant contribution to reducing project risk. The process of applying the criteria during the early design stages generates a high degree of certainty that the project's quality and performance targets will be reached upon completion. The certification process promotes the integrated consideration of all aspects of the development process from initial concept to completion on site. This leads to greater transparency and better-defined processes and

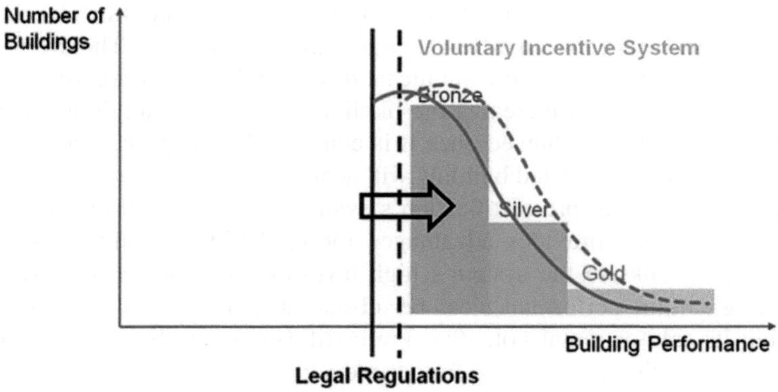

Fig. 23.2 Generating the upward drift (*Source*: DGNB 2013)

contributes to quality-assurance during the design, construction and commissioning stages.

As a result of these aspects, the DGNB Certificate denotes greater quality and workmanship. The focus on user comfort and flexible and adaptable design contributes to increased user-friendliness and improved rental rates. In capturing these qualities and providing an independently recognised and assured means of documenting investors, owners and users commitment to sustainability, certification can also serve as a marketing tool.

Certification makes the high quality of a building tangible for owners and users. Moreover, it signals a performance-enhancing work environment as well as high user satisfaction. The certificate thereby increases the building's sale and rental potential as it celebrates its holistic high quality for owners and users whilst helping reduce energy consumption and costs during operation.

23.6 Achieving Best Practice More Efficiently

The early integration of criteria for the evaluation of a buildings performance in terms of sustainability offers the opportunity to identify the most cost-effective approaches to achieving the best overall result. For this reason, pre-certification allows investors and building owners to optimise their projects during the planning stage. Including an audit for pre-certification formulates a clear target for achieving sustainable construction and goes hand in hand with an integrated approach to the building's design and construction. Pre-certification is an excellent instrument to increase transparency, ensure clear planning and construction processes, improve risk management, and increase the building's overall quality through integral quality assurance in the process of compiling the audit trail. For pre-certification, all main sustainability criteria must be addressed statements of intent or targets at an early stage in the design process.

Pre-certification thereby supports decision-making whilst alerting stakeholder's attention to trade-offs between different targets and requirements. Therefore, it is an important means by which to communicate design and construction objectives. In addition, pre-certification increases the likelihood that a building's intended performance goals will be achieved once it is completed. This process also makes it more likely that the completed building will achieve certification without problems and that the delivery of pre-certification's evaluation results will thus be assured. Pre-certification also provides advantages for marketing a building still under construction. Thanks to the system's high level of transparency and credibility, the building's future performance can be substantiated as early as the design stage, increasing its sale or rental potential. Pre-certification can thereby also increase security for financing projects and further reduce project risk.

23.7 International Development

One of the DGNB system's greatest strengths is its high degree of flexibility, and this makes it very well suited for adaptation and use in other markets. The system can not only be adapted to meet changing technical and social requirements, but it can also be made to match individually varying national or regional particularities. These aspects might include the climate, structural and legal requirements, or cultural factors. These qualities allow for a very rapid globalisation of the DGNB certification system.

Where a suitable partner organisation has been identified in a country, the DGNB can work in tandem with this organisation to align the certification system even more closely with local requirements and the prevailing building culture. This allows buildings all over the world to be evaluated based on the same standard and using the same approach, thus ensuring comparability, transparency and security for investors, building owners and users alike.

In June 2009, only half a year after the first DGNB certificates were awarded, the Town Town building in Vienna received the first certificate for a building outside Germany. Since then, a fully adapted Austrian version of the DGNB certification system has been developed and is implemented by the DGNB's partner organisation in Austria, the *Österreichische Gesellschaft für Nachhaltige Immobilienwirtschaft*, or ÖGNI. The implementation of the operational system in a specific country can also be carried out by the local partner, following the establishment of a corresponding contract of cooperation. Accordingly, nationally adapted versions of the DGNB system have been developed and implemented with the *Bulgarian Green Building Council* in Bulgaria, with the *Danish Green Building Council* in Denmark, with the *Swiss Green Building Council* in Switzerland, and with the *Thai Association for Sustainable Construction* in Thailand.

To date, fewer than 100 projects have gained certificates and pre-certificates according to the DGNB and its nationally adapted versions (DGNB System 2013). This figure also includes a number of projects certified directly according to the

DGNB's international system, which is based on current European standards and building regulations. Thanks to this system, it is possible to certify buildings and urban districts in any location worldwide. Where required, an authorised DGNB auditor working on a specific pilot project can liaise with the DGNB office to adapt the system to local requirements in three clearly defined steps.

23.8 Conclusions and Recommendations

The first key insight gained from the development and implementation of the DGNB certification system, and touched upon in Sect. 23.2, is the fact that economic sustainability, as captured by a lifetime cost analysis, is key to establishing a broad mainstream consensus in support of sustainable design and construction practices.

A second key aspect, noted in Sect. 23.5, is the focus on performance-related outcomes rather than specific inputs as a crucial factor for the adaptability of the certification tool, whether in terms of varying building uses, national regulatory contexts, or the climatic parameters of the location.

A third point, from Sect. 23.3, is the relationship between voluntary standards, such as certificates, and legal regulations, with voluntary standards requiring and rewarding a continual 'upward drift' in building performance, and lending competitive advantage to building practices which outperform the legal requirements.

A fourth point, made in Sect. 23.6 relates to the early integration of certification into the design development process. This early integration is crucial in terms of reducing the marginal cost of improved building performance in terms of their long-term sustainability.

In the long run, the knowledge gained from the consistent application of criteria-based evaluation tools can help implement carbon-neutral business practices and organise the building process increasingly in terms of material lifecycles. One key investment that will make it possible to master future challenges is to integrate competence in sustainable building into tertiary education. This is where not only prospects for the future are created, but also important bridges between real-world experience and tertiary education will be built. For the emerging generation of building and real estate experts, sustainable building will be a 'given'. Now is the time to lay the foundations in our institutes of higher education.

References

Barthauer M (2007) Ökologische Nachhaltigkeit von Büroimmobilien. Jones Lang LaSalle. http://www.joneslanglasalle.com/ResearchLevel1/JLL_Germany_%C3%B6kologische%20nachhaltigkeit%20von%20b%C3%BCroimmobilien.pdf. Last accessed 05 July 2013

DGNB – Deutsche Gesellschaft für Nachhaltiges Bauen (2012) Excellence defined. Sustainable building with a systems approach. DGNB, Stuttgart

DGNB – Deutsche Gesellschaft für Nachhaltiges Bauen (2013) Making sustainability measurable. DGNB presentation

DGNB System (2013) DGNB pre-certified and certified projects. http://www.dgnb-system.de/en/projects/. Last accessed 05 July 2013

Henzelmann T, Büchele R, Engel M (2010) Nachhaltigkeit im Immobilienmanagement: Kurzfassung der Studie. Roland Berger Strategy Consultants. http://www.rolandberger.at/media/pdf/Roland_Berger_Nachhaltigkeit_im_Immobilienmanagement_20100413.pdf. Last accessed 05 July 2013

Jones Lang LaSalle (2013) How will 'sustainability' affect offices? http://www.joneslanglasalle.eu/EMEA/EN-GB/Pages/offices-2020-sustainability.aspx. Last accessed 05 July 2013

Part V
Challenges and Future Trends

Whilst featuring as a concluding topic throughout all chapters within this publication, this section deals explicitly with voluntary standards with an eye to the future. Hindsight is commonly referred to as always being "20/20", in contrast with the confusion and unknown hazards which trouble those predicting future trends. The following chapters are anticipatory of voluntary standards developments, based on the benefits of hindsight gleaned from lessons learned previously and our current experience within VSS application.

In the not-too-distant past, consumer consensus on 'corporate social responsibility' (CSR) was a curiosity at best and a blatant example of 'green-washing' at its worst. Over time businesses have realised the importance of implementing CSR policies to improve consumer confidence and maintain market share, whilst also securing future supply of materials to their business. Chapter 24 investigates how this switch in corporate strategy has pushed corporations to externally source standards for operation which will aid in achieving targets amidst the age of the 'informed' consumer. The challenges of integrating VSS into current operations is discussed with the future of impact measurement featuring high on the agenda for corporations looking to better understand the impacts throughout their procurement and supply chains processes.

Chapter 25 explores how VSS can be applied within the urban and community development environment, in an effort to make cities and communities resilient and sustainable. Where previously such issues have been dealt with through a heavily bureaucratic and technological approach, the case is made for utilising VSS for sustainable communities and cities to meet the goals associated with attaining a liveable and ecologically sensitive community. The relevant components of the British Standards Institute's guidance for community sustainable development (BS8904) are pondered along with the emerging 'Transition Movement' within the UK. Topics range from risk and resilience in sustainable community development to setting the conditions to encourage informal social contracts and moral ownership within such endeavours.

Chapter 26 asks, "What links a small-scale cocoa farmer in Africa to the compact and distinguishable 'eco-labels' on products throughout EU supermarket shelves?". Are we safe in the assumption that claims related to our informed purchase are actually benefiting such a farmer, or has this human-being been relegated to acting simply as a unique selling point at their own cost? With the popularity of eco-labels within VSS, there have arisen conflicts and problems over the market- and consumer-orientated characteristics of eco-labels, while neglecting the suppliers' perspective. The discussion around eco-labelling is made here whilst keeping the producer in focus in relation to the allocation of burden upon them, their potential to contribute, and empowerment when involved in decision making.

Chapter 27 discusses how the impact of a trade policy tool like Generalised System of Preferences (GSP) could go beyond simply improving legislative frameworks and lead to an actual change in production conditions in emerging and developing countries. The focus lies in the promotion of compliance with social and environmental standards in producer states. As demonstrated by GSP+ (an enhanced development of GSP), the EU can link the award of tariff preferences to certain development targets and sustainability criteria. The idea lies in a linkage of the GSP with the provision of evidence regarding sustainable production, which would allow tariff preferences for sustainably produced goods. This chapter examines whether sustainable business activities can be promoted through a system of tariff preferences and whether state-level recognition of certification systems would be practicable in the case of the EU's own GSP.

In closing this book, Chap. 28 takes the form of a political 'call-to-arms' in realizing and promoting the use of VSS, based upon its utility in various circumstances where legal standards and governments have failed to promptly address a problem with appropriate and meaningful action. As an emerging governance feature within both public and private sectors, there has been lacklustre rate of growth in VSS uptake and implementation, which is disappointing considering the demand for change from 'mainstream' consumption choices. After outlaying the political setting of VSS, voluntary standards are presented as a learning experience and various stakeholder views are voiced. Following on from these perspectives are examples of strategic relevance, particularly those related to supply chain patterns. In the concluding segment, the inherent limits of VSS are presented along with the debate as to what is considered actually considered 'voluntary'.

Chapter 24
Corporate Social Responsibility and the Role of Voluntary Sustainability Standards

Daniele Giovannucci, Oliver von Hagen, and Joseph Wozniak

24.1 Introduction

Corporate Social Responsibility (CSR) is becoming a standard feature particularly for large and consumer-oriented firms. What started in the late 1960s as something closer to charity or philanthropy has evolved dramatically in recent years. Yet, as actualisation of the CSR concept is increasingly explored and becoming better-defined, there is limited understanding of how to operationalise CSR and how to manage it for desirable results at the ground level. This gap is particularly salient in the purchasing relationships with producers in developing countries. Voluntary Sustainability Standards (VSS) such as Organic, Fairtrade, Rainforest Alliance, Forest Stewardship Council, Ethical Tea Partnership, GlobalG.A.P., and UTZ Certified present an important step in this process but, like many tools, require some learning.

The business environment is radically altering. Vague concepts of sustainability and CSR are giving way to specific and auditable standards. More firms are now employing what Kolk (2005) calls "a cascade of codes of conduct". Clear definitions are useful and it is worth noting the distinction between codes of conduct and a VSS. Codes of conduct can be internally developed or externally. They are a set of practice guidelines characterised by flexible implementation rules that tend to lack enforcement mechanisms and may not have audit or reporting criteria. VSS are here defined as the independent and publicly determined standards that have, as primary criteria of compliance, multiple aspects of sustainability defined as specific social,

D. Giovannucci (✉)
The Committee on Sustainability Assessment (COSA), 1006 South 9 Street, Philadelphia, PA, USA
e-mail: DG@theCOSA.org

O. von Hagen • J. Wozniak
The International Trade Centre (ITC), 54-56, rue de Montbrillant, 1212 Geneva, Switzerland
e-mail: hagen@intracen.org; wozniak@intracen.org

C. Schmitz-Hoffmann et al. (eds.), *Voluntary Standard Systems*, Natural Resource Management in Transition 1, DOI 10.1007/978-3-642-35716-9_24,

environmental and economic guidelines that feature transparent auditing and more credible (typically external) third-party enforcement mechanisms.

The general research and discussion on CSR to date does not adequately address the recent evolution of both CSR approaches and their interface with a broadening range of external standards or VSS. Part of the recent story is that VSS have become prominently intertwined and increasingly integrated with CSR into the strategy of many firms. This chapter explores the roles and challenges of VSS within the objectives of a firm and its CSR strategy and contributes a practical understanding grounded in the authors' combined experience in both the private sector and in the public sector.

The overall purpose of the chapter is to elucidate several key areas of understanding:

- How and why the VSS have come to prominence for CSR applications.
- How critical shifts in corporate strategy, driven in part by better consumer understanding and greatly expanded levels of communication and supply chain transparency are leading corporations to look externally for operating standards that not only help them reach their goals but that also confer social legitimacy.
- Most VSS were not designed as corporate tools, and their integration into procurement and corporate supply chains while often successful, can also be a challenge.
- Impact measurement will be an integral part of the next evolution of CSR as firms and public agencies move toward more effective use of VSS tools and a greater understanding of how to measure and manage their impacts.

The next section of the chapter starts by exploring the roots of CSR and its relationship to VSS. It covers their common and divergent objectives and also how firms have put them into practice, discussing notable successes and how even large firms can sometimes get it wrong. Section 24.2 also outlines the distinctions between public standards and private or corporate standards. Section 24.3 illustrates the rise of the main VSS for food and agriculture and how pervasive standards have become in terms of numbers and market share. Section 24.4 describes new approaches to understanding how VSS can serve the specific objectives of their stakeholders, including producers, consumers, and firms. Section 24.5 highlights the main conclusions and offers recommendations for enhancing the symbiosis and effectiveness of the corporate relationship with these voluntary standards.

24.2 A Search for Credibility: The Roots of CSR

About a century ago, most people lived in proximity to a town or village wherein most of their and their neighbours' actions were known—like it or not—to all of the community. Community members were interdependent on each other personally for most food and services. If the grain miller cheated or was fair, the results were usually evident and the corresponding consequences were obvious and direct. For much of the world, daily needs were locally met and everyone lived at a scale where self-regulation was plausible within the community.

The guidelines for day-to-day interactions and transactions were locally established and enforced as part of the social code that found its legitimacy in custom and in the daily presence of institutions that included religious and temporal authority. As such, the social norms were adapted to the needs of the community and were easily understood and usually followed. But something had already begun to shift.

By the late nineteenth century the twin factors of specialisation and economies of scale had already emerged from military theory and, along with a burst of diverse technologies, were driving the engines of the Industrial Revolution. The basic fuels necessary for this revolution were labour and capital. Capital came from the increased scope and freedom of the corporation[1] while labour came from the migration of growing and relatively poor rural communities often living under feudal conditions. The ensuing concentration of labour, capital, and outputs fed the rapid expansion of cities and the deracination of many smaller communities. The social fibres that had held families and communities together and ensured a certain level of shared well-being began to unravel with the many threads gathering in urban areas. The speed of growth and considerable scope of these developments quite literally re-created town and urban communities making them more transitory and heterogeneous. The novel diversity undoubtedly offered considerable benefits but came at a price.

The social conventions of mutually agreed upon limits or boundaries for the purpose of longer-term and common benefit—what today might be called sustainability—began to lose their power. Even religious authority for most fast-urbanising society gradually devolved toward a level of some disregard.[2] The definition and influence of ethics and morality were migrating from their source of localised legitimacy and now the closest expression occurred within state or governmental control. While such coalesced power had existed since ancient times, it evolved in recent centuries to supersede local societal controls in new and more complete ways. The apex of the power of the state in this regard may have occurred in the late twentieth century. More recently, in the current age of global capitalism, governmental or political boundaries have begun to dissolve as the corporation has taken on new and more powerful roles and, in some cases, corporate influence may even surpass that of the state (Glasbergen 2011). These shifts of power, in a relatively short period of time, mean that it is no longer clear who decides social legitimacy and ethics and on what basis.

[1] See, for example, the 1856 UK Joint Stock Companies Act that served as a template for similar company laws in the US and other nations.

[2] There are clear exceptions, especially among the more fundamental segments of Christianity, Judaism, and Islam, but these stand in contrast to mainstream life of society in most cities.

Box 24.1: New Limits

Problems created by booms and busts in supply and demand, which are due to economic, political or speculative reasons, are actually dwarfed in comparison to those created by environmental limits. Because when it's gone, it's gone; meaning "no resources = end of business". *(A. Ionescu-Somers 2012)*

This presents an understandable concern particularly when it is abundantly clear that our technology now permits a scale of human activity or intervention that can rapidly and profoundly alter our way of life. It can be enormously positive or it can put not just a single community but society as a whole at risk. In just a few decades, the stakes have become formidable. The threats to food and agriculture range from climate change and depletion of natural resources to population explosion and chronic malnutrition. Some of the key issues for business revolve around scarcity of basic non-renewable resources including water and arable land leading to higher commodity prices and protectionism. This is particularly relevant in the realm of agriculture and ecology where the evidence is stark in every region of the world. A few examples include:

- a decades-long and possibly irreversible decline of many major fresh water sources in key US farm regions;
- persistently high prices for multiple agricultural commodities with resulting civil unrest and export bans in a number of countries;
- a rapid removal of the forests in the biodiversity-rich areas of South America, Malaysia, and Indonesia for more soy, timber and palm oil;
- the reduction of one of the world's great rivers to a toxic trickle, nearly destroying the sea and fisheries that it once fed, due primarily to Central Asia's cotton farming practices.

From these challenges, new opportunities have emerged. Some leading corporations have come to understand that responsible stewardship is necessary to ensure their own longevity in terms of both resources and public opinion. But putting this understanding into action has not been easy, particularly for publicly held firms where many shareholders focus more on short term profit than long-term success. In 2003, the CEO of Starbucks Corporation, one of the world's most popular beverage brands, noted that while it made sense for the business to invest even more in the sustainability and the long-term well-being of coffee farmers (Starbucks is a major buyer of coffee from dozens of developing countries), the pressure to deliver positive quarterly financial reports made that very difficult.[3] The resistance to enduring viability for firms may thus come from their own shareholders who are often relatively anonymous and unaccountable to the firm, the community, or the environment and whose private gain can therefore easily compromise public and

[3] Personal discussions between Daniele Giovannucci and Starbucks CEO Orin Smith.

corporate good without personal consequence. However, it is untenable to put all of the blame at the feet of shareholders; the firm's leadership clearly has a say and can also be responsible. Paul Polman, CEO of consumer goods giant Unilever, responded to the demands of short-sighted analysts and shareholders by ordering his company managers to stop delivering quarterly results to the financial markets, thus instilling a longer-term view of the company's success factors.[4] Fuller and Jensen (2010) concur and suggest that it is necessary for leaders to make more socially responsible, value-focused decisions.

In recent decades, business thinking is evolving—at least in branded food and consumer goods sectors—beyond the sclerotic grip of short-sighted corporate theories (see, for example, Friedman 1970) that were better suited to an age of robber barons than they are to today's emerging need to cooperate as much as to compete with regard to our finite resources. Harvard Business School professors Porter and Kramer (2006), for example, make a strong case for the value of CSR as a source of long-term competitive advantage. One of the world's top business school deans, INSEAD's Dipak Jain, firmly champions the emerging recognition of the multi-faceted value of a 'purpose driven' executive.[5] Carroll and Shabana (2010) review the value of CSR from a business perspective. Well-known financial scholar, SSRN Chairman, and Harvard Professor Emeritus Michael Jensen posits that: "A firm cannot maximise value if it ignores the interests of its stakeholders." (Jensen 2001) Stakeholders, he states, include not only financial claimants or customers, but also employees, communities, governmental, and the environment.

The business environment is shifting. Vague concepts of sustainability and CSR are being replaced by better defined and more transparent standards that consumers increasingly expect of the brands they choose. Measuring and reporting are increasingly valuable. Various multi-stakeholder initiatives reinforce the principle that corporations must be transparent. This is especially relevant in terms of their support for human rights—including those related to discrimination, labour, water, and food. Such initiatives include: the UN Global Compact, OECD Guidelines on Multinational Enterprises, the ILO Tripartite Declaration on Multinational Enterprises and Social Policy, the European Union Strategy for Corporate Social Responsibility, the International Finance Corporation's Performance Standards, and the UN Guiding Principles on Business and Human Rights (WBCSD 2010).

More firms, particularly dynamic multinationals are now exploring and employing VSS. The opportunity is particularly interesting for those first movers that want to capture the benefit of such market positioning which appeals to the 'heart space' of consumers and can contribute to brand loyalty in unique ways.

By 2009, Mars, one of the world's largest privately-held food companies, announced that it would source 100 % VSS certified cocoa by 2020. In 2010, global giant Unilever launched an innovative and public 10-year Sustainable Living Plan to address environmental, social and economic factors and to halve the negative

---

[4] http://www.guardian.co.uk/sustainable-business/unilever-ceo-paul-polman-interview.

[5] Personal communications between Giovannucci and Jain beginning January 2011.

environmental impacts of their products. Companies—especially brand-owning firms—are increasingly held responsible for the social and environmental performance of their supply chains (BBC News 2010; Muradian and Pelupessy 2005). Working conditions and environmentally unfriendly practices are among the major issues facing these companies. While many are trusted for their products or services, very few are trusted to be socially fair or to be good stewards of our natural resources. Meanwhile, governments, whether trusted or not, are barely able to keep up with the fast-paced change of the business world. In their World Bank report, Fox et al. (2002) note that governments—whose clear primary purpose is the common good—nevertheless struggle to effectively find policy options that foster productive and responsible corporate activity. Increasingly prominent social concerns mean that corporations are now being called upon to be more conscious of their impacts. The calls are coming from increasingly conscious consumers who have ever more information available to them (Fig. 24.1).

For many firms the interest in VSS goes beyond social legitimacy to addressing even more demanding challenges, affecting the viability of the company itself. Four relatively recent phenomena are influencing the increasing adoption of VSS by all sorts of firms[6]:

1. A *consumer environment* characterised by strong interest in personal health and concern about the social and environmental conditions in the place of origin.
2. A concentrated and more competitive *business environment* requiring new methods of differentiation, more agile reputational risk management, and more sophisticated supply chain management where greater efficiencies in costs and logistics are only the beginning.
3. A *regulatory environment* with new and import rules and greater food safety requirements such as traceability while also struggling to keep up with fast-moving global trade developments.
4. Social *communications advances* that are global in scale, exposing corporations and individuals to greater levels of scrutiny that can alter reputations in a matter of hours and even offer tangible proof of civil or criminal responsibility regarding food safety, labour violations, and environmental impacts.

The combined pressure from consumers and civil or non-governmental organisations (NGOs) in tandem with increased corporate awareness led to the evolution of both the term and the practice of Corporate Social Responsibility (CSR). Dahlsrud in his review of CSR definitions (2008) notes that while these are generally congruent there is ample confusion in terms of how they are applied in specific contexts.

What was, until the late 1990s, a merely philanthropic corporate expression has today progressed to the integration of social and environmental 'good practices' into day-to-day business operations (Porter et al. 2007). While laudable from a

[6]Based on similar ideas elaborated in Giovannucci (2008) and also Giovannucci and Purcell (2008).

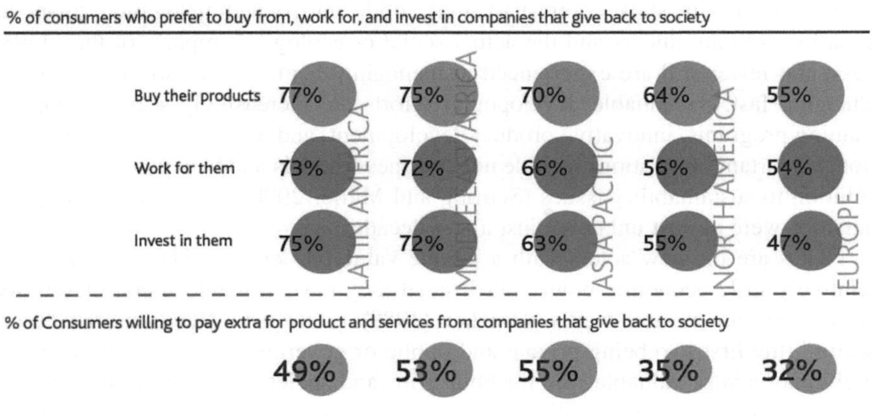

Fig. 24.1 Consumers' preference for companies that "give back to society" (*Source*: Nielsen Global Survey of Corporate Citizenship 2012)

humanistic perspective, there is nevertheless little evidence that this charitable approach makes much of a difference in the long term sustainability of a business or the environment in which it operates—and few firms are yet good at doing this (Economist 2008). This is because the contributions are often short-lived, relatively modest to the scale of the challenges, and do little to alter the actual business operations where corporate impact can be greatest. On this latter point, altering corporate messaging is easy but companies' incentive systems have often not been sufficiently oriented to encourage or reward desired behaviour (Lorne and Dilling 2012). In the work of the Committee on Sustainability Assessment (COSA) in many developing countries, this disconnect between the intention of senior management and the reality of distant line operations is often evident regardless of the firm's size or CSR orientation.

Addressing sustainability as an integral part of business means treating sustainability as a core operational issue that is no different than inventory, cycle time, cost of materials, and logistics. It means going beyond saying that a firm has "sustainability in its corporate DNA" to actually reworking a supply chain's structures and incentives so that it actually can behave in a socially and environmentally responsible manner. The work of Andersen and Skjoett-Larsen (2009) looking at one of the world's most successful supply chains states that integration of all staff is critical for the success of any CSR approach. Strategically involving active inputs from both management and workers in a firm is not a new concept in corporate theory and evidence for its value dates back several decades to the work of business pundits W. Edwards Deming and Peter Drucker (Deming, 2000; Drucker, 1989). Drucker is noted for his related comment about the difficulty of making tough choices: "Management is doing things right; leadership is doing the right things". Pronouncements from top management are necessary but hardly sufficient as prescriptions to generate a sustainable enterprise.

Most firms—in stark contrast to their subtle understanding of their financial situation—do not understand the actual social or ecological impacts of their business, and fewer still are experienced in managing them. For leading firms, this is changing fast. Sustainable development efforts are increasingly seen in supplier training programs, innovative product development, and new logistics for distribution. Important innovations include new business models and new partnerships as a solution to sustainability issues (Seuring and Müller 2008). Corporate-NGO partnerships were almost unknown just a few decades ago.

NGOs are the new actors with a unique value proposition. They tend to have relatively little economic power but instead they have valuable social credibility among consumers and media. Drucker (1989) referred to NGOs as the "third sector" (the first two being private and public or government) that would increasingly play a very valuable role for both firms and society. It is NGOs that fill the vacuum of trust as representatives of credible social and ethical positions. Because of this, NGOs are serving to create a certain level of social legitimacy for corporations and have increasingly become as a trusted conduit between firms and the perceived desires of the individual or community.

Via different forms of public–private partnerships NGOs have evolved VSS to provide the normative framework that corporations use for social legitimacy and essentially constitute a social contract whose compliance is assured by independent certification (Giovannucci and Ponte 2005). Meanwhile the state, rather than being directly involved, tends to focus on basic guarantees or regulations such as contract rules and food safety. There are however, many more roles that can be played by the public sector to facilitate and encourage CSR as a powerful complementary tool for public policy. Fox et al. in their World Bank report (2002) present an array of viable options for more active and supportive public sector participation in this process.

24.2.1 The Objectives of Standards and How Firms Use Them

The many different VSS, although often lumped together, are not at all alike. Yet, they do as a whole tend to deal with the areas not functionally addressed by most firms and global trading structures. Figure 24.2 shows the average number and type of sustainability criteria covered by some of the fastest-growing and more prevalent sustainability standards: Fairtrade International and Fair Trade USA, Forest Stewardship Council, Rainforest Alliance and UTZ Certified.

Mayer and Gereffi (2010) and Jaffee et al. (2011) are among the many scholars recently reporting on the proliferation of businesses adopting standards and codes of conduct and the array of relationships that they have with standards. It is likely true that many firms have a simply transactional relationship with VSS in which they purchase products that are certified to a particular standard in order to fulfil a procurement necessity. These are often followers in the CSR arena. They can

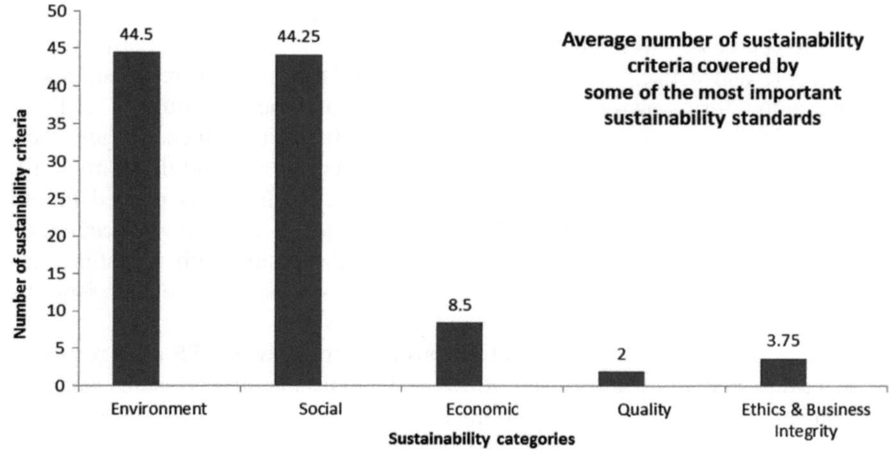

Fig. 24.2 Sustainability criteria covered by major VSS [*Source*: Association Materials Management, Purchasing and Logistics (2010) Standards Map, ITC]

nevertheless serve to influence standards, especially if they are large. The Wal-Mart choice to have organic versions of their most popular products resulted in the considerably greater availability of organic cereals from major mainstream suppliers that had not invested in such VSS prior to the 2006 Wal-Mart announcement.

There is often a dynamic tension between firms and VSS. Some standards require change in the firm's practices or costs and may not meet all of a firm's needs. Firms, particularly large ones, can try to influence VSS and some go so far as to create their own standards. Yet some firms engage VSS as functional tools of change and integrate them to become a *de facto* part of the firm's CSR 'strategy'. In recent years, a number of successful companies have evolved a range of ever more intimate and interesting relationships with VSS.

Most of the VSS were not designed as corporate tools, and their integration into procurement and corporate supply chains can be challenging. The VSS organisations and many of the businesses they work with have fundamentally different origins, different values or intentions, and different operating models.[7] They also may have varying types and levels of experience in particular areas as well as very different levels of resources to pursue their objectives and to collaborate. Fortunately, many VSS are built on working partnership models that can open space for cooperation. Some even learn from the firms they partner with.

In some cases, efforts to integrate VSS into business have led to unexpected outcomes. When Starbucks declared their position as the leading buyer of Fairtrade coffee in the early 2000s, it was attacked by consumer and student activists who

[7] Notable exceptions exist including the certified B-Corporations that use the power of business to solve social and environmental problems and meet high levels of sustainability criteria. See: www.bcorporation.net.

accused it of exploiting the Fairtrade name while only a very small percentage of its total coffee purchases were Fairtrade certified. Senior executives were surprised by the outcome and a likely result could be the firm's subsequent reticence about making claims for its own private standard: Coffee and Farmer Equity (C.A.F.E.) practices. Nestle, the world's largest food company faced its own challenges when it launched a small test of a Fairtrade certified product in England that garnered a mix of negative and positive reviews. On one side, the firm was praised for its efforts and for venturing to support a VSS while, on the other side, it was accused of creating only window dressing and pandering to the public with a gesture that actually represented only a tiny fraction of its business. Clearly, there are lessons to be learned about the relationship between CSR and VSS.

Mayer and Gereffi (2010) note that the push to engage with VSS are a response to increasing and more globalised social and environmental pressures and the inadequacy of governmental institutions in addressing these pressures. However, there are clear limits to what VSS can be expected to accomplish. They hypothesise that the effectiveness of such forms of private governance depends on four factors:

1. The structure of the value chain in which production takes place;
2. The extent to which demand for a firm's products relies on its brand identity;
3. The possibilities for collective action by consumers, workers, or other activists;
4. The extent to which commercial interests of lead firms align with social and environmental concerns.

Mayer and Gereffi's hypotheses suggest that VSS as a form of private governance will only flourish in certain circumstances and need to reflect the interests of multiple stakeholders to succeed.

Firms that are practice leaders in CSR tend to take an active stance in regard to their supply chains and elect to partner with standards to evolve their procurement and even leverage standards to evolve their corporate persona. The world's largest banana brand turned around a dismal public reputation and low profitability partly as a result of its close partnership with Rainforest Alliance and adoption of its sustainability standards (Taylor and Scharlin 2004).

Other firms have moved in the same direction. A number of large brands such as Sara Lee, Mars, and Tchibo and global retailers such as Ahold, IKEA, and Rewe work closely with UTZ Certified and have all significantly grown their business with the UTZ Certified label from year to year especially in coffee, cocoa and tea.

Access to higher value markets is one reason for producers to participate, but the requirements can be daunting and even constitute barriers to entry for smaller and poor producers. Yet, rates of expansion among farmers continue to be remarkable. The Dutch Sustainable Trade Initiative (IDH) expects that 22 % of total worldwide exported tea will be certified by 2015. Considering that in 2007 about 1 % was certified, the growth is impressive.

The approaches adopted by firms tend to depend on whether they are brand owners, consumer-facing or in the business-to-business markets. Some take bold initiatives. A leading U.S. brand, Ben and Jerry's Ice Cream, has recently overhauled its procurement to align the global sourcing of more than 3,000 ingredients with the company's mission and core values. The firm's Values-Led Sourcing initiative includes a commitment to source Fairtrade certified ingredients for its entire global flavour portfolio by the end of 2013 (Alvarez et al. 2011).

Consumer products giant Unilever's collaboration with Rainforest Alliance was the product of its decision to invest in its current suppliers' capacity rather than seeking new sources. As a major buyer of tea, it actively engaged local NGO partners to train small and large scale tea farmers and supported them to become Rainforest Alliance certified, Unilever thus established a measure of supply security and likely a first mover advantage in tea. Competitors such as Tata, Tetley, and Twinning's followed and soon after also started purchasing and selling certified tea.

Two of the world's leading chocolate brands have made commitments to fully source from suppliers meeting the VSS of global NGOs (Cadbury with Fair Trade and Mars with several VSS[8]). Similar examples of large-scale corporate commitments include: Mondelez's claim to sustainably source all its European coffee by 2015,[9] Nestles' commitment to only source sustainable palm oil and Unilever's promise to source 100 % of agricultural raw materials sustainably by 2020.[10]

Some initiatives go beyond the firm level. Efforts such as the Keystone Field to Market, SAI Platform, and Sustainable Food Lab are platforms generated primarily by the private sector taking a strategic opportunity approach to VSS and the challenges of sustainability as a pre-competitive issue and working to advance industry-wide behaviour in a collaborative way. A survey of 254 senior leaders in procurement and supply chain management indicates the current and emerging rationale for their adoption of sustainability criteria or VSS in their procurement (see Fig. 24.3). What the VSS bring to companies and brands is not only some assurance of functional benefits such as traceability and better practices but also the goodwill of a public that is increasingly aware of such standards and that generally trusts the NGOs that manage them.

24.2.2 The Distinction Between Public and Private or Corporate Standards

Some corporations have elected to create their own standards either independently or as part of associations. A number of labels are propagated by individual firms and it is not clear whether they have an impact in terms of global trade since they are often internal standards or sometimes they can be primarily marketing-oriented efforts. Supermarkets often create their own labels as a distinctive communication to their consumers. Association or industry-wide standards have broader effects. Most are business-to-business standards such as GlobalG.A.P., the Round Table for

---

[8] http://www.mars.com/global/press-center/press-list/news-releases.aspx?SiteId=94&Id=1482.

[9] http://www.mondelezinternational.com/DeliciousWorld/sustainability/coffeemadehappy.aspx.

[10] http://www.unilever.co.uk/sustainable-living/sustainablesourcing/.

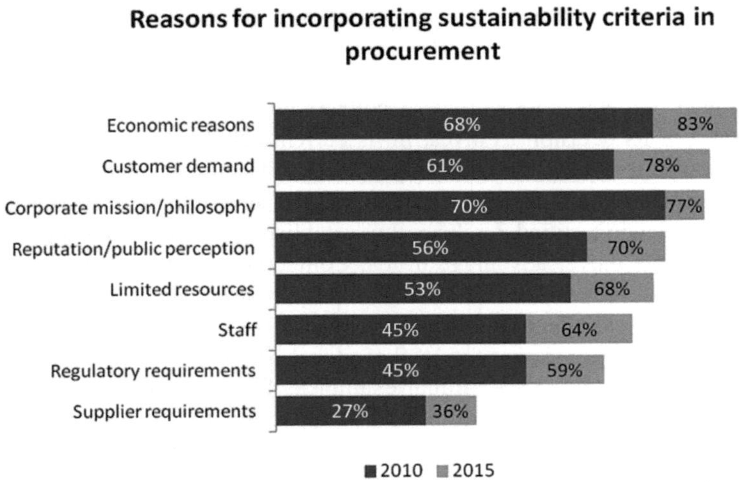

Fig. 24.3 Reasons for sustainability criteria in procurement (*Source*: Roland Berger Strategy Consultants 2010)

Sustainable Palm Oil (RSPO), and Ethical Tea Partnership (ETP) that raise awareness and establish minimum guidelines. GlobalG.A.P. has become so widely used that it is now routinely incorporated into other standards and in some sectors and markets e.g. fruit and vegetables to the EU, it is becoming a *de facto* business requirement for some segments of trade.

There are distinctions between consumer-facing VSS and B2B standards or codes of conduct. The latter are typically more concerned with quality, food safety, and traceability than with more comprehensive aspects of sustainability and they have not needed to prioritise transparency and independent audits. While they provide a useful base, most of the B2B standards are modest on social and environmental requirements when compared to the consumer-oriented standards and set the bar for compliance at a fairly low level.

Most, but certainly not all, standards and verification programmes that are established exclusively within the corporate arena are often excluded from discussions of VSS because they tend to differ from the salient values of VSS in several ways:

1. They are often imposed on producers and supply chains and rarely include the serious input of producers in their design;
2. The lack of independent oversight or third-party certification suggests that the private firms that control them can alter, dilute, or simply not fully apply the standard at their prerogative;
3. When lacking adequate support or remuneration for sustainable production practices, they can serve as significant barriers to entry for producers;
4. They are rarely transparent and if they lack accountability that engages consumers, they are limited as a market mechanism that drives sustainability.

Some firms do not avoid the temptation to launch their own standard. For those, it is often a cost with little measurable benefit. Even giants such as Wal-Mart, Nestle, Unilever, Kraft Foods and Mars have elected to not create such efforts. Their research suggests, and some have stated, that consumers do not want them to compete in this space and prefer them to align with a VSS as a more accepted arbiter of sustainability.

24.3 The Growth and Pervasiveness of Standards

In 1967, in a remote area of Chiapas, Mexico the first VSS certification (Organic Demeter) by an independent third party was granted to a coffee farm. Organic is, by several measures, the grandfather of agricultural VSS (Giovannucci and Koekoek 2003). Fair Trade later emerged to be standardised in the 1980s and also began with coffee.[11] Today both of these seminal VSS are globally recognised multi-billion dollar segments that have spread to nearly every type of agricultural product from cocoa to cheese to cotton. By 2011, global sales of Fair Trade topped US$6 billion[12] and Organic is estimated to have topped US$60 billion, both more than tripling their value in a single decade.

The 1990s and early 2000s saw the seeding of several new VSS for food and agriculture including the standards associated with the Rainforest Alliance and with UTZ Certified[13] that offered related objectives but somewhat different sustainability theories. In the most traded commodities these have provided arguably less challenging requirements in some areas and more business-friendly approaches. As a result, their growth rates have skyrocketed particularly as large mainstream firms engage more actively with them. While no VSS approaches the global range of products, depth of market awareness, or global recognition that Organic has achieved, both Rainforest Alliance and UTZ Certified are growing much faster than Organic or Fair Trade.

Among the major global food firms in terms of consumer brands, there has been a remarkable consensus on the commitment to certified products with notable exceptions. Kraft Foods has pledged that all of its coffee brands in Europe will use fully certified sustainable sourcing by 2015. In 2011, 28 % of the tea purchased for all Unilever brands was sourced from Rainforest Alliance Certified farms and it plans to have 100 % of its tea certified by 2020. Mars has made a public commitment to certifiably source 100 % of the cocoa, coffee and tea for all of its global brands by 2020. Hershey has made a similar commitment. Starbucks has about

[11] First with the Max Havelaar label and a more formalised Fair Trade system launched in the Netherlands in 1988 but had been functioning informally, as had organics, for decades prior.

[12] Reuters Article accessed Nov 1, 2012 online: http://uk.reuters.com/article/2012/07/16/uk-fairtrade-softs-idUKBRE86F19P20120716.

[13] Originally started in coffee as Utz Kapeh.

90 % of its coffee certified to its own standard. Chiquita banana, the world's largest banana brand, is fully certified by a VSS. Cadbury's top selling confectionary products are also fully VSS certified. IDH has committed to having fruit and vegetable imports into the Netherlands being 100 % certified by 2020.

There are indications of further uptake at the mass-market levels. When Wal-Mart, the world's largest retailer demanded organic versions of their most popular products, major suppliers first said it would be extremely difficult but most had them on the shelves within 12 months. Global retail food-service giant McDonalds is already applying better standards to its poultry supply chains and exploring the effects of VSS for some of its liquid products, particularly in its fast-growing coffee business.

Coffee has been the leading commodity to apply different VSS. Trend indications are also coming from different products such as tea (noted above), seafood, and cotton. VSS certification for coffee, the world's most valuable export crop, and for bananas, the most important fruit in global trade have both seen substantial growth in the past decade and these multi-billion dollar markets expect similar tendencies in the future (see Fig. 24.4).

An emerging trend is the formulation of new national sustainability standards such as China's Green Food, Indonesia Sustainable Palm Oil, Brazil's Certifica Minas Café, and the Sustainability Initiative of South Africa (SIZA). These are now emerging as local alternatives but because these domestic standards tend to be less restrictive and less credible to markets, they are not commonly useful for international trade. They may have some relevance for national domestic markets especially in light of an increasing sentiment, particularly, but not only, within the BRICS (Brazil, Russia, India, China and South Africa) countries that some international standards may be too expensive or burdensome to implement or lack certain domestic market relevance. Perhaps the most common complaint is that the benefit of international standards accrues to brands and to traders but not to producers themselves. National standards can be a step toward international VSS, but if they do not serve to compensate producers or improve their conditions, they may be imposing yet another layer of burden on farmers.

24.4 Do Standards Serve the Firm's CSR Objectives?

24.4.1 It Is Important to Have Objective Assessment of VSS

While there are a number of guidelines or frameworks for social accounting, environmental reporting, and even 'auditing', most are self-reported and only a few use independently verified measures to ensure clarity or comparability.[14] Until

[14] For example: ISEAL Alliance Impacts Code, AccountAbility AA1000, Fair Labour Association Workplace Code of Conduct, Fair Wear Foundation, Global Reporting Initiative, Carbon

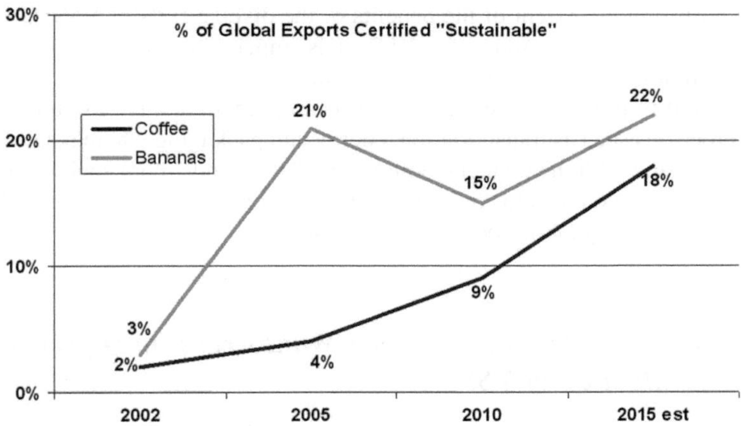

Fig. 24.4 Growth of sustainability certifications for two major export crops [*Source*: Daniele Giovannucci, for coffee. For banana: FAO, COMTRADE, Rainforest Alliance, ACP-EU Technical Centre for Agricultural and Rural Cooperation, FLO, Agritrade.cta, Forschungsinstitut für biologischen Landbau (FiBL)]. N.B. percent of exports (green coffee and bananas) certified by independent third parties as complying with VSS. Estimates for 2015 are not linear projections from the current data; instead, they reflect calculations based on the stated commitments and expectations of leading buyer firms or their representatives and related trade experts

recently, there have been no reliable and globally comparable metrics to understand the *actual* impacts of the VSS as distinct from their stated objectives.

> In an increasingly performance-oriented society, metrics matter. What we measure affects what we do. If we have the wrong metrics, we will strive for the wrong things. - *Mismeasuring Our Lives* (2010)
> *by Joseph Stiglitz, Amartya Sen, and Jean-Paul Fitoussi*

24.4.2 VSS Align with CSR, but Are More Useful When Understood Objectively

The evidence and experience to date indicates that VSS align well with CSR concepts. They can serve corporations as an already formulated and pre-vetted approach. Market-driven solutions are promoted by many as the ideal ways to drive sustainable practices and VSS or certifications have become the mechanism of choice (Hartmann 2011). However, little is known about the actual impacts of VSS, including the effects on productivity and risk. There is still little scientific literature on how effective VSS are as a tool to further a firm's CSR objectives in the food and agriculture sector. Recently, concerns have begun to emerge about the direct and

Disclosure Project, SA8000 (Social Accountability International), ISO 14000 and 23000, and United Nations Global Compact.

indirect costs and the extent of the benefits of the diverse sorts of VSS. Given the scope and scale of the markets for VSS, it is imperative that firms using them comprehend how they work and how to use them.

Sustainability in agriculture may evolve from predominantly environmentally-related processes such as transportation, energy and packaging where results have a clear relationship with the financial bottom line[15] to include more socially-oriented choices whose economic value may at the moment appear less obvious especially in a world with growing labour pools and less stability.

24.4.3 Firms and Public Agencies Want Access to Objective Evaluation of VSS

Having sound information on impacts is becoming a priority for firms and also for investors. JPMorgan's Impact Investor Survey (Saltuk et al. 2013) tracks a fast-growing business segment by polling investors who committed US$8 billion to impact investments in 2012. The majority of respondents report that they seek market rate financial returns but want to have positive social or environmental impact as well. The survey found that 70 % of respondents hold that standardised impact metrics are 'important' or 'very important' to the development of their industry.

The question of impacts is a significant one for governments and policymakers as well because these standards are not only part of fast-growing, multi-billion dollar market segments, they are also being adopted by millions of producers. Until recently, most of the publicly available measurements of the effects of the VSS were either very specific case studies of one point in time or anecdotal assessments. The resulting lack of time-series data or data that is comparable across countries or regions allows only a fragmented understanding of these VSS or certifications and an inadequate evaluation of their impacts. This lack of clarity hinders the ability to move efficiently toward sustainability.

24.4.4 The Importance of Common International Standards

In the ever more complex situations of global production and trade, good business runs on good data. It follows that succeeding at sustainability requires the same: an understanding of not only costs and benefits but also of the results or of particular investment or operational choices. To effectively improve sustainability, we need to understand it much better than we currently do. The answer does not lie only in scientific experimentation. Like any successful business, effective sustainability

[15] Association of Materials Management, Purchasing and Logistics (2010).

relies on the day-to-day application of sound management feedback loops. To extend the business analogy: having clear and consistent standards (i.e. International Financial Reporting Standards) is vital for efficient business controls. The same is true regarding a business's CSR practices, but the challenge is that the field of sustainability has not had clearly defined metrics for its intrinsic social, economic, and environmental dimensions. That is quickly changing.

24.4.5 COSA and ITC Provide Tools for Objective Evaluation and Understanding

Two complementary and mutually supportive initiatives are contributing a critical new and transparent understanding of sustainability. The global partnerships of the Committee on Sustainability Assessment (COSA) have developed innovative ways to understand the myriad of possible impacts to sustainability at the ground level with producers, organisations and communities. The International Trade Centre's Standards Map is part of its pioneering initiative: Trade for Sustainable Development (T4SD) and provides a unique way to understand the distinct features of the most important VSS on a single platform.

24.4.6 ITC's Tools

With the proliferation of sustainability labels—436 available in 2011—it is important for both consumers and firms to distinguish what is trustworthy and to have access to neutral information for understanding them. The T4SD's Standards Map provides independent and credible information on the relative features, requirements and compliance policies of the most important VSS as well as audit protocols and retailer codes of conduct.[16] T4SD is also developing diagnostic and self-assessment tools that can help producers and companies make better decisions on the implementation of standards.

Box 24.2: Understanding the Basics of VSS: A New Map

Standards Map is the new International Trade Centre market analysis tool on voluntary sustainability standards. It provides information on more than

(continued)

[16] The that houses the T4SD and the Standards Map initiatives is an agency created to provide independent technical advisory services under the auspices of the United Nations and the World Trade Organization.

Box 24.2 (continued)

130 standards and allows users to compare VSS on diverse social, environmental, economic, and quality criteria (among others). The tool offers geo-graphic and product-related scopes, as well as up-to-date coverage of compliance policies and requirements for implementation.

www.standardsmap.org

24.4.7 COSA's Tools

The new tools developed by the Committee on Sustainability Assessment complement the Standards Map with a standardised approach to getting information about the actual effects of such standards—going beyond the written or paper standard to ascertain what happens in practice. The ability to scientifically measure the results of VSS—and the result of any approaches to improve sustainability—paves the way for better management so as to achieve corporate objectives as well as the wider societal or ecological benefits to which the VSS are intrinsically dedicated.

Beginning in 2006, the Committee on Sustainability Assessment, a non-profit consortium, set out to alter the knowledge gap by formulating a consistent and reliable metrology based on exhaustive scientific review of methods and multi-stakeholder consensus on the most important key indicators to measure. The result is a set of neutral, state-of-the-art assessment tools to generate science-based information on the social, economic and environmental impacts of agricultural practices. These are captured year to year and because the methods and indicators are standardised, the resulting information can, for the first time, be compared across time and borders. As COSA partner institutions add thousands of data sets each year, they will be able to more acutely discern trends and patterns as well as determine what approaches work for sustainability and which do not.

Appropriate to its public beginnings under the umbrella of the International Institute for Sustainable Development and the United Nations Conference on Trade and Development,[17] COSA gleans expert input from a global array of scientists, producer groups, private firms, NGOs, and development agencies. Ensuring balance among the diverse needs of stakeholders has gained it widespread acceptance and recognition. COSA focuses on developing countries and has already been tested and applied in 12 countries (Fig. 24.5).

[17] COSA and its projects have had the support of multiple research and development agencies; since 2009 these include the Swiss State Secretariat for Economic Cooperation (SECO), International Institute for Sustainable Development, Ford Foundation, ENTWINED International Research Consortium, International Finance Corporation, NORAD, Solidaridad, and the InterAmerican Development Bank Multilateral Investment Fund (list is not complete).

Fig. 24.5 Map of COSA operations 2010–2012

24.4.8 COSA Indicator Groups

COSA's consistent methods and comparable metrics facilitate more structured learning and enhance the ability to test almost any investment or project interventions. COSA's broad set of more than 130 indicators offers diverse insights and access to new ways of understanding the impacts of various efforts. These can be used selectively as needed. For example, one set of indicators can provide total costs and net income; another set can identify basic risk factors, while another can offer insight into training and gender. COSA indicators help to discern efficiencies such as the relative use of labour for the net income achieved or the amount of inputs such as fertilisers or pesticides used relative to yields. The correlations to vital factors such as food security, education levels and good governance are also available in order to understand the less direct effects of the selected practices (including VSS practices or any other approach). The main categories, within which multiple specific indicators exist, are shown in Table 24.1.

24.4.9 How COSA Surmounts the Challenge of Reliable Data in Developing Countries

Getting consistent data is a challenge and COSA believes that this challenge should be met at two distinct levels. First, the most basic data can be gathered by companies or co-operatives in the course of their work and used for real-time decision making. This is a basic monitoring function and is no different than basic bookkeeping functions for those who aspire to be more sustainable. Second, each country needs the basic scientific capacity to do occasional in-depth impact assessment (similar to a financial audit) to improve or refine what is gathered at the

Table 24.1 Major categories of COSA standardised indicators

Economic	Social	Environmental
Revenue	Health and safety	Conservation
Costs	Living conditions	Water Quantity and Quality
Income	Labor rights	Waste management
Diversification	Education	Input management
Information	Gender	Soil health
Credit	Food security	Biodiversity
Volatility	Participation	Carbon sequestration
Vulnerability	Transparency	Climate risk mitigation
Business development	Investing in capacity	
Differentiation	Social Situation	
Efficiency		
Governance		
Producer economic, social, and environmental perception		

business level and thus add to the quality and relevance of the data and also to the scientific understanding of the country's sustainability. With this critical function in place, scaling up can occur with much more confidence. COSA integrates both of these functions into its operations. It trains firms and farmer groups to conduct credible monitoring using standard tools and it also partners with top-notch institutions in developing countries so they can achieve world-class capacity to measure.

As a non-profit research organisation, COSA fosters such global networks to provide reliable information permitting stakeholders to make better and more informed choices so that they can be drivers of sustainability. Monitoring and Impact Assessment are therefore used as tools for learning that enable better decisions in the service of economic, social and environmental sustainability. The resulting aggregated information is then collated and will be publicly available via COSA's own network as well as via its partnerships with agencies such as the International Trade Centre.

24.4.10 Using Data for Better Decision Making and CSR Strategy

The early data is already providing interesting results and some examples are outlined below. It is important to note that these are preliminary and intended primarily to be illustrative of the range of knowledge available from the COSA efforts, rather than to be interpreted as concrete or definitive conclusions. COSA work is still at the beginning stages, and as data and efforts expand, this increased knowledge will allow for much more reliable conclusions.

For example, COSA data will easily offer a broad sectoral understanding as illustrated in the work of COSA's Colombian partner CRECE. They wanted to

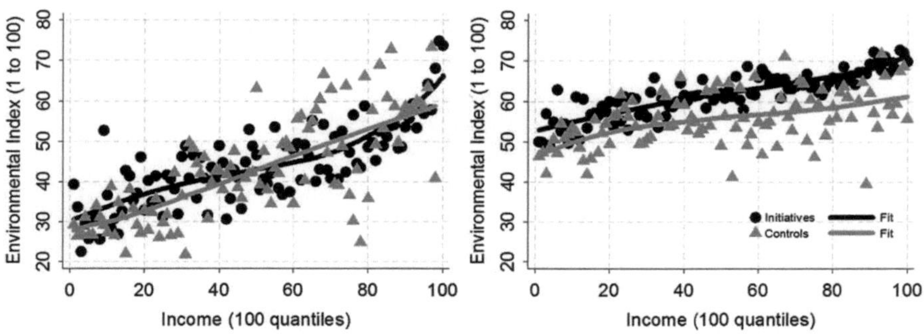

Fig. 24.6 Relating environmental indicators to income, years 1 and 4 (*Source*: CRECE and COSA)

know if certification really produced significantly better environmental results and whether good environmental results were correlated to better income; in other words, if improved environmental performance on the farm was related to simply having greater earnings. For several years CRECE collected COSA information from thousands of producers applying seven different VSS. CRECE calculated the outcomes for a basket of COSA environmental indicators and compared those to a matched control group (grey triangles in Fig. 24.6) that did not use any VSS and then mapped the results to the income levels of the 3,298 producers being measured. The results, indicated in Fig. 24.6, strongly suggest that the VSS (dark circles) develop significant environmental benefits and that higher income does not necessarily correlate to better environmental outcomes. With such an understanding, investors and policymakers would be better informed to select or design projects and investments. For example, they would be able to better consider that developmental interventions interested in positive environmental results should not expect to achieve those simply by addressing economic issue or increasing incomes.

Other data, when gathered in consistent manner, also begin to produce comparable results that can be tracked from year to year. Figure 24.7 shows an important social aspect of health and safety as measured by the difference in the use of one or more good safety practices for the application of pesticides, herbicides, herbicides and other agrochemicals among VSS certified coffee farmers in Colombia and VSS certified cocoa farmers in Cote d'Ivoire. The graph shows the percent difference from the beginning of certification (Y1) when compared to very similar conventional farmers in the same respective regions. The Colombian farmers started with a higher level of achievement and show less relative improvement but both clearly suggest substantive change related to the certification process. In addition, the data shows the change over time and Y2 notes the percentage difference of applying one or more methods seen in certified producers compared to conventional producers (control). The consistency, especially if repeated in other assessments that COSA and its partners perform could lead to a useful understanding of the important impacts of such VSS, even across different cropping systems.

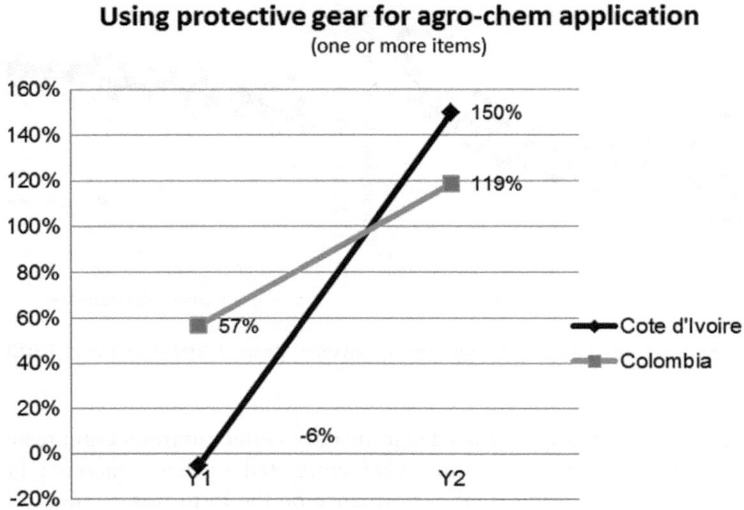

Fig. 24.7 Farms using protective gear for agrochemical applications (*Source*: COSA and CRECE)

Figures 24.7 and 24.8 are simple examples of what we could learn about the application of specific social or environmental measures across countries when the measurements and methods are consistent. Both indicate the percentage difference between VSS farmers and similar but conventional (not VSS certified) farmers used as control groups in each country.

COSA research is also results-oriented and can provide a sort of cost-benefit analysis. It can observe and report different types of outcomes simultaneously so that users can determine the appropriate level of trade-offs. This can be used to evaluate projects, assess the effectiveness of technical support, or compare different investments. It can even be used to calculate all the key costs and reveal actual producer net income—not just price premiums or revenue. Accordingly, such information can serve not only farmers and their organisations but also corporate managers who want to achieve better results with their farmer-suppliers. Clearly, this would also serve governments or development agencies that need to understand what works from country to country when it is clear that the same approach cannot always be applied elsewhere in a cookie-cutter approach. Figure 24.8 indicates how the average net income of producers applying a specific VSS (organic in this case) can be quite different when applied in different countries or conditions. One conclusion emerging from the initial COSA efforts is that it is difficult to generalise about results such as income advantages for VSS given the widely divergent conditions and contexts of developing country agriculture. By having consistent measures, we can assess the differences.

It is worth noting that these results are indicative of specific crops and regions. Because, relative to the hundreds of thousands (even millions) producers in a

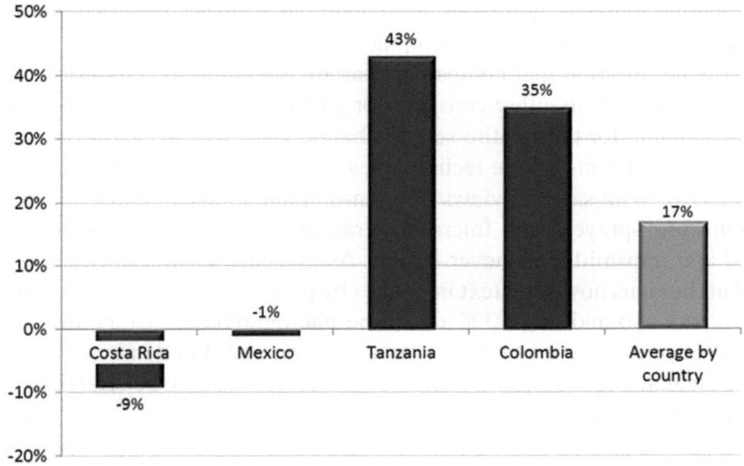

Fig. 24.8 Net income comparison of certified organic coffee farmers vs. conventional coffee farmers [*Source*: Committee on Sustainability Assessment (COSA)]

country, the sampling (n) for much of the data presented here is a very modest proportion. The findings should not be extrapolated to make any firm assumptions about a country as a whole or about any particular VSS on the whole since conditions and applications can affect results. Nevertheless, one begins to see the enormous potential of the information coming from consistent methods and growing in scale over the coming years.

24.5 Conclusions and Recommendations

We are witnessing a co-evolution of CSR and VSS as they help serve each other's objectives. Both are relatively recent phenomena and both are still at the early stages of finding the means to be truly effective. VSS are demonstrating the potential to be very useful market-oriented mechanisms for achieving a number of the CSR goals that firms pursue; these range from traceability to safe working conditions to environmental stewardship. They are, however, far from perfect.

The VSS do not always meet their stated objectives. They can sometimes impose substantial requirements on producers for only modest returns. VSS, quite simply, are not the single complete answer to sustainability; instead, they are a useful tool. It must be remembered that the VSS are not well-resourced, globally-ubiquitous standards regulators. Instead they are typically well-meaning and activism-oriented NGOs that simply cannot be expected to resolve all of the complex issues of sustainability for entire supply chains. Many have a few dozen or a few hundred staff to cover global operations for many products and typically receive modest sums considering the scope of their work. As market mechanisms, they may nevertheless be as effective as large sums of development aid.

There is no question that companies that are operating in responsible environments, particularly with either consumer or government influence, will be increasingly accountable for their entire supply chains. They will be held responsible via new and sometimes disruptive technologies that are difficult to manage or control (Gorbis 2013). With satellite views of palm-oil plantations, internet-linked sensors on cotton crop sprayers and micro-cameras in cocoa farms, everything can be recorded and transmitted as never before. As evidence of this shift, consider that 60 % of all humans now send text messages by phone systems (mobile) that did not exist 25 years ago and that 10 % of all the photographs in history were taken in 2011.[18] We are in the era of hyper-communication and 'big data'.

There are many approaches to sustainability and in order to be effective, we need to better understand what does and what does not work. We can already see successful pilot approaches that are offering useful insights and practical tools for managing and advancing sustainability. The leading research institutions, firms, and development agencies that gather under COSA are working toward performance management that is much more agile and effective by being integrated into smart self-monitoring systems at the ground level that are themselves linked to reliable means of impact assessment or confirmation.

In the coming decades, the social and environmental processes that companies must manage will come to be as understandable as the economic processes they now manage, if not as controllable. Information, in many forms, but increasingly directly from the source, will help producers and firms to more efficiently advance their sustainability. The need for comparable credible data that can be verified will be fundamental. It will aid our understanding of sustainability and of how VSS can play an active role in CSR.

These combined interests will be best served by their articulation into a more seamless understanding of the landscape-scale complexity of our production systems. The advent of an age of clarity—where vast information is vetted—will help us to better understand the interdependence of resources, including human resources, and to better manage our systemic choices as companies and as a society.

References

Alvarez G, Malnight T, von Hagen O (2011) Ben & Jerry's: inside the pint – values-led sourcing and linked prosperity. IMD Business School, Lausanne

Andersen M, Skjoett-Larsen T (2009) Corporate social responsibility in global supply chains. Supply Chain Manage 14(2):75–86

[18] Fortune (2012), pp. 69–75.

Association of Materials Management, Purchasing and Logistics (2010) Sustainability as a competitive factor. GDP 4, vol 1 (original in German: Bundesverband Materialwirtschaft, Einkauf und Logistik (2010): Nachhaltigkeit als Wettbewerbsfaktor, BIP 4, 2010, 1 Jahrgang) Published: 2010/03/24 Dutch Sustainable Trade Initiative. www.idhsustainabletrade.com

BBC News (2010) Cocoa's bitter child labour ties, 24 March 2010. http://news.bbc.co.uk/panorama/low/front_page/newsid_8583000/8583499.stm. Accessed 28 Aug 2013

Carroll AB, Shabana KM (2010) The business case for corporate social responsibility: a review of concepts, research and practice. Int J Manage Rev 12(1):85–105

Dahlsrud A (2008) How corporate social responsibility is defined: an analysis of 37 definitions. Corp Soc Responsib Environ Manage 15(1):1–13

Deming WE (2000) The new economics for industry, government, education, 2nd edn. MIT Press, Cambridge

Drucker PF (1989) What business can learn from nonprofits. Harv Bus Rev 67(4):88–93

Economist (2008) Special report: corporate social responsibility. The Economist Magazine, 17 January 2008

Fortune (2012) What data says about us. Asia-Pacific Edition, pp. 69–75, 24 September 2012

Fox T, Ward H, Howard B (2002) Public sector roles in strengthening corporate social responsibility: a baseline study. World Bank, Washington

Friedman M (1970) The social responsibility of business is to increase its profits. The New York Times Magazine, 13 September, 1970. Retrieved 14 Sept 2012

Fuller J, Jensen M (2010) Just say no to Wall Street: putting a stop to the earnings game. J Appl Corp Finance 22(1):59–63

Giovannucci D, Purcell T (2008) Standards and agricultural trade in Asia. Asian Development Bank Institute, Tokyo

Giovannucci D, Koekoek FJ (2003) The state of sustainable coffee: a study of twelve major markets. International Coffee Organization/International Institute for Sustainable Development/United Nations Conference on Trade and Development, London/Winnipeg/Geneva

Giovannucci D, Ponte S (2005) Standards as a new form of social contract? Sustainability initiatives in the coffee industry. Food Policy 30(3):284–301

Glasbergen P (2011) Mechanisms of private meta-governance: an analysis of global private governance for sustainable development. Int J Strateg Bus Alliances 2(3):189–206

Gorbis M (2013) The nature of the future: dispatches from the socialstructed world. Simon & Schuster, New York

Hartmann M (2011) Corporate social responsibility in the food sector. Eur Rev Agric Econ 38 (3):297–324

ITC – International Trade Centre (2013) Standards map. www.standardsmap.org. Accessed 28 Aug 2013

Ionescu-Somers A (2012) "Going, going...": the long term sustainability impacts of short term focus. Eur Bus Rev 84–88

Jaffee S, Henson S, Rios L (2011) Making the grade: smallholder farmers, emerging standards, and development assistance programs in Africa (a research program synthesis). World Bank, Washington

Jensen M (2001) Value maximization, stakeholder theory, and the corporate objective function. J Appl Corp Finance 14(3):8–21

Kolk A (2005) Corporate social responsibility in the coffee sector: the dynamics of MNC responses and code development. Eur Manage J 23(2):228–236

Lorne FT, Dilling P (2012) Creating values for sustainability: stakeholders engagement, incentive alignment, and value currency. Econ Res Int 2012:9 pages, Article ID 142910. doi:10.1155/2012/142910

Mayer F, Gereffi G (2010) Regulation and economic globalization: prospects and limits of private governance. Bus Polit 12(3):1–25

Muradian R, Pelupessy W (2005) Governing the coffee chain: the role of voluntary regulatory systems. World Dev 33(12):2029–2044

Nielsen (2012) The global, socially-conscious consumer. The Nielsen Company, Diemen
Porter ME, Kramer MR (2006) Strategy and society: the link between competitive advantage and corporate social responsibility. Harv Bus Rev 84(12):78–92
Porter M, Kramer M, Zadek S (2007) Redefining corporate social responsibility. Harv Bus Rev
Roland Berger Strategy Consultants (2010) BME – Bundesverband Materialwirtschaft, Einkauf und Logistik e. V. Online at: http://www.bme.de/fileadmin/bilder/BME_Roland_Berger_Sustainability.pdf
Saltuk Y, Bouri A, Mudaliar A, Pease M (2013) Perspectives on progress: the Impact Investor Survey. JPMorgan Chase & Co. and the Global Impact Investing Network, New York
Seuring S, Müller M (2008) From a literature review to a conceptual framework for sustainable supply chain management. J Clean Prod 16(15):1699–1710
Taylor JG, Scharlin P (2004) Smart alliance – how a global corporation and environmental activists transformed a tarnished brand. Yale University Press, New Haven
The guardian website: http://www.guardian.co.uk/sustainable-business/unilever-ceo-paul-polman-interview. Accessed 28 Aug 2013
World Business Council for Sustainable Development (WBCSD) (2010) Vision 2050. Online at: www.wbcsd.org

Chapter 25
Voluntary Standards and Approaches for Sustainable Communities

John Blewitt

25.1 Introduction

This chapter will explore the ways in which voluntary standards for sustainability can shape urban and community development. It is argued that a flexible approach to identifying, implementing and amending standards for sustainable cities and communities will optimise both democratic participation and social learning while recognising that technological, bureaucratic and other interventions, although important, are by no means sufficient for ensuring liveable and ecologically sensitive communities to grow. Two specific guidance schemes exemplify this contention: first, the guidance for community sustainable development (BS8904) recently published by the British Standards Institute; and second, the stages, 'ingredients' and principles that have emerged from within the Transition Movement in the UK and elsewhere. Moving on, Sect. 25.2 outlines the demographic, social, political as well as environment context of urban growth and development in the first half of this century. Section 25.3 will interrogate the concept of risk and resilience as it is applied and developed in sustainable community development making key reference to the debates on risk and vulnerability and the ways in which the community based Transition Movement practically engage with them. Section 25.4 examines the relationship between sustainable community and liveability, particularly as this pertains to health. Section 25.5 shows how the design, moral ownership and commitment to voluntary standards may secure a sense of obligation sometimes understood as being an informal social contract. Section 25.6 examines the ideas, suggestions and prescriptions of BS8904. Finally, Section 25.7 offers some tentative conclusions and recommendations the main one being, the necessity for standards to be a stimulus for continuing social learning and creativity in practice

J. Blewitt (✉)
Aston Business School, Aston University, Aston Triangle, Birmingham B4 7ET, UK
e-mail: J.D.BLEWITT@aston.ac.uk

C. Schmitz-Hoffmann et al. (eds.), *Voluntary Standard Systems*, Natural Resource Management in Transition 1, DOI 10.1007/978-3-642-35716-9_25,

avoiding the restrictions and limitations imposed of a managerialist culture that sees a standard as a box that must be ticked.

25.2 An Urbanising World

Over half the world's population live in urban environments and the United Nations anticipates that 9.3 billion people will inhabit this planet by 2050 with 6.3 billion people living in urban areas by 2050. Thus, urban areas are expected to absorb all the population growth in the next 40 years as well as drawing in many people from rural areas (United Nations 2012). Over 180,000 people join the planet's urban population every day. This demographic increase is placing immense strains on the planet's already stressed ecosystems and urban environments. The city, at all levels of spatial organisation, is under huge pressure to reduce its ecological footprint as well as being more socially and economically just (Rees and Wackernagel 1996). In fact, if cities and communities are not effectively sustainable by the end of this century, the prospects for improving the quality and standard of life for most people will be severely impaired. This means that cities will have to consume less, be more energy efficient and implement processes of renewable energy generation—bioenergy, solar, wind and arguably nuclear. Global demographics and population movement will also inevitably mean the construction of more cities and the expansion of existing ones as is currently the case in China. More land is being used for urban development every year. Having said that, there is also a counter trend of urban decline, the hollowing out of older industrial urban areas, which is likely to continue in some regions. Many spaces in rustbelt cities like Detroit may find themselves returned to agriculture and horticulture. Population density is also likely to increase in many places which bring both costs and benefits. Although dense compact cities are perceived as offering significant advantages and opportunities for realising economic and energy efficiency targets, increased urban density also produces other challenges. These may relate to water and food security, housing provision, the availability of meaningful work, crime prevention, transport and accessibility, educational opportunity and fair income distribution. The ongoing maintenance of peace and social harmony among differing groups who may entertain deep suspicions of, or dislike each other, also compromises achievement of social inclusion and social cohesion as overarching policy goals. On the other hand, a compact urban environment may lead to opportunities for enhanced sociability through community action, the development of social capital and economic enterprise. However, a distinction needs to be made between the idea of a compact city, one where proximity to amenities radiating from the urban core is privileged to a densely populated one that entails a simple concentration of dwelling units irrespective of proximity and accessibility to services and amenities. Densely populated cities, districts and neighbourhoods therefore offer both threats and opportunities to human social well-being and the potential for human society to fashion a non-exploitative and respectful relationship with the Earth.

At the heart of this concern for the wellbeing of urban dwellers is the notion of 'the right to the city'. This idea was first coined by the urban sociologist Henri Lefebvre in the late 1960s and further developed by urban planner Peter Marcuse, geographer David Harvey and many others (Brenner et al. 2012). It has since taken root in the public policy and the global human rights discourse. As the UN Habitat report, *State of the World's Cities 2010/2011*: *Bridging the Urban Divide* (UN Habitat 2008, p. 123) stated,

> The right to the city should not be viewed as a new legalistic instrument, but rather as an expression of the deep yearnings of urban dwellers to see their multiple human rights become more effective in urban areas. In this perspective, the right to the city serves as a bulwark against the exclusionary types of development, the selective benefit-sharing and the marginalisation and discrimination that is rampant in cities today. The right to the city provides an adequate platform for action as well as a framework for human rights enforcement.

However, many people, particularly in the developing world have not fully benefited from the 'urban advantage', do not live in decent accommodation, do not participate in decision making, do not live in healthy and environmentally friendly places and are unable therefore to exercise their full rights to urban citizenship. There is a problem of fairness, equity and equality. Thus underpinning all this and informing much of the work in the development of voluntary standards for sustainable cities and sustainable communities is therefore the concept of environmental justice. For Agyeman et al. (2002, p. 78),

> Sustainability (...) cannot be simply a 'green' or 'environmental' concern, important though 'environmental' aspects of sustainability are. A truly sustainable society is one where wider questions of social needs and welfare, and economic opportunity are integrally related to environmental limits imposed by supporting ecosystems. (...) The basis for this view is that sustainability implies a more careful use of scarce resources and, in all probability, a change to the high-consumption lifestyles experienced by the affluent and aspired to by others.

This will entail long term significant shifts in human behaviour, mindsets and capabilities that must address the many natural and anthropogenically created uncertainties and risks that our dominant mode of essentially urban economic development and growth has produced.

25.3 Risk, Resilience and Climate Change

A fundamental concern facing all of us is climate change and its anticipated impact on human well being and the general state of the planet. Average global temperature increase, sea level rise and increasing unpredictability and extremes in weather patterns are now universally accepted by national and city governments, business organisations, international agencies and the general public. Despite failures to reach satisfactory and legally binding global agreements regarding targeted reductions in global greenhouse gas emissions and a discernible reluctance from 'the international community' to move swiftly away from carbon fuelled economic development, considerable efforts are being exerted to address, adapt to or mitigate the effects of climate change in urban environments. The swiftly growing 'Transition Town' movement is one bottom-up example of urban and rural communities working together to develop and then implement carbon descent plans, local food production schemes and new tools for community based conviviality. The aim is to restore resilience and ecological responsibility to a culture that had been carelessly and mindlessly destructive during the age of abundance and affluence. Despite criticisms of the Transition movement as failing to adequately address broader issues of political power, vested financial interests and economic inequality, the movement does offer a fresh perspective on community self reliance, resilience, environmental responsibility and respect for others within and beyond immediate social spheres and geographic locales. Indeed, *The Transition Handbook* is aptly subtitled 'from oil dependency to local resilience' and offers heuristic guidance to communities of various descriptions as to how to embark on this change process (Hopkins 2008). The free edit version of the book, published on the Internet under a creative commons copyright, combines with the accompanying wiki, networked meetings and growing use of social media to ensure the Transition movement remains co-operative, collaborative and dynamic. Significantly, the movement is reluctant to prescribe a set series of standardised steps, procedures and actions for transition. However, although Hopkins (2008, p. 98) does offer his own "Twelve Steps to Transition", he provides only a rough charcoal drawing rather than a finely etched engraving of the transition process,

> [The steps] don't take you from A-Z, rather from A-C, which is as far as we've got with this model so far. These steps don't necessarily follow each other logically in the order they are set out here; every Transition initiative weaves a different way through the Steps, as you will see. These Twelve Steps are still evolving, in part shaped by your experience of using them. There may end up being as few as six or more than fifty!

These Steps to Transition have recently been modified enabling groups to decide what issues and actions to engage with first and how. This enables each group to draw on their own local ecological knowledge thereby creating both a heuristic and iterative approach to sustainable community developments. Transition has moved from adopting a metaphor of steps (and ladders) to one of recipes and 'ingredients' suggesting a creativity tailored to specific tastes and needs (Hopkins 2011). Although more top down perhaps, local government bodies such as the ICLEI (Local Governments for Sustainability) have been closely involved in nurturing and implementing local sustainability initiatives. In 1994 the ICLEI launched the Local Agenda 21 Model Communities Programme and more recently this umbrella group, representing over 500 municipal authorities throughout the world, announced the start of its Resilient Communities & Cities Initiative which focuses on developing tools for disaster risk management, training and capacity building. Both mitigation measures, designed to reduce carbon emissions, and adaptation measures designed to reduce vulnerability and limit the effects of climate change, are part of the policy and practice mix. So apart from the more traditional planning concerns that deal with floods and public health, climate adaption measures often overlap with those concerned with environmental sustainability such as those designed to protect the viability of ecosystem services, improve urban green space, foster urban agriculture, and promote improvements in green building design and construction, and urban transport infrastructure. Additionally, climate adaptation addresses issues relating to sustaining urban economic vitality, supply chains and attending to the material needs of poor and vulnerable populations. The ICLEI (2011, p. 1), like the Transition movement, places resilience at the centre of its deliberations and its practice defining the concept as referring to "the capacity and ability of a community to withstand stress, survive, adapt, bounce back from a crisis or disaster and rapidly move on. Resilience needs to be understood as the societal benefit of collective efforts to build collective capacity and the ability to withstand stress".

For many people, building urban and community resilience requires an integrated ecosystems approach which, as the World Bank report, *Cities and Climate Change: An Urgent Agenda* (World Bank 2010, pp. 11–12) states include a number of key actions, namely:

> (i) robust decision making (incorporating broader-based cost and benefit assessments that include societal values, ecosystem services, risks, and longer time horizons); (ii) buttressing of key infrastructure (e.g. increased robustness of water and power supply systems); (iii) social inclusion (ecosystems abhor extremes, for example, pronounced differences between rich and poor); (iv) urban risk assessments; (v) emergency preparedness (practice, know where the risks are likely, make this information public); (vi) partnerships with other cities, agencies, and governments; (vii) greater adaptive capacity through buildings and critical infrastructure to withstand increased climate variability, for example, metros; (viii) reduced social tensions; (ix) where practicable, and cost effective, streamlining of key services and infrastructure; and (x) protection and integration of key ecosystem services.

It has been apparent for some time that the nature and magnitude of the risks being confronted are of a different order from the more natural and predictable risks of earlier centuries. Ulrich Beck's (Beck 1992) 'risk society' is one confronting both environmental disturbances and climate turbulence with threats to social, cultural and national identity caused by the decline and in some cases collapse of traditional values, norms and customs. The idea and practice of fashioning sustainable urban environments and communities is confronted with many unknowns compounded by the problems of fashioning meaningful work and securing consistent employment in both developed as well as developing countries, shifts in and resistance to changes in gender roles and expectations, changes to the family life and structure, to class consciousness and loss of geographical, and perhaps even spiritual, rootedness brought on by increasing levels of capital and labour migration. New technological and scientific developments such as nanotechnology, genetic engineering, nuclear power and synthetic biology are only adding to a sense of disorientation and dislocation. Newman et al. (2009), see individuals, cities and communities as being essentially similar in that fear destroys resilience and exacerbates risks whereas hope builds strength and confidence. For hope to be realised, human society needs to mend its ways and learn from the past. For instance, building social-ecological resilience requires that we understand ecosystems and it is not only professionals, the core scientific disciplines and academic experts who have the requisite knowledge and understanding. The tacit environmental knowledge of local people must be properly respected and taken into account by policy makers and practitioners. As Folke et al. (2002) suggest, ecological ignorance undermines resilience and the still all too common assumption that human society is separate from nature rather than part of it continues to be an underlying cause of the vulnerability of human social systems. In this a clear connection needs to be made between social resilience and social vulnerability and economic resilience and vulnerability. Environmental change can impact seriously on a group or community's livelihood and the resilience of social, economic and political institutions. Markets may be disrupted or destroyed, the availability of key resources may be compromised and, as a result, economic well being can be threatened, crime increase and people of working age may lose their jobs and/or migrate to other locations. As Adger (2000, p. 361) writes, "*the centrality of social resilience to sustainable development remains a critical question*".

25.4 Sustainable Community

Community is at the human heart of urban sustainable development, social and ecological resilience. At a moment when globalisation is seen as a fact of life as well as a desirable consequence of neoliberal economic development, there has been a return to community and to the local. Environmentalists have, for many years, articulated the idea 'think global, act local' and problems of social exclusion, alienation and disaffection have seen a flurry of policy activity informed by particular understandings of communitarianism, social capital and social cohesion. There has also been a spatial turn in political, economic and social thinking. A spatially based critical politics of consumption could build understanding and awareness of the global ramifications of the way we live, play and work offering important ethical and material lessons for those wishing to develop more sustainable lifestyles. The belief that a sense or spirit of place is key to health and well being, feelings of belonging and a capacity and willingness to care for self and others (human and non humans) has grown considerably in recent years too. However, Doreen Massey has critically dissected the meaning of space and warned against an excessive attachment to localism and/or globalism. The local, she says, can never be simply walled off from the global but must somehow be weaved in to the changing global environment in ways that are distinctly advantageous for local economies and local communities (Massey 2004). Globalisation is made in local places; the global is constructed out of the local and vice versa. Globalisation is really an abstraction for ultimately it is constituted by a complex network of actors and relationships, phenomenological experiences and place based groups and actions that increasingly transcend one locality. These local places are not, and should not be perceived as being, victims of globalisation but rather as sources of social reproduction and cultural innovation. In this way, a global sense of place implies that local communities and cultures are a product of relations that extend far beyond their specific boundaries. We know that there are many transboundary flows . . . of information, capital, people, pollution, commodities, etc., and we know that climate change is a global phenomena and we know that cities do not move. They may grow or shrink and although various 'imagined communities' and social identities may be quite cosmopolitan they will invariably be physically located in specific spaces and places.

With many towns and cities hosting a wide array of different cultural and ethnic groups, many opportunities to examine the place of locality, economic opportunity and social and environmental (in)justice within a globalised world exist. Groups and individuals tend to frame their behaviour according to context or 'sites of practice' thereby exposing themselves to a number of contradictions and conflicts. Whether or not to fly to that wonderful ecotourist holiday destination on the other side of the world may be just one example albeit only for the relatively well off (Barr et al. 2011). An understanding of the role of place in public health is also of considerable importance and here four aspects of the built environment are of critical importance: nature contact, public space, buildings and urban form. Nature contact may reduce stress and enhance work performance. Green open space together with what is known as 'green infrastructure' such as woodlands and wetlands, etc., help cool urban environments, contribute to flood protection and the provision of clean air and water (Marton-Lefevre 2012). Green building design and construction may not only be energy efficient reducing the seriously high rate of greenhouse gas emissions of urban areas but also have direct health impacts on their inhabitants. Public spaces are important areas for physical activity, the free exercise of sociability, the building of social capital and fostering of individual and collective mental well being. Sustainable urban form, good transport and accessibility, proximity to leisure amenities and work, address issues of poverty related ill health, educational under achievement, drug use, crime and disorder (Frumkin 2003). Local food systems, specifically urban and peri-urban agriculture, can help integrate more sustainable food diets with the management of natural resources and ecosystem systems and build rural and urban connectivities that are of central importance to urban resilience (Custot et al. 2012). With a focus on the local and with activities cognizant of the global but resolutely emerging from the community or the neighbourhood, it is possible that sustainable development can resonate with the needs, desires and life experiences of groups and individuals. As Bridger and Luloff (2001, p. 461) write:

> by focusing on sustainability at the local level, changes can be seen and felt more immediately. Further, discussions of a "sustainable society" or a "sustainable world" are relatively meaningless to most people since they require levels of abstraction not relevant in their daily lives. The community, in contrast, is more conceptually manageable. After all, the consequences of environmental degradation are most keenly felt and the results of intervention most noticeable in one's own backyard (. . .). To the extent that successful intervention becomes a tangible aspect of local life, we increase the likelihood that sustainability will acquire the widespread legitimacy that has thus far proved elusive.

Community economic development may also go some way to addressing issues of social and economic inequality which are themselves prime indicators of relatively low levels of wellness and quality of life and high levels of deprivation and ill health. Unfortunately, community based economic activity alone is not sufficient to transcend the systemic nature of these inequalities although they may stimulate political action that may usher in a new social contract that may shape more sustainable ways of living, being and creating wealth (Wilkinson and Pickett 2010).

25.5 Sustainability Standards as a Social Contract

Voluntary (and sometimes compulsory) sustainability standards, codes and indicators have become increasingly important management tools, guides and learning devices in sustainable community and urban development. Indicators, for example, are tools which provide information for community members that may inspire action, lead to constructive deliberation and better decision making. In this way, indicators can empower both citizens and local government officials although they often are a cause for discussion and sometimes even dispute. Similarly standards, particularly those with a voluntary status, play an important role in reshaping knowledge, understanding, social awareness, ethical perspectives, normative frameworks and cognitive mindsets. If they become established they may help to create a form of social contract within civil society between businesses, community groups, local government, professional bodies, campaign organisations, education and research institutions and so on. Although not legally binding, in time they establish a set of expectations, cultural proclivities, (pre)-dispositions and structuring frameworks that in effect form a type of habitus. It is also worth recalling that for the political philosopher Jean-Jacques Rousseau, the social contract meant that a community recognised having a collective good which is not the same as private interests and through this collective good, civil liberty and social progress could be achieved. Of course, the concept of community is one that has been subject to almost as much debate as sustainable development but it nonetheless has a considerable degree of public acceptability. It is a term, a presumed reality, to be applauded, protected, developed and aspired to particularly at a time when risk, uncertainty, insecurity and competitiveness dominate (Bauman 2001). The conceptual relationship between habitus and community is therefore quite important, but complex. For Bourdieu (2005), habitus should not be considered in isolation but must be used in relation to the notion of 'field' which is a dynamic space of tensions, contradictions, conflicts and struggles in which various actors seek to make adjustments according to their own skills, understandings, interests and needs. The concepts of community and habitus are both relational, engaging both structure and agency. They have a spatial dimension, a political aspect and have implications for both governance and governmentality i.e. those organised practices including the various mentalities, calculations, analyses, reflections, techniques, powers, apparatuses and rationalities that shape the way we create, administer and manage sustainable lifestyles, social practices, communities and cities (Dean 1999). For the OECD voluntary standards for sustainability help guide consumer behaviour in more pro-environmental directions, they foster corporate responsibility, raise public awareness, establish grounds for new sustainable learning experiences, shift NGO campaigns away from being purely oppositional, inform future governmental standards for sustainability and establish new or enhanced certification regimes (Salmon 2002).

Processes of developing standards, like those of sustainability indicators, are most effectively accomplished when undertaken in a democratic and participative manner. They need to be sensitive to place and enable future learning, revision and refinement. If applied thoughtfully and critically, voluntary standards for sustainability may become a form of distributed intelligence (Innes and Booher 2000). They may also enable the emergence of a more developmental and generative approach to design, construction, management and community engagement. Raymond Cole (2012) has analysed the development of building codes and practices distinguishing between those he considers to be largely technical and confined to the actual building itself (LEED, BREEAM, the UK Code for Sustainable Homes); those he considers to be sustainable which has more of a relational dimension being sensitive to the wider built, natural and social environments such as Arup's Sustainable Project Assessment Routine; and, those he considers to be regenerative which go way beyond eco efficiency and stable state sustainability measures. Cole writes (2012, p. 47):

> Regenerative design thereby requires a fundamental re-conceptualization of the act of building design primarily in terms of imagining, formulating and enabling its role within a larger context. It would therefore seem appropriate that the representation of regenerative design in support tools should reflect this interplay. (. . .) Regenerative design prioritizes the understanding and engaging in the unique qualities of place and continues the Bioregionalist commitment to developing communities integrated with their surrounding ecosystems.

In contrast, the UK Code for Sustainable Homes (2008), closely linked to current Building Regulations, simply establish basic performance measures which are known to reduce environmental impacts and can be objectively assessed, evaluated, delivered and verified. Each criterion carry a certain number of credits which in sum inform the rating awarded to the building. A certificate is then issued with anything from one to six stars and this can then be appended to the building. The UK Housing Corporation, a government QUANGO that funds and regulates housing associations, includes the initial iteration of the Code for Sustainable Homes (see Table 25.1) as an element in its *Design and Quality Standards* (Housing Corporation 2007) which sets out its expectations and recommendations for all new affordable homes, registered social landlords and housing associations that receive the Social Housing Grant. New affordable homes must at least attain Code Three.

There are now a wide range of tools and voluntary standards available to help promote urban sustainability and sustainable community development but lessons from their use are not always learnt or applied in a serious, rigorous or consistent manner by those using them. This may be because a genuinely inclusive vision of

Table 25.1 Summary of environmental impact categories and issues (*Source*: Department for Communities and Local Government 2008, p. 10)

Categories	Issues
Energy and CO_2 emissions	Dwelling emission rate (M)
	Building fabric
	Internal lighting
	Drying space
	Energy labelled white goods
	External lighting
	Low or Zero Carbon (LZC) technologies
	Cycle storage
	Home office
Water	Internal water use (M)
	External water use
Materials	Environmental impact of materials (M)
	Responsible sourcing of materials—building elements
	Responsible sourcing of materials—finishing elements
Surface water run-off	Management of surface water run-off from developments (M)
	Flood risk
Waste	Storage of non-recyclable waste and recyclable household waste (M)
	Construction waste management (M)
	Composting
Pollution	Global Warming Potential (GWP) of insulants
	NOx emissions
Health and wellbeing	Daylighting
	Sound insulation
	Private space
	Lifetime homes (M)
Management	Home user guide
	Considerate constructors scheme
	Construction site impacts
	Security
Ecology	Ecological value of site
	Ecological enhancement
	Protection of ecological features
	Change in ecological value of site
	Building footprint

Note: (M) denotes issues with mandatory elements

an urban sustainable community has not been created; or because equitable and communicative partnerships between community stakeholders, developers and local governments were not properly formed; or because either concerns for land and property market valuations remained prominent; or because the material realities of local communities were not understood or effectively articulated. Collaboration is not the same as either consensus or agreement and where this is lacking the new social contract, and in time habitus, will struggle to take shape and become a spatially embedded reality (Deakin 2011). Islands of sustainability, green buildings and eco home developments may emerge because they can but in terms of urban strategy something is lacking. Clearly a growing 'toolification' and 'normalisation' of sustainability within urban sustainable development is evident as various tools, indicators and standards increasingly come to define and manage sustainability. However problems persist particularly when there is limited knowledge about sustainability or when these tools are applied too rigidly. Even when they have been integrated into policy frameworks, strategies and action plans, they may not be sufficiently or sensitively adapted to either place or circumstance (Jensen and Elle 2007).

25.6 BSI 8904: Sustainable Communities

In the United Kingdom, during the 13 years of 'New' Labour Government from 1997 to 2010, a significant amount of attention was given to developing frameworks, processes and strategies for developing sustainable communities. A Sustainable Communities Plan was published in 2003 (Department for Communities and Local Government 2003) together with a set of regional variations which was reviewed by Professor Anne Power of the London School of Economics for the Sustainable Development Commission in 2004 (Power 2004). A Sustainable Communities Act was later passed in 2007 and amended in 2010 establishing a statutory framework whereby local councils in England could work with community groups to devise proposals aimed at improving sustainable economic, environmental and social wellbeing. The Act provides a checklist of potential community issues ranging from the use of local waste to community health, jobs and organic horticulture that ought to be considered. The Commission saw sustainable communities as being defined by a set of aims, tools and measures. These included (Power 2004, p. 5),

The Three Aims

> *a healthy environment* involves minimal ecological impact, minimal waste or pollution and maximum recycling, protection and enhancement of the natural environment, wildlife and biodiversity, so that all may enjoy environmental benefits such as greenery, careful planning for physical and social wellbeing, space to walk, cycle, meet, play, and relax.
>
> *a prosperous economy* generates wealth and long-term investment without destroying the natural and social capital on which all economies ultimately depend; minimises resource use and environmental impact; develops new skills through education and training; meets basic needs, through local jobs and services.
>
> *social well-being* arises from a sense of security, belonging, familiarity, support, neighbourliness, cohesion and integration of different social groups, based on respect for different cultures, traditions and backgrounds.

The Four Essential Measures (or Building Blocks)

> *planning, design, density and layout* will influence the shape of a community, the level of services and the way people interact with each other and their environment, e.g. low density sprawl makes public transport and local shops unviable; higher densities support shops, buses, neighbourhood schools and a sense of community.
>
> *minimising energy use and environmental impact* contributes to sustainability, helps combat global warming and encourages 'long-term stewardship of' communities; e.g. recycling buildings helps to reduce resource use and encourages care and low impact approaches.
>
> *a viable local economy and services* provide the rationale and underpinning for community development and survival; e.g. loss of manufacturing has made many traditional urban communities unviable and requires a major economic shift and new uses for existing infrastructure if they are to flourish again. They also require transport links to wider job markets, and education and training for new skills.
>
> *community organisation and neighbourhood management* are essential to social networks and urban viability, ensuring well maintained, secure conditions which are the prerequisite of stable, long-term, participative and cohesive communities; e.g. regeneration companies, local housing companies and neighbourhood management organisations can transform basic street conditions, community safety and security, social contact and youth engagement, by acting as a local conduit for decisions, co-ordinating supervision and frontline service delivery.

Interestingly, although the 'resilience' concept frequently appears in many approaches to sustainable community development, it is not explicitly referred to in 'New' Labour's Sustainable Communities Plan, the Sustainable Development Commission Review or the Sustainable Communities Act. Nor did it feature explicitly in the guidance provided for the Conservative-Liberal Democrat Coalition Government's Localism Act of 2011 which is ostensibly designed to give back power from central government to local authorities and for local communities, "*giving them the freedom and flexibility to achieve their own ambitions*" (Department of Communities and Local Government 2012). Even so, given the contexts in which the Acts operate, resilience remains of key underlying importance albeit implicitly.

In January 2012 the British Standards Institute (BSI), the UK's National Standards Body (NSB), published its own standards for sustainable communities. BSI standards are designed for voluntary use and are not regulations. NSB standards are produced 'to make life simpler' and, as the NSB states, to increase reliability and effectiveness by providing a bridge between expert knowledge and experience. Essentially, standards are an agreed, repeatable way of doing something. A published document is invariably produced containing a technical specification which is expected to be used consistently as a rule, guideline, or definition. Based on the 8900 series of sustainable development management standards, the *BS 8904—Guidance for Sustainable Community Development* outlines 'a step by step' process by which sustainability may be embedded into the everyday lives of local communities. BS 8904 addresses issues of cost efficiency by seeking to reduce environmental impacts and to improve social relations broadly understood in terms of social cohesion and inclusivity. It also claims to lay out the grounds for a robust economic resilience which will mitigate risks relating to health, shelter and food.

National government departments, local authorities, higher education institutions, community-building organisations, planning officers, representatives from consumer groups, the National Health Service and a number of independent experts on sustainable development were involved in the iterative development process that lasted over a year. The process took into account related developments by the International Organisation for Standardization (ISO) and the European Committee for Standardization (CEN) but BS 8904 is the first standard of its kind and has been presented by the BSI as a valuable contribution to the professional 'toolkit' for sustainability. These standards are claimed to ensure quality as it would do for any other product, service or management system.

The drafting of BS 8904 was informed by five key 'principles':

1. That users would use the standards to identify the community it aims to serve as well as possible benefits and desirable outcomes.
2. That the embedding of sustainable development in everyday community life would be continually evolving and challenging.
3. That the process of building sustainable communities could be either grass roots or local authority led.
4. That some communities may wish to apply some form of verification to their achievements although the BSI is not empowered to recommend any specific auditor or certification system.
5. That a 'maturity matrix' would be important for assessing future progress, clarifying next steps and identifying future actions and the linking of sustainability principles with practice.

These principles were then extrapolated for ease of comprehension in tabular form (see Table 25.2).

Where academics may find the concepts of community and sustainable development difficult to pin down, the Guidance document sees the concept 'sustainable communities' as making sustainable development 'tangible' for the sustainable development process which may become rooted in a specific place or 'community of interest'. Such communities may be of any size and dimension but to benefit fully would most likely already possess some degree of social cohesiveness that would foster empowerment, participation, ownership, engagement, flexibility, adaptability and resilience. A flourishing local economy, enhanced quality of life, reduced ecological footprint and a greater degree of social and intergenerational equity than presently exists is all part of this standards package. Self reliance, self sufficiency and an ability to overcome vulnerability are additional qualities a sustainable community will need to develop to successfully deal with future uncertainties and BS 8904 will help achieve this in eight clear steps:

1. People coming together to agree core principles such as mutuality, sense of place, connectedness, resilience, etc.
2. Involving others and engaging stakeholders—individuals, community groups, and local organisations.

Table 25.2 Sustainable development maturity matrix (illustration/extract, *Source*: BSI 2012, pp. 17–18)

Principles of sustainable development	Key issues	Start up	Gaining momentum	Self-sustaining commitment	Leadership and innovation
Engagement and inclusivity	Facilitating as many as possible using good governance and democratic principles. Good decision making and governance will embrace peer review to challenge and steer progress	Informal and formal communications open to all interested parties in accessible format	Good organisational governance along accepted principles of local democracy	Full accountability for decision making and records kept on open access. Wider evaluation of sustainability e.g. community health and happiness	Engaging in national and international projects on sustainability. Relationships that will help perform critical friend and peer review roles
Environmental limits	Enhancing biodiversity, stewardship of natural resources	Identifying what parts of the environment are important for the community	Start projects that illustrate the history of the local natural environment through old maps. Start annual back garden wildlife surveys	Name local nature reserves and identify volunteer wardens	Build a partnership with a voluntary organisation and the local council to protect and enhance natural resources
Resilience and adaptability	Embedding collaboration so that it continues and adapts in times of pressure or complacency	Increasing diversity of job opportunities, protecting local services and reducing risk, e.g. flood control	Developing mixed energy sources in a policy that is owned by all	Developing renewable energy schemes locally e.g. HEP, ground water, wind, solar	Share energy sources between neighbours or community and economies of scale. Co-housing becomes a norm

3. Defining key issues including services such as health and lifelong learning, energy conservation measures, sustainable farming, and retrofitting homes.
4. Identifying community capability including existing human and material resources, assets as well as mapping out potential risks and hazards.
5. Planning and selecting options with the application of sustainable development principles, SMART (i.e. specific, measurable, achievable, relevant and time-bound) objectives and pointers for building confidence.
6. Executing the plan including resource allocation and budgeting.
7. Evaluating and analysing agreed measures, outcomes and reviewing the continuing relevance of the community's vision, values and purpose in order to identify and implement any necessary future changes.
8. Learning and building community capability by acquiring new knowledge, skills and dispositions whereby future issues may be successfully addressed.

25.7 Conclusion

At the time of writing it is difficult to estimate the extent to which BS 8904 will be taken up by communities and local authorities. It clearly has some correspondence with the aims and purposes of the Transition Movement although standardisation is not something that always fits easily with sustainability practitioners at local level. It also has some resonance with organisations such as The Young Foundation in the UK whose work on building resilient communities has involved creating a community action toolkit and a Wellbeing and Resilience Measure (WARM) (Young Foundation 2008). In claiming to measure life satisfaction by capturing information on how well, or otherwise, a community is faring by mapping local assets such as self efficacy and resilience as well as vulnerabilities, WARM is largely about informing local decision-making. There are also other standards, measures, guides, toolkits and indicators in the sustainable and community development market place that individuals and groups may select. They all offer variations on a familiar theme and in some ways it is hard to choose between them. In fact, as guides they are probably best used heuristically and should be adapted to time and place. This means that although there may be an overlap in categories, concepts, steps, actions, advice, guidance and putative strategies on offer, it is really up to local communities to work things out for themselves. Indeed, this is what the Transition Movement in its many manifestations in the UK and elsewhere is actually doing. For Scott Cato and Hillier (2010) Transition towns and communities encompass the development of sustainable local economies and renewable energy capabilities but most importantly offer spaces for experimentation in sustainable living and opportunities. Only through experimentation will an alternative sustainable reality emerge.

However, the drive for standardisation continues. In October 2011 the French certification body AFNOR proposed a new CEN Technical Committee to develop a series of European Standards on Sustainable Development in Communities. The ISO was also busy throughout 2012 working on its own 'Guidance for Communities

Sustainable Development and Resilience' building on its earlier guidance work on incorporating sustainability in the development of standards. These community standards, as one would hope, focus on action that will be meaningful and are credibly within the power of individuals, community groups and organisations to do something about. Discussion of climate change or global warming rarely dominate these community based approaches as they so often do in the more city wide agendas and strategies or the mentalities of international bodies. However, this is not to say that climate change is of little regard for those wishing to build more sustainable communities, it is rather that social and economic issues weigh at least equally or more heavily. Anthropogenic climate change affects everyone and its effects are evident at all spatial scales. Low carbon initiatives are frequently referenced but the issues at community and neighbourhood levels are usually those that are perceived as immediately relevant to and fall clearly within the bounds of individual and community efficacy. One major task for sustainability practitioners, educators and others is therefore to connect the local with the global in ways which Massey (2003) suggested. Voluntary sustainability standards may help to some extent but awareness and recognition also comes with intuition and reflection on one's lived experience and the trying out of new and different thinking. Managerialist frameworks and toolkits sometimes invite some thinking 'out of the box' but this may be compromised by having to record any consequent outcomes of in the box. Thus box ticking, although sometimes derided, remains a too common experience. It is important to go beyond this and perhaps even standards and standardisation though they may indeed be a help, a guide and a support to get things moving. If there is one key recommendation emerging from this discussion then it is one shared with Rob Hopkins. It is important for the processes of sustainable community development to be creative and to empower. Standards and recipes are guides we need to fashion sympathetically to culture, heritage and ecology, to taste, preference and fulfilment. Think of Transition like cooking; and like cooking building sustainable communities requires some order, guidance and some clear stages. However, as writes Hopkins (2011, p. 90),

> There are all kinds of amazing ingredients we can assemble in order to make, say, a cake, and the creation of every cake will be unique, reflecting his or her abilities and culture, and the local resources available.

References

Adger NW (2000) Social and ecological resilience: are they related? Prog Hum Geogr 24(3):347–364

Agyeman J, Bullard RD, Evans B (2002) Exploring the nexus: bringing together sustainability, environmental justice and equity. Space Polity 6(1):77–90

Barr S, Gilg A, Shaw G (2011) Helping people make better choices: exploring the behaviour change agenda for environmental sustainability. Appl Geogr 31(2):712–720

Bauman Z (2001) Community: seeking safety in an insecure world. Polity Press, Cambridge

Beck U (1992) The risk society: towards a new modernity. Sage, London

Bourdieu P (2005) Habitus. In: Hillier J, Rooksby E (eds) Habitus: a sense of place. Ashgate, Aldershot, pp 43–49
Brenner N, Marcuse P, Mayer M (2012) Cities for people, not for profit. Routledge, London
Bridger JC, Luloff AE (2001) Building the sustainable community: is social capital the answer? Sociol Inq 71(4):458–472
British Standards Institute (2012) BS 8904 – guidance for sustainable community development. BSI, London
Cole RJ (2012) Transitioning from green to regenerative design. Build Res Inf 40(1):39–53
Custot J, Dubbeling M, Getz-Escudero A, Padgham J, Tuts R, Wabbes S (2012) Resilient food systems for resilient cities. In: Otto-Zimmerman K (ed) Resilient cities 2. ICLEI/Springer, Bonn, pp 125–137
Deakin M (2011) Meeting the challenge of learning from what works in the development of sustainable communities. Sustain Cities Soc 1(4):244–251
Dean M (1999) Governmentality: power and rule in modern society. Sage, London
Department for Communities and Local Government (2003) Sustainable communities: building for the future. TSO, London
Department for Communities and Local Government (2008) The code for sustainable homes: setting the standard in sustainability for new homes. TSO, London
Department of Communities and Local Government (2012) Historic powers in the Localism Act pass down to communities. Public announcement. https://www.gov.uk/government/news/historic-powers-in-the-localism-act-pass-down-to-communities–2. Accessed 31 Aug 2013
Folke C, Carpenter S, Elmqvist T, Gunderson L, Holling CS, Walker B (2002) Resilience and sustainable development: building adaptive capacity in a world of transformations. Ambio 31 (5):437–440
Frumkin H (2003) Healthy places: exploring the evidence. Am J Public Health 93(9):1451–1456
Hopkins R (2008) The transition town handbook: from oil dependency to local resilience. Free edit version. http://www.transitie.be/userfiles//transition-handbook(1).pdf. Accessed 31 Aug 2013
Hopkins R (2011) The Transition companion. Transition/Green Books, Totnes
Housing Corporation (2007) Design and quality standards. Housing Corporation, London
ICLEI Briefing Sheet (2011) Towards urban resilience. May. http://www.iclei.org/fileadmin/user_upload/documents/Global/News_Items/Image_Documents_web_news_11/Briefing_Sheet_Urban_Resilience_20110616.pdf. Accessed 16 Aug 2012
Innes JE, Booher DE (2000) Indicators for sustainable communities: a strategy building on complexity theory and distributed intelligence. Plan Theor Pract 1(2):173–186
Jensen JO, Elle M (2007) Exploring the use of tools for urban sustainability in European cities. Indoor Built Environ 16(3):235–247
Marton-Lefevre J (2012) Nature at the heart of urban design for resilience. In: Otto-Zimmerman K (ed) Resilient cities 2. ICLEI/Springer, Bonn, pp 113–118
Massey D (2003) The responsibilities of place. Local Econ 19(2):97–101
Massey D (2004) Geographies of responsibility. Geogr Ann Ser B 86(1):5–18
Newman P, Beatley T, Boyer H (2009) Resilient cities: responding to peak oil and climate change. Island Press, Washington
Power A (2004) Sustainable communities and sustainable development: a review of the sustainable communities plan. ESRC/CASE/SDC, London
Rees W, Wackernagel M (1996) Urban ecological footprints: why cities cannot be sustainable – and why they are a key to sustainability. Environ Impact Assess Rev 16(4–6):223–248
Salmon G (2002) Voluntary Sustainability Standards and Labels (VSSLs): the case for fostering them. Background paper for the OECD round table on sustainable development
Scott Cato M, Hillier J (2010) How could we study climate-related social innovation? Applying Deleuzean philosophy to transition towns. Environ Polit 19(6):869–887
UN Habitat (2008) State of the world's cities 2010/2011: bridging the urban divide. Earthscan, London

United Nations (2012) World urbanization prospects: the 2011 revision. United Nations, New York. http://esa.un.org/unup/pdf/WUP2011_Highlights.pdf. Accessed 31 Aug 2013

Wilkinson R, Pickett K (2010) The spirit level: why equality is better for everyone. Penguin Books, London

World Bank (2010) Cities and climate change: an urgent agenda. The International Bank for Reconstruction and Development/The World Bank, Washington

Young Foundation (2008) Local wellbeing: can we measure it? http://youngfoundation.org/wp-content/uploads/2013/02/Local-Wellbeing-Can-we-Measure-it-September-2008.pdf. Accessed 31 Aug 2013

Chapter 26
Small-Scale Farmers' Involvement in Ecolabelling: Limitations and Conflicts

Carolin Möller, David Smyth, and Michael Schmidt

26.1 Introduction

In a globalised market characterised by socio-economic cooperation, trading behaviour has become a significant and globally impacting concept. Especially within the current era of environmentally aware consumers, the focus upon 'green' orientated decision-making has come to the fore as a potential tool for market operations. Besides formal government implemented legal standards, ecolabels have emerged in various guises throughout the globe, aimed towards encouraging consumption structures with consideration for sustainability issues. The idea behind voluntary environmental standards hinges upon the inclusion of a broad and representative range of stakeholders participating within the production chain of a particular product.

This chapter sets out to discuss the position of small-scale farmers and 'environmentally-friendly' markets in regards to ecolabels. The core of this chapter revolves around the question of how small-scale farmers, such as those broadly represented by farms in Sub-Saharan Africa or Southeast Asia, are able to benefit from the ecolabelling tool. This includes a detailed discussion on the potential of farmers regarding their participation in empowering decision-making and long lasting conceptual agreement processes between all involved stakeholders, while also being concerned with environmental standards. The importance of this discussion is given based on the market- and consumer-orientated characteristics of ecolabels, while often neglecting the suppliers' perspectives. This discussion, however, will contribute to the need of holistic and efficient socio-environmental health while understanding shared stakeholders' responsibility and questioning the burden allocation of production and consumption patterns. Recently emerged

C. Möller (✉) • D. Smyth • M. Schmidt
Department of Environmental Planning, Brandenburg University of Technology (BTU), P.O. Box 101344, 03013 Cottbus, Germany
e-mail: moeller.carolin@googlemail.com; david_smyth@live.co.uk; michael.schmidt@tu-cottbus.de

C. Schmitz-Hoffmann et al. (eds.), *Voluntary Standard Systems*, Natural Resource Management in Transition 1, DOI 10.1007/978-3-642-35716-9_26,

conflicts between market demand in Europe and production capacity in the West African base serve to demonstrate some of the restrictions associated with ecolabels.

26.2 Ecolabelling Processes

The established 'Ecolabels' are meant to represent holistic, environmentally-friendly performance that refer to the production as well as the life-cycle categories of goods or services provided to recognised consumer groups. By analysing the internal structures of ecolabelling, the top-down attitude of this tool becomes visible. This attitude is based on power shifting towards consumers and suppliers represented by standards which are defined and determined by exporting dominantly to westernised countries (UNEP 2012). Ecolabels that are officially accepted, in contrast to a self-styled environmental symbol or claim statement developed by a manufacturer or service provider, are only awarded to products or services by an impartial third party if they meet the established environmental criteria as stated by the International Organisation for Standardisation (GEN 2004).

Ecolabelling refers to the provision of information to consumers about the relative environmental quality and impact of a product and/or service. Nowadays, a large variety of ecolabels occur in various sectors including: electronics manufacture, food production, cosmetics, and tourism on the global trade market (GEN 2004). Some commonly known ecolabels which address their respective product sectors include 'Energy Star' (electrical products), 'Forest Stewardship Council' (wood products), 'Green IT' (Information Technology), Marine Stewardship Council (fish) and 'Bio' (food).

The main guiding principle of ecolabels addresses the consumer lifestyle and creates awareness for the established standards it represents. This consumer-orientated layout generates the objectives, the vision of the processes and the image of products furnished with an easily identified ecolabel. However, the common market demand principles are the guidelines for ecolabel implementation. Its creation can be attributed to the growing global awareness and concerns for environmental protection. The demand for these market tools is driven by an environmentally-oriented consuming public and impacts both business and governmental decision-making. The consumers' demand for environmental protection induces a 'rebound' effect on the market behaviour by motivating businesses to align their decisions with maintaining environmental well-being.

The concept of ecolabels originates from Europe, being legally established on 23 February 1992. It is defined by the Council Regulation (EEC) No. 880/92 on a Community eco-label award scheme with the intentions to, "promote the design, production, marketing and use of products which have a reduced environmental impact during their entire life cycle, provide consumers with better information on the environmental impact of products" (GEN 2004).

Nevertheless, this ecolabelling scheme excludes market sectors related to food, drinks and pharmaceuticals from the product categories. The concept of 'Life Cycle Assessment' (LCA) has become a basis for awarding ecolabels in all sectors. Conclusively, these tools should aim to integrate all involved agents and stages of a production and supply chain such as one defined by the United Nations Conference on Trade and Development (UNCTAD) Secretary. The life cycle of products is divided into distinct stages of pre-production, production, packaging and distribution, utilisation and disposal. Each stage is analysed for potential environmental impacts to be classified into fields such as waste relevance, noise generation, air and water contamination, soil pollution and degradation, effects on ecosystems, consumption of natural resources and energy consumption. Ecolabels address environmental impact from various perspectives, with approaches varying between being based on the targeting of a single sector of the products and services, individual characteristics, or a collective set of attributes.

There are three types of ecolabel concepts which are characterised as follows (GEN 2004):

Type I

- A voluntary, multiple-criteria based, third party program that awards a license which authorises the use of environmental labels on products *indicating overall environmental preference* of a product within a product category based on life cycle considerations.

Type II

- Informative *environmental self-declaration* claims.

Type III

- Voluntary programs that provide *quantified environmental data on a product* under pre-set categories of parameters set by a qualified third party and based on the life cycle assessment which is verified by them or another qualified third party.

These conceptual approaches for assessing products and services are not only found in the frameworks provided by European contributors but also by various 'in-country' representatives (i.e. those located within or geographically adjacent to the producing nation) such as Eco Mark Africa (EMA). EMA is a third party certifying body based within the African continent that aims to credibly verify the sustainability profiles of African products; to furnish them with ecologically-friendly market identifications for global, national, and/or regional markets and enhancing trade access. Thus, EMA and other similar ecolabel contributors provide general market-based incentives for a shift towards sustainable production patterns, for example, between African and European markets in order to meet the sustainable consumption trends of various consumer groups. Like other impartial ecolabel contributors, EMA adheres to the economic principle of non-discrimination which attempts to combine socio-economic development with global trading

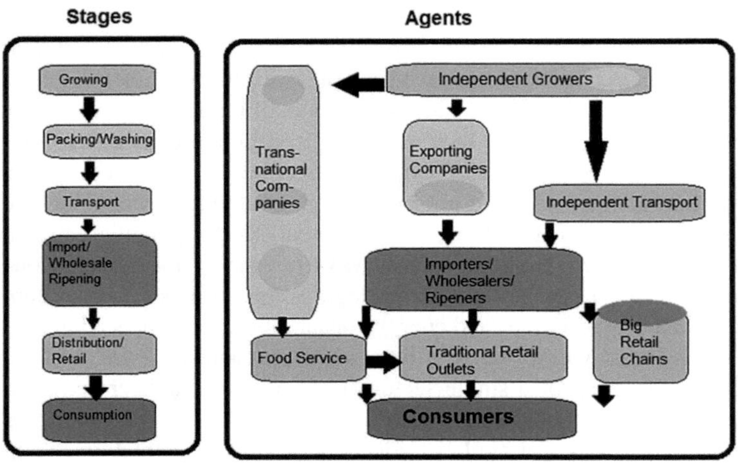

Fig. 26.1 Production and supply chain stage and agents (modified by C. Möller from UNCTAD Secretary)

structures (AEM and GIZ 2011). Under the supervision of the African Ecolabelling Mechanism (AEM), a pan-African political structure and technical framework cooperative, EMA focuses on capacity building in alignment with the needs of smallholders as the backbone of the African economy. In Fig. 26.1, a general production chain is depicted which highlights the narrow range of possible access and interference points for small-scale farmer involvement. It also demonstrates the dominance of large-scale producers and retailers which are all aimed towards satisfying consumer demands.

26.3 Affected Stakeholder Environment

The potential transformation of environmental concern and protection into a tangible market advantage is strongly linked to the various stakeholder groups involved along the production and consumption chain. This transformation is possible by accessing alternative product and service markets as defined by the environmentally conscious consumer groups.

The market process of ecolabelling is based upon the existing market system and therefore, defined by trends and ideas on the free market economy. It is even better defined as being 'consumer-centred' as it meets environmental concerns and consumer willingness to adapt personal consumption patterns to become more convenient. These patterns act as drivers for sustainability.

To address the involvement of small-scale farmers, one must consider the constraints of their participatory involvement given the market-orientated character of ecolabelling. An initial definition of involved stakeholders and small-scale farmers also has to be given. The involvement strategies of small-scale farmers in ecolabelling can be approached from the perspectives of two different assumptions. On one hand, the focus is set on the mere persuasion of the farmers to adapt to the concept and to the related standards for their production processes and their products, depending on the label type. Conversely, it is possible to provide small-scale farmers with the incentives to create their own design for the processes, essentially, aiming for participation throughout the entire decision making processes.

The target group in this chapter's discussion are small-scale farmers as actors in a production chain. Even so, there is no universally established definition of small-scale farming; a general idea categorises it as farming practises falling between subsistence agriculture and large cash crop production. Small-scale farmers mostly operate and serve local, domestic or regional markets. These farmers are generally well represented in regions in Southeast Asia and Sub-Saharan Africa (Pye-Smith 2011). By trying to include them into ecolabelling processes, ambient stakeholders along the production and supply chain must enunciate and understand their positions. The operating environment is composed of the central government that represents the enabling of institutional and legal frameworks, monitoring and evaluation processes (FAO 2008). The relevant governmental institutions are represented by economic and environmental institutions. They operate and interact in national and international legal frameworks which are highly relevant for assessing the entire production chain. They are necessary in various levels of regional, national and objective-related decision making. The positions and interventions of topic related non-governmental organisations (NGO), private industry, consumers and other consulting third parties such as the previously mentioned EMA are equally significant. Industries and consultancies represent a demanding and standardised environment. NGOs often take-over the coordination of supportive tools such as discussion, implementation and realisation processes. This enables the integration of all stakeholder positions, especially those of the farmers themselves. Nevertheless, the farmers are also involved in decision making for the ecolabelling process allowing for their own capacity building (FAO 2008).

A core feature of branding a product with an ecolabel is subject to the decisive control and power distribution among the stakeholders. Consequently, due to the nature of ecolabelling, branding is primarily adjusted and defined by consumer interests and lifestyle as well as retailer profits and outcomes. By looking at the power distribution, stakeholder interests and consequences, all actors have to be aware of and accept the significance of contributions from small-scale farmers as well as the cooperation of the economic sector on various levels on a globally interacting market.

The view on the contributions of small-scale farmers to the global product trade and their impact on the production chain must be enriched. Society should accept their overarching responsibility for the well-being of the greater community. This should be based on food security, sovereignty, and socio-economic pillars while satisfying the demands of various consumer groups and related foreign markets. These farmers enjoy an extended share of agricultural production benefits from monetary and employment assessments, especially in developing regions such as Sub-Saharan Africa and Southeast Asia.

The geographical, economical and participating or decisive distance between the stakeholders within the processes of ecolabelling must be considered. The concern should therefore emphasise: What is being decided? Who is deciding? To what extent? At which stage of production? All of these initial and stimulating questions demand a scrutinising view on the consumer position as the main driving force for ecolabels.

26.4 Necessity of Small Scale Farmers' Involvement

Noting the previously mentioned characteristics and concepts of ecolabels within various markets, the importance of the involvement and the participation of small-scale farmers in the process can be concluded. In regards to the capacity of involving small-scale farmers into the processes of ecolabelling, a range of strict considerations are essential for apprehending the potential difficulties related to this matter while also shifting the focus to the need for farmer participation. Their involvement is essential within the decision-making processes to develop different mechanisms. Their participation should also be analysed to provide valid sustainable development purposes while integrating environmental health and socio-economic security on a local and global scale.

To find appropriate solutions for involving these farmers, the necessity of their participation, whether it is an active or a passive involvement, has to be conceived. Noteworthy arguments can be found by reasoning that from the free economic market point-of-view, ecolabels are perceived as mere market-oriented tools and may potentially bring financial benefit to the involved actors.

Participation from the concerned farmers group can be substantiated by access to new eco-labelled markets such as organic markets, ethnic markets, bio-markets, 'Lifestyles of Health and Sustainability' (LOHAS), and 'Corporate Responsibility' (CR) markets which leads to a guarantee of price premiums for the certified products in certified high-value markets (Chaturvedi and Nagpal 2002). Nevertheless, a price premium does not necessarily imply or guarantee the self-sustaining financial management of a farm.

The idea could be converted based on future-oriented gains while focusing on long lasting business infrastructures. Addressing these intensively consumer-determined markets means aligning to current consumer demands and trends. Consequently, this requires the satisfaction of both market and consumer demand while maintaining farm operations and livelihood. Other market-oriented arguments for the establishment and satisfaction of the ecolabel market would be pressurised by competiveness, where the only means of viability is based on the ability to compete. Conventionally, small-scale farmers are often excluded from certain markets due to asset shortages. This exclusion impedes the necessary investment to allow their entry into these markets. Basic market operation should be considered, enabling land tenureship at a minimum of 2 ha for farmers that are individually active as a market stakeholder. This could shift subsistence farmers into the range of a more active, economically feasible small-scale farmer that serves domestic, national, or even foreign markets (UNEP 2012). The market drives towards implementing these two approaches in integrating small-scale farmers as passive adapters or as active designers of the ecolabelling processes. Farmer participation seems to be the primary argument while environmental protection is an additional concern. This is due to the fact that ecolabelling is purely a market tool, using environmental standards as an orientation point for marketing. Attempts to integrate the farmers into decision making processes or convincing them to adapt ecolabels to their products can be accomplished. It can then be guided and shaped by both the farmers' needs or the market needs. Measures should be taken to ensure their livelihood is at least maintained or becomes more sustainable.

In regards to the farmers' actions, even a hint of neglect would cause disturbance in the production and supply chain of product flow from the farmers towards the consumer. Much more importantly, it would create a breakdown of the sustainable livelihood of the local and regional socio-economic network. The necessity of the farmers' involvement is given by handing over a voluntarily used financial opportunity and should not depend on consumer driven demand due to the highly flexible nature of consumer choices.

26.5 Integration Procedures in Ecolabel Markets

The integration of the farmer group into ecolabel markets is nestled in the concept of private certification and/or governmental regulation. Both approaches carry inherent advantages as well as particular constraints. Private certification is a multi-stakeholder, flexible, broadly accepted and international approach. However, it is merely voluntary and often lacks governmental legitimacy or fundamental support. On the contrary, governmental regulations provide state legitimacy, enforceability, and applicability but its' development and implementation is often very languid. Therefore, the integration of small-scale farmers fluctuates between arguments for mandatory government-associated measures and voluntary private-sector related measures (FAO 2008). The general quality infrastructure is defined by the combination of these two approaches for integration. The regular approach stands for basic safety through its technical requirements and legal methodology, whereas quality and standard information on the production infrastructure is introduced via the voluntary concept (Jaffee et al. 2011). As an essential part of sustainable management, providing space for innovation should be assured by using a combination of these approaches. Generally, a value chain is embraced with the starting point at the production of a good or service and the end point at certification. The position of the farmers and their production along the chain can be found right at the beginning. Their role makes a proportionally smaller contribution when compared to the decision making share of other contributors in regards to exporting. Exporting is often within the core trading system in ecolabel incentives when addressing western consumption tendencies. Therefore, the debate over the various kinds of integration methods into ecolabelling processes and markets should begin with some questions. The following questions emerge when assessing the distribution among the production and supply chain:

- How can small scale farmers change the entire chain by implementing on-farm standards?
- To which extent can environmental impacts of these farmers be compared and balanced with the environmental impacts of each step of the production and supply chain?
- How negatively or positively are the dependencies of producers on immediate chain neighbours affecting the ability to individually implement successful environmental and social standards on single farms?
- What are conclusively honest and effective incentives for small scale farmers adapting environmental standards of ecolabels, while being tightly embedded into the chain concept?

However, the general participation approach of small-scale farmers in the high value product markets must to be guided by certain landmarks, especially those centralised by the FAO (2008). The essential and interwoven key issues are a set of legal frameworks and regulation mechanisms, effective food control systems (regional food security), given certification and laboratory infrastructure, specific physical infrastructure, open-business development environment and established extension services. Firstly, a legal framework and its regulation mechanisms are related to policy recommendations in general. They address the poor integration of small-scale farmers into market systems while promoting the sustainability concept among all stakeholders nationwide (AEM and GIZ 2011). The government has to balance the relationship between environment maintenance and trade concerns. Governmental support for farmers can be realised through investment in research and easy implementation tools of the assessment results, and furthermore, matters of real land and device ownership.

An ecolabel is not related to an individual product but rather to a chain of production processes. Governmental management and integration of all stakeholders is essential to the manufacture of a product. Governmental management is especially necessary for the coordination and traceability of the entire chain that can be in turn, performed by the government and also supported by self-regulating mechanisms or third parties. An example of government participation involving small-scale farmers occurs in food safety legislation, where a national framework is integrated with international food trade requirements. Food safety laws are related to the need for an effective food supply to address health and safety for domestic and foreign consumers as well as for the local producers themselves (Grote 2002). Furthermore, the certification and laboratory infrastructure includes certifying bodies and auditing services in which the farmers' contributions should not be excluded or underestimated.

Traditional knowledge and experience passed down through the generations are inevitable concepts, which cannot be distanced from the sustainability approach. AEM and Africert are only two examples of local ecolabel infrastructure with a guaranteed political ownership over ecolabel mechanisms. AEM and Africert are also able to represent small shareholder certifications while counteracting the European top-down approach of certification by including their own internal control systems (AEM and GIZ 2011). This is closely connected with the necessity for attainable and physical infrastructure such as pack houses, transport, storage, water supply and irrigation techniques, reliable electricity access and land ownership. The infrastructure in such agriculture represents an example of governmental support and control while balancing trade, health, and environment using a holistic assessment and mitigation of environmental and health risks.

The next important landmark in involving small-scale farmers is an appropriate business development environment defined by improvements in production processing efficiency, financial support such as loans and micro-financing, and promotion of entrepreneurship skills and spirit. The last aspect embracing extension services is essential for establishing sustainability via ecolabel measures. Extension services include general capacity building, stakeholder training, information and education transparency for all stakeholders along the food chain. These services, discussed in detail in a following chapter, should be defined, promoted and accepted as the foundational tools for implementing the concept of sustainability, the core aim of ecolabelling.

Extension services as a landmark for sustainable participation of small-scale farmers into ecolabelling processes are promising. Some options for extension services include forming farmer associations, managing interests, handing over support farmers, addressing the entrepreneurship spirit, and setting up 'business-to-business' (B2B) structures. External activists, but even better internal activists, contribute significantly to development. These activists also create awareness through informational and educational advertising on the improvement of farming practices and conditions. Campaigning can also be used as a tool that relates environmental protection to direct benefits for farmers such as improved health, livelihood, nutrition, wealth, and sustainable farm management.

In general, creating awareness along the production chain and with the end consumer helps to bridge the gap between consumers and producers making ecolabel ideas transparent and possible. Transforming awareness into practical implementation is done mostly through fundamental capacity building and soft skills in finance administration, technology and/or management (FAO 2008). More specifically, capacity building depends on micro-financing, understanding farm equipment, adjusting payment fees and scaling for ecolabelling processes. For example, cost-sharing grants can be distributed and local, rural, and agricultural banks can be involved.

Additionally, the shift from the common top-down approach by third party non-local decision makers towards a bottom-up involvement process can bring deep-rooted, long-lasting benefits. This includes real participation, especially through active decision-making and feasibility statements from the farmers on integrated environmental farm management. It should not be dependent on foreign third parties that tend towards impulsive green consumption habits. Again in this case, the relevance and character of traditional knowledge should not be given any less value than the newly emerged sustainability approaches. Both tactics are based on their ability to perform sustainably and conduct environmentally friendly agriculture practices.

26.6 Benefits and Constraints Behind the Ecolabelling Concept for Small-Scale Farmers

Incentives for the targeted farmers should be established on a level where benefits and constraints are taken into consideration and are balanced holistically. The previously discussed operating framework of ecolabelling as summarised by the FAO in 2008 also includes intersectional provisions of certain on-farm tools, materialistic as well as informative. All of this is guided by environmental objectives such as encouraging the efficient management of renewable resources to ensure their availability to future generations; promoting the efficient use of non-renewable resources, including fossil fuels; facilitating the reduction, reuse and recycling of industrial, commercial and consumer waste; encouraging the protection of ecosystems and species diversity; and encouraging the proper management of chemicals in products (GEN 2004). These principles and tools act as support for other measures such as record keeping, technical agricultural practice (crop management) and further documentation at the farm level, especially through self-regulation. Capacity building through the topic-specific manuals and forms available from the WTO also assists in promoting sanitary and phytosanitary quality management. The goal related benefits, including general and specific health issues, are an omnipresent concern for small-scale farmers. Further support for benefits is found within integrated agricultural management. For example, this includes more crops per drop, mineral fertilisers and pesticides savings, integrated crop and seed management, recycling methods like composting, and holistic sustainable management of resources that the farmers depend on. These resources can include soil, man-power, plants, climate conditions, and time management. The idea of ecolabels carries a slight spirit of biodynamic farming while allowing for a smaller workload with good yield and balancing different farm elements. The efficient farming measures, besides sanitary and phytosanitary health, aim to increase economic and monetary benefits (Chaturvedi and Nagpal 2002). Namely, this could occur through lower investment costs from the application of artificial fertilisers and pesticides, as well as decreases in water and electricity costs through reduced consumption. Ecolabelling, as its vision indicates, stimulates long-term planning by protecting the earth which farmers depend on; mitigating health risks; striving to create a sustainable business with economic security for families; and establishing long-term, trustworthy business relationships with neighbouring actors in the production and supply chain (FAO 2008).

Constraints are derived partially from monetary aspects but also from market management forces. The monetary aspects are directly and indirectly represented by certification costs. Ecolabel adaption costs are diverted into initial investment costs (fixed) and recurrent costs (certification interval). These costs vary based on existing infrastructure, technical and business services, extent of capacity building, individual or group certification realisation and occurrences of local certification bodies and laboratories. For feasibility, all investments have to be covered by price premiums for ecolabel investment pay offs. However, a price premium does not conclusively include profit premium for farmers (UNEP 2012). Yet, supportive measures, according to the profit issue, can act as cost sharing along the production and supply chain while balancing costs and responsibilities to ensure long-term sustainability.

Additionally, small-scale farming has the potential to promote a better balance between urban and rural areas by closing the nutrient cycle. This vibrant and dynamic relationship between rural and urban activities could bring further benefits through its socio-economic focus, helping to inhibit urban migration through rural employment and meeting urban food security via well-developed services. Strengthening and guaranteeing structures benefiting rural farming would bring stable and secure family planning due to the fact that approximately 2.5 billion individuals are involved in agriculture as farmers or workers and at least 1.5 billion live and work on small family-run farms (Quan 2011). Based on the informative aspect, ecolabel products have the potential to lessen the gap between small-scale producers and end consumers. Small-scale farmers can be detached from the market-burden to support the shift towards sustainable consumption and production patterns, especially by promoting the socio-economic producers through fair trade characteristics (Janisch 2007).

26.7 The West African Situation

Ecolabel trends have prevailed in relation to small-scale farmer conditions. Pascal Gbenou emphasised that the problem of small-scale farmers is "not in accessing European markets but to be able to feed the local people and themselves" (Pascal Gbenou, 2012, personal communication). He stresses that, "Nothing should be disturbing, considering high ecolabel requirements, the capacities of local producers tackling food sovereignty and satisfying local demand and creating a reliable food security for stressed countries and societies. Meeting the needs of western demand often goes along with constraint and harsh competition, as well as delayage [buzz words]" (Pascal Gbenou, 2012, personal communication).

In general, Africa often experiences large amounts of food importation coming mainly from outside of Africa. These import structures are accompanied by relatively low prices that cause and feed competitive product situations for small-scale farmers who implement sustainable farm approaches. Currently, approximately 25 % of African food demand is imported with an increasing trend due to the inability to meet essential food requirements. The reasons for food shortages and whether or not this can be overcome should be questioned while considering the impacts of imports and local markets' non-competitiveness. In 2008, West Africa was afflicted by a severe food crisis bringing the "...question of the implications of West Africa's dependency on global food markets" into focus (Pascal Gbenou, 2012, personal communication). At present, approximately 40 % of West African rice is imported. A share of 20 % of this rice is traded on the international market. The rice-growing capacity in West Africa is substantial with Nigeria having the highest potential at 2.4 million hectares. Studies and research on 'Systems of Rice Intensification' (SRI) have become the main focus of local and experienced agricultural stakeholders, enhancing the current 1.5 and 3.6 t paddy rice cultivation (numbers referring to Liberia and Senegal relatively). Even so, the production costs are comparable to world market cultivators due to the ability to improve productivity and process capacity. Due to high market costs, local farmers are hampered in participating sustainably in local and regional businesses.

Nevertheless, numbers show that the change in global rice prices has reached West Africa more slowly and to a lesser extent (first quarter of 2008: international rice prices tripled, doubled in Senegal, increased by half in Mali and Benin). Food insecurity in urban areas is becoming more severe and more frequent, drastically affecting household food security (expenses for urban households in 2008 increased from 20 to 25 % for rice). Nevertheless, West African nations are struggling for consistent and reliable food security while on-going export trading still occurs, as seen in Beninese and Belgian supermarkets. To strengthen local farmer capacity for self-sufficiency in rice production, long-term measures have been adapted to meet the needs of local markets to possibly establish and continue trading without cutting back on local food security (Aker et al. 2011). These are the essential steps towards balancing global production and consumption. However, market-related infrastructural circumstances and climatic stresses manifest and imply that West Africa's dependency on international markets are impeding the farmers from maintaining their position as the backbone of the African economy and holding responsibility for food security through small-scale cash crops and beyond-subsistence agriculture (WAC 2011). Recent topics—"energy in lieu of food"; food sovereignty related to land tenure; integration of gender roles in sustainability; and the accumulation of land and water resource grabbing via Foreign Direct Investment (FDI)—are questioning and scrutinising the potential and benefits of ecolabels for small-scale farmers applied under these conditions.

26.8 Conclusions and Recommendations

The initial question for this paper is based on the potential of involving small-scale farmers into ecolabelling processes and in accordance with the farmers' interests. This entails marking their products with a broadly known and accepted eco-certification. After discussing certain arguments and necessities pertaining to participation, it can be concluded that all of the involved aspects have to be balanced. Ecolabels occur as a regulated governmental or voluntary private market-influential policy tool. Above all, ecolabels are aligned to the consumers' conscience and personally designed needs. The consumer-orientated aspect urges for a debate over potential discrimination and disadvantages for the farmers who join the ecolabel market. The developed and expected environmental standards and feasibility should be assessed to avoid environment-related non-tariff barriers (ETB) and therefore, economic and livelihood maintaining discrimination (Chaturvedi and Nagpal 2002). Market access for small-scale farmers should be available without imposing potential constraints or generating mismanagement. The financial burden for small-scale farmers in joining ecolabel parties represents a pressure on agricultural performance and therefore, provides general poverty alleviation, as is often promoted (UNEP 2012). Even more competitive pressure is being injected into farm management by making these farmers strongly dependent on the consumer tendencies that are based on market dynamics and demand (Chaturvedi and Nagpal 2002).

The concept of ecological agricultural production is included in the ecolabelling standards and its implementation can be supported beyond the idea of providing commodities for the global market. Many small-scale farmers in Southeast Asia and Africa practice these standards but do not want to spend the recurring certification costs or are unable to pay for it. If the focus is set primarily on addressing the ecolabel market for western countries, Africa and Southeast Asia's backbone, small-scale agriculture, might be threatened and may not perform sustainably (UNEP 2012).

The distance to the end-market and the contradiction between trade and environmental needs are obstacles for honest and holistically beneficial concept adaption. Small-scale farmers should be actively engaged in minimising the gap between trade demand and environmental health. Albeit, these farmers can benefit from the ecolabel market without being exposed to the potential burdens of merging under the umbrella of farmers associations, making various farm measures easier through the sharing of finances and equipment (Le Courtois et al. 2011). This can also regulate homogenously applied sustainable agriculture by shifting small-scale, independent land segmentation towards well-organised, sustainable small-scale agriculture beyond mere subsistence performance. Concepts such as holistically applied infrastructural concepts to water, waste, fertiliser, financing and delivery infrastructure should also be considered. Using ecolabels can boost domestic markets and be a platform for innovation. Establishing and guaranteeing environmental standards can contribute positively to local food security and health while promoting sustainable consumption within affected regions.

In closing, current ecolabelling concepts are quite one-sided and should be questioned as to their intentions and focuses. Small-scale farmers should not be a means to satisfy only the western market. The necessity of integrating social and economic benefits for producers in the market system should be naturally initiated by an enriching cooperation of fair-trade and ecolabels (Le Courtois et al. 2011). Furthermore, the idea of defining small-scale farmers as suppliers for western, non-domestic markets has to be reconsidered and shifted towards domestic and regional enhancement. Even so, developing countries hosting a majority of small-scale farmers are able to gain from market liberalisation and market access opportunities, especially in agricultural trade. Domestic sustainable development should be promoted through capacity building and therefore, extension services should be supplied. Ecolabels as of today are not intended to focus on the pure sustainable livelihood condition of small-scale farmers. However, transforming this market tool into a concept considering regional wealth and food security for the small-scale farmers' area, as well as applying the concept to all involved stakeholders along the production chain, could lead to an efficient and beneficial condition for consumers, intermediate actors, and all types of producers.

Acknowledgment This assessment work is based on highly detailed communication between the authors and the experienced and exemplary organic farmer Pascal Gbenou, S.A.I.N., Benin, as well as direct consultation with Fabian Mauermann, Tamera, Portugal. The accessory legwork of Henry Mensah and Michael Acquah supported the findings in the conclusion.

References

AEM and GIZ – African Eco-labelling Mechanism and Gesellschaft für Internationale Zusammenarbeit GmbH (2011) EcoMark Africa: final draft report of the first sector working group meeting. http://www.ecomarkafrica.com/downloads/EMA_SWG_Report.pdf. Accessed 22 Sept 2013

Aker JC, Block S, Ramachandran V, Timmer CP (2011) West African experience with the world rice crisis, 2007–2008. CGD Working Paper 242. Center for global development, Washington, D.C. http://www.cgdev.org/files/1424823_file_Aker_Ramachandran_Block_Timmer_West_Africa_FINAL.pdf. Accessed 22 Sept 2013

Chaturvedi S, Nagpal G (2002) WTO and product related environmental standards: emerging issues and policy options before India. Econ Polit Wkly 38(1):66–74

FAO – Food and Agriculture Organization of the United Nations (2008) Supporting farmer compliance with private standards: agrifood systems brief. Rural infrastructure and agro-industries division, FAO. http://www.enaca.org/modules/library/permalink.php?id=oai:enaca.org:1208420021. Accessed 22 Sept 2013

GEN – Global Ecolabelling Network (2004) Information paper: introduction to eco-labelling. http://www.globalecolabelling.net/docs/documents/intro_to_ecolabelling.pdf. Accessed 22 Sept 2013

Grote U (2002) Environmental and food safety standards and international trade: concerns and challenges for developing countries. Center for Development Research ZEF, Bonn

Jaffee S, Henson S, Rios LD (2011) Making the grade: smallholder farmers, emerging standards, and development assistance programs in Africa. A research program synthesis. Report no. 62324-AFR. Economic and sector work, The World Bank, Washington, DC. http://siteresources.worldbank.org/INTARD/Resources/Making_the_Grade_ePDF2.pdf. Accessed 22 Sept 2013
Janisch C (2007) Background assessment and survey of existing initiatives related to eco-labelling in the African region. http://www.unep.org/roa/docs/pdf/RegionalAssessmentReport.pdf. Accessed 22 Sept 2013
Le Courtois E, Galvez-Nogales E, Santacoloma P, Tartanac F (2011) Enhancing farmers' access to markets for certified products: a comparative analysis using a business model approach. Agricultural management, marketing and finance working document no. 28. FAO, Rome. http://www.fao.org/docrep/016/k9849e/k9849e.pdf. Accessed 22 Sept 2013
Pye-Smith C (2011) Farming's climate-smart future. Placing agriculture at the heart of climate-change policy. Technical Centre for Agricultural and Rural Cooperation AC P-EU (CTA)
Quan J (2011) A future for small-scale farming. Science review: SR25. Foresight project on global food and farming futures. The Government office for science (UK), London
UNEP – United Nations Environmental Programme (2012) Ecolabeling – as a potential marketing tool for African products: an overview of opportunities and challenges. http://www.unep.fr/scp/marrakech/consultations/regional/pdf/MTF_CoopAfrica_EcolabellingBrochureDOC.pdf. Accessed 22 Sept 2013
WAC – West African Challenges (2011) West African Challenges. Sahel and West Africa. Club Secretary SWAC. OECD N°02

Chapter 27
Tariff Preferences for Sustainable Products: A Summary

Philipp Schukat, Jenny Rust, and Julia Baumhauer

27.1 Introduction

With the introduction of the Generalised System of Preferences (GSP), the European Union (EU) developed a trade policy tool which allows easier access to the EU economic zone for emerging and developing countries through granting tariff preferences for the importation of industrial and agricultural products. The aim of the regulation is to support developing countries in their efforts towards poverty reduction, good governance and sustainable development (BMWI 2013; CARIS 2010, p. 21). This ambitious goal is to be achieved, among others, through the so-called GSP+ which grants a selection of countries additional tariff preferences if they meet certain criteria. One of the preconditions is the ratification of fundamental international conventions covering human rights and labour standards, the appropriate use of environmental resources and good governance. However, although the regulation has had an impact on the legislative framework in partner countries, when it comes down to driving changes at the operational level and to improving the actual conditions of production in those countries; the approach still reveals weaknesses (CARIS 2010, p. 10).

P. Schukat (✉)
Social and Environmental Standards Programme, Deutsche Gesellschaft für Internationale Zusammenarbeit GmbH (GIZ), Eschborn, Germany
e-mail: philipp.schukat@giz.de

J. Rust
Social and Environmental Standards Programme, Deutsche Gesellschaft für Internationale Zusammenarbeit GmbH (GIZ), Eschborn, Germany

Como Consult GmbH, Winterstraße 4-8, 22765 Hamburg, Germany
e-mail: jenny.rust@como-consult.de

J. Baumhauer
Give Something Back To Berlin, Lausitzer Str. 10, 10999 Berlin, Germany
e-mail: j-baumhauer@web.de

C. Schmitz-Hoffmann et al. (eds.), *Voluntary Standard Systems*, Natural Resource Management in Transition 1, DOI 10.1007/978-3-642-35716-9_27,

This chapter will discuss the question of how the impact of a trade policy tool like GSP could go beyond the sole improvement of legislative frameworks and lead to an actual change in production conditions in emerging and developing countries. The focus lies in the promotion of compliance with social and environmental standards in producer states. As demonstrated by GSP+, the EU can link the award of tariff preferences to certain development targets and sustainability criteria. The idea lies in a linkage of the GSP with the provision of evidence regarding sustainable production, which would allow tariff preferences for sustainably produced goods. This chapter examines whether sustainable business activities can be promoted through a system of tariff preferences and whether state-level recognition of certification systems would be practicable in the case of the EU's own GSP. The chapter comprises the results of a study published in 2012 by the Deutsche Gesellschaft für Internationale Zusammenarbeit (GIZ), which deals with the potential role of sustainability standards in generalised preference systems based on the European Union model (GSP).[1] The study (GIZ 2012), as well as its summary, are merely offered as an outline and would need to be fleshed out in greater detail should such a system actually be adopted.

We begin with taking a closer look at the previous strategy for the promotion of sustainable development under the GSP. It suggests to further support the efforts of the private sector through a system of state recognition for certification systems and the extension of the existing trade policy tool for sustainable development. Section 27.3 will discuss strengths and weaknesses of state-level recognition of private certification systems and outlines the framework conditions for possible implementation. Section 27.4 drafts requirements and principles for the realisation of such a state-level recognition system for sustainable goods and points out the potential opportunities the approach implies towards shaping sustainable conditions of production in the countries of origin. The chapter ends with conclusions and recommendations in Sect. 27.5 highlighting the effects on business activities in producer states as well as the transferability of the approach to other contexts.

27.2 The EU's Generalised System of Preferences (GSP)

The European Union's Generalised System of Preferences (GSP) aims to promote trade with emerging and developing countries by awarding tariff preferences that facilitate access to the EU market. The origins of the GSP date back to the early

[1] The study 'Tariff preferences for sustainable products: An examination of the potential role of sustainability standards in generalised preference systems based on the European Union model (GSP)' had been published in 2012 by The Deutsche Gesellschaft für Internationale Zusammenarbeit (GIZ) on behalf of the German Federal Ministry for Economic Cooperation and Development (BMZ).

1970s. It is based on a series of recommendations of the *United Nations Conference on Trade and Development* (*UNCTAD*) that developing countries should be awarded preferential treatment to help boost their share of international trade. The general objective of the GSP is to increase and diversify exports from emerging and developing countries.

The current GSP is sub-divided into three 'regimes':

- the standard GSP (currently 176 beneficiary countries);
- the GSP+ (*Special incentive arrangement for sustainable development and good governance*), which grants additional preferences to a select group of developing countries[2] subject to evidence that they have ratified and implemented 27 international conventions and whose share of total EU imports of a specific product from all GSP countries does not exceed 1 %[3];
- the *Everything but Arms Programme* (EBA) for the 50 least developed countries (LDCs). Under EBA rules, products from the least developed countries (with the exception of arms and munitions) may be imported into the EU without quotas or customs tariffs.

A 'graduation' system (from which countries in the EBA programme are exempted) is also in place to ensure that countries whose exports in one or more product groups exceed 15 % (12.5 % for textiles) of total imports into the EU are no longer awarded preferential treatment as they are clearly able to survive in normal competition. By contrast, in response to changing trade flows, a parallel system of 'de-graduation' means that tariff preferences can be awarded to GSP countries previously excluded from beneficial treatment.

To ensure that the GSP can be regularly updated in response to changes in the system of multilateral trade, a decision was taken to implement it in 10-year cycles (current guidelines 2006–2015). The system is implemented in the form of consecutive regulations, each of which remains in force for 3 years. Regulations are updated on the basis of key trading data. Countries that wish to benefit from the special incentive rules of the GSP+, in addition to the GSP, must submit a formal application with evidence that they meet the criteria.

The current GSP regulation came into force in 2009 and on this basis, was to expire at the end of 2011. However, in late May 2011, the EU Commission decided to extend the existing regulation up to the end of 2013 to allow more time for the current review process. The new GSP regulation is expected to come into force in 2014. At present the GSP covers approximately 7,000 dutiable goods and 176 countries qualified for the GSP[4] (UNCTAD 2011).

As the interim report produced by the Centre for the Analysis of Regional Integration at Sussex demonstrates, trading under the GSP is predominantly

---

[2] The number of countries that currently benefit from preferential treatment under the GSP+ is 16.

[3] To be raised to 2 % in line with current EU proposals for reform of the GSP (CARIS 2011, p. 1).

[4] The number of countries that actually benefit is lower, as some have signed up to bilateral or multilateral trade agreements and already make use of these.

restricted to just a few sectors, including "live animals, vegetable products, processed foodstuffs, textiles and clothing" (CARIS 2010, p. 8). On this basis, the comprised study focuses on agriculture, fisheries, flowers and the textiles and clothing sector.[5]

27.3 Co-Regulation as an Instrument for Promoting Social and Environmental Standards

A system of state recognition for sustainability standards would allow private-sector certification systems to furnish evidence to governments that certain policies (e.g. laws) are being implemented. Any such recognition would be based on the sustainability and system requirements defined in legislation. The use of a private-sector instrument by the state is known as co-regulation (see Chap. 7 for a detailed discussion of the co-regulation concept). This innovative approach makes it possible to employ systems developed by the private sector as a complementary tool for the implementation of government policies. One example of this kind of 'co-regulation' is the recognition system adopted by the EU to help implement its Renewable Energy Directive (RED).

Advantages of co-regulation and a system of state recognition for sustainability standards:

- The recognition of already established private-sector certification systems would safe a great deal of development time.
- Accordingly the costs to the state regulatory body would be limited to initial recognition and the subsequent monitoring. The costs of the initial and ongoing development are with the private certification system.
- The increasing range of private certification systems can better meet the differing sectoral and regional demands and allows a beneficial competition between those systems in terms of providing the most efficient method of implementing the state's minimum requirements. A 'race to the bottom' would be prevented.
- Based on internationally recognised and WTO-compliant ISO quality standards, operational audits are performed by accredited private certification companies. These release the state regulator from auditing business operations in sovereign non-EU states.
- Sensitive judgments on policy in non-EU states can be mainly avoided. Assessments are only being made on the performance of private social and environmental standards systems.
- Certification can directly promote or sanction against any failure in the implementation of sustainability criteria at individual company level. This can be done

[5] The sectors examined in this study were chosen in the light of research (using data from the International Trade Centre's (ITC) Intracen database) into the main products imported into the EU from a selection of GSP countries.

without either granting or withdrawing tariff preferences in a blunt fashion across all companies of one state, as it could theoretically happen under the terms of the current GSP+ which could punish states that ignore the designated sustainability criteria through the suspension of tariff preferences.

- Co-regulation gives the private sector the flexibility it needs to develop the most efficient certification system possible and integrate a system of value chain management. In turn, the state needs only to focus on monitoring the effectiveness of those systems.

However, it is important to give attention to potential weaknesses of the approach, which would have to be considered in the development of a system of state recognition for sustainability standards:

- The initial and ongoing development of private-sector certification systems is quite cost and time consuming, since it involves wide-ranging discussions and integrating processes of different interest groups. However there is already a wide range of certification systems in existence, which could be used in a system of state recognition and tariff preferences. Nevertheless the long development process for certification systems needs to be considered by the state regulator when drawing up any schedules.
- Private-sector certification systems cannot guarantee 100 % accomplishment nor can they promise 100 % sustainability. However, tariff preferences for sustainable products could be a strong incentive for sustainable production and development (cf. ISO 17021:2006).

27.3.1 General Requirements for a State Use of Private-Sector Certification Systems

Through a system of state recognition, the regulatory body sets a framework for minimum requirements. This results in questions about the implementation as well as how to deal with deviations. Additionally there is a need for regulations on accessibility and the provision of information. Proposals on how to deal with some of those issues are submitted below.

- Over-compliance of minimum requirements will be recognised, whereas partial compliance requires mechanisms, which for instance allow provisional recognition with integrated deadlines for compliance with outstanding criteria (phasing in) or partial approvals, e.g. for a specific product group (EU 2010).
- Most certification systems are committed to a process of continuous optimisation, e.g. in terms of user-friendliness and effectiveness. Consequently, the certification systems are obliged to notify the regulator of any modifications that affect the requirements before they take effect (EU 2010).
- As a matter of principle, responsibility for compliance with the standard lies with the producer (ISO 17021:2006; 4.4.1). In the event of misuse of tariff

preferences the guiding principle applied should be that of 'good faith' in view of the importers. However the extent of any liability of the certification system itself, or the certification office, needs to be clarified and it has to be decided which requirements can be applied for a sufficient proof of compliance.[6]

- The exclusion of the respective producer could act as an effective deterrent against any kind of misuse. However, it is also in the interests of certification systems to protect their own trade-marks. Most of the certification systems therefore already employ a number of mechanisms to protect their name and 'tried-and-tested' procedures to deal with misuse are in place.
- It has to be clearly allocated that the legal authority lies with the state regulatory body. Certification systems in turn act as a complementary instrument whose expertise will help to manage global supply chains.
- The usage of certification systems has to be equally accessible for all companies including small and medium-sized enterprises (SMEs). This has been made possible by developing numerous adaptations, such as e.g. group certification schemes for small farmers.

27.4 EU/GSP Recognition of Certification Systems

The following section gives an example of how state recognition of certification systems could be integrated into the EU GSP Regulation. The idea is to incorporate the recognition system into the GSP's legal framework, which would determine the sustainability and system requirements. This would require the establishment of internationally recognised sustainability criteria and system requirements by the EU. The first part of this section will therefore look at which sustainability requirements could be applied, what stages of production would have to comply with them and which system requirements would need to be in place. The second part focuses on the issues surrounding the implementation of a system of state recognition linked to the GSP. It looks at already existing certification systems for the sectors covered by the GSP and offers recommendations on how clearing of imported goods through customs, under a system of state recognition for private certification systems, and tariff preferences for sustainable products could be regulated.

[6] cf. FSC—Trademark Assurance (FSC 2013) or MSC—Report ecolabel misuse (MSC 2013).

27.4.1 Sustainability and System Requirements

Legitimacy of Sustainability Criteria

In defining sustainability criteria and system requirements which would build the basis for a state recognition of certification systems as part of an EU tariff preferences system, it is essential that the developed criteria are internationally recognised and their legitimacy is guaranteed. Therefore it is recommended, that the principles and criteria underlying the EU's sustainability requirements must relate to international conventions and guidelines that have been recognised by a large number of states at different stages of development and that have been developed in participatory and transparent processes (Charnovitz et al. 2008). International standards could be for example International Labour Organization (ILO) or environmental conventions such as the UN Framework Convention on Climate Change (UNFCCC) or the Convention on Biological Diversity (CBD) or guidelines like the FAO Codes of Good Agricultural Practices (GAP).

Stages of Production to Comply with Sustainability Requirements

In our globalised economy, value chains now tend to cross a number of national borders. Cotton from Mali is exported to China, where it is processed into fabrics that are made into shirts in Bangladesh. These are then sold in the EU. However, the GSP grants different tariff preferences to different non-EU states. This problem was addressed through the introduction of 'rules of origin', the aim being to determine at what point a product should be regarded as having been made in a certain country. A simplified explanation is that the current rules are based on the principle that for a country to be regarded as the producer, it must have added sufficient value to the product. So under the rules of origin, it is not possible for China to have the shirts ironed in Bangladesh simply in order to benefit from that country's tariff preferences.

Applied to a tariff preference system for sustainably produced goods this means that the sustainability requirements would have to be observed by those companies identified in the existing GSP that are regarded as 'defining the origin' on the grounds that sufficient value has been added.

For many agricultural products, the rules of origin stipulate that they have to be cultivated within the country. The sustainability requirements would then apply to the cultivation of the product. The same rules would apply to agricultural products and their subsequent processing in the same country.

Credibility Through System Requirements

As well as minimum social and environmental standards, a system of state-level recognition would also need to define a set of quality requirements covering implementation of the standards system. This would avoid the risk of arbitrary implementation and ensure the application of clearly structured, objective and transparent procedures to guarantee credibility and independence of certification systems.

One of the key elements of a credible system that wishes to obtain state recognition should be a set of methods of regulating accreditation and certification processes within those systems and methods of auditing the value chain. This could include for example, an impartial compliance audit, i.e. compliance with the corresponding standards would be checked by an independent third party.

Additional requirements should be that a complaints mechanism is in place, which guarantees the transparent and objective processing of complaints.

Another important factor is the monitoring of custody chains. Chain-of-custody controls are intended to ensure, for example, that a tonne of sustainably produced flowers from Kenya is not then classified as 3 tonnes of supposedly sustainable flowers on arrival in the Netherlands as a result of a fraudulent declaration. Any organisation that is involved in trading or processing certified goods needs to be monitored accordingly. Specific monitoring arrangements are generally set out in the chain-of-custody requirements of established certification systems and used to record evidence of sustainability. In terms of a tariff preference system for sustainably produced goods, sustainability requirements must be observed by the companies that determine a product's origin. Consequently, the chain of custody only needs to be monitored from these companies onwards.

27.4.2 Practical Implementation Issues

Existing Certification Systems

State-level recognition would need to be based on credible and established certification systems (see Fig. 27.1). A large number of certification systems have been developed and used in recent years for those sectors of particular relevance to the GSP, i.e. agriculture, textiles and clothing, flowers and fisheries. In terms of content, most of the existing certification systems already cover the relevant minimum requirements fully or to a great extent, so there would be no need to set up completely new certification systems. Furthermore, EU recognition of these certification systems would promote further harmonisation and establish a minimum standard for sustainability.

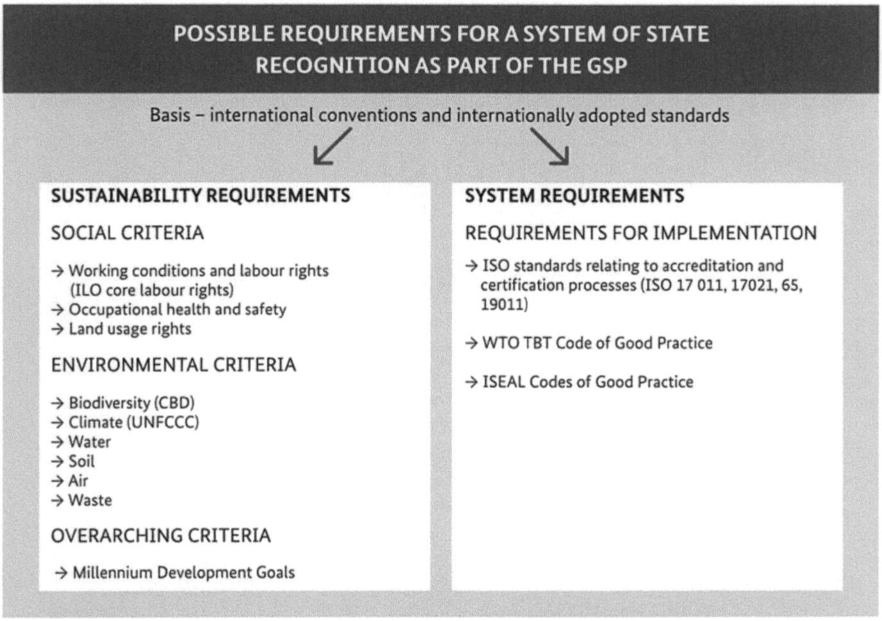

Fig. 27.1 Possible requirements for a system of state recognition as part of the GSP (*Source*: GIZ 2012)

Technical Issues: One Sustainability Certificate for All Systems

Individual certification systems would be authorised to issue a single, uniform sustainability certificate for customs/GSP purposes. Instead of multiple certificates issued by different systems, the importing state's customs officials would simply have to deal with the new EU sustainability certificate. A clearance check on one additional document (e.g. a sustainability certificate) would not entail any significant increase in costs for the importer or the customs authority itself.

This conclusion is primarily based on the fact that the GSP regulation applies across the EU and does not have to be transposed into national law. Accordingly, apart from the customs authority, there would be no need for a further official body to check the certificates or other documentation. The systems would be recognised at EU level, and the EU would itself determine how to deal with any misuse, e.g. counterfeit certificates.

Cost Sharing

Companies that wish to take advantage of the GSP would be expected to provide evidence of sustainable business management using a recognised system of standards. There would be support for the efforts of governments in GSP countries to

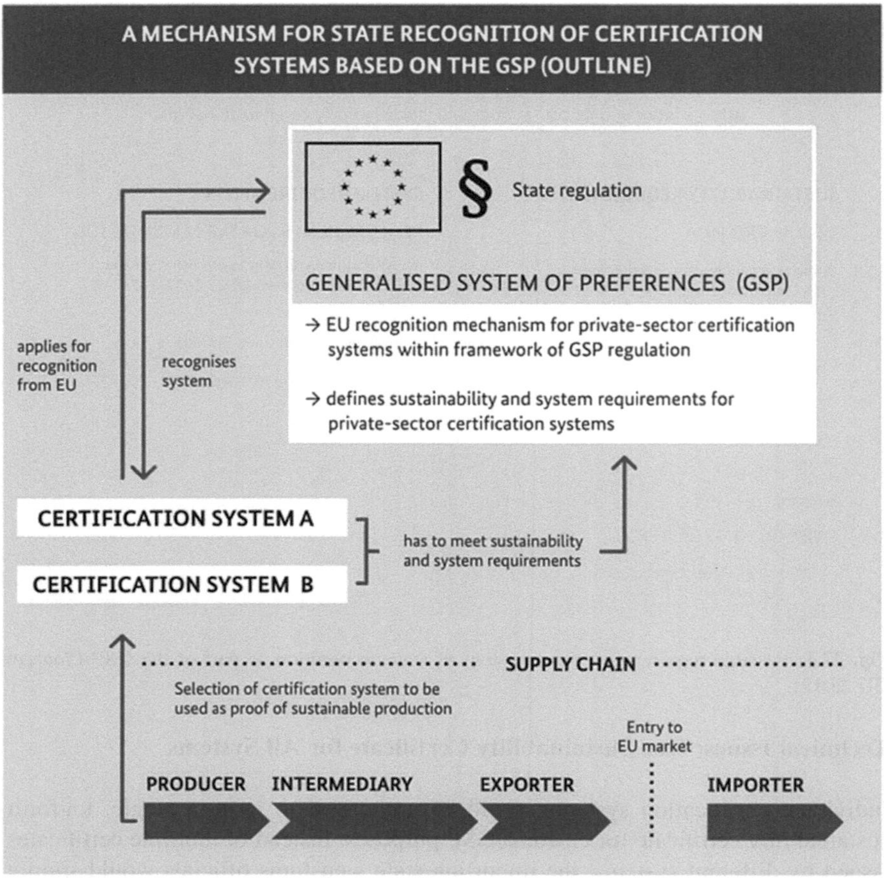

Fig. 27.2 A mechanism for state recognition of certification systems based on the GSP (*Source*: GIZ 2012)

monitor and enforce minimum social and environmental standards at the production stage. While the EU would bear the cost of providing a financial incentive in the form of preferential tariffs, a large part of the corresponding monitoring costs would be paid for through the value chain or by consumers. Information on exporters is already collected and stored as part of the registration process, so the only additional dataset required would be that containing the stipulated evidence of sustainability.

The following Fig. 27.2 draws an example of a potential mechanism for state recognition of certification systems.

27.5 Conclusions and Recommendations

In recent decades there has been a considerable increase in demand, especially in industrialised countries, for products that have demonstrably been manufactured in compliance with social and environmental criteria. Increasingly, companies from many different sectors that import widely on the global market from emerging and developing countries and public sector bodies in various European Union (EU) countries have committed themselves to only sourcing products that have demonstrably been manufactured using sustainable methods.

As we work towards a global system of sustainable development and trade, it is important not only to generate demand on one side of the market but also to give producer states (in particular developing countries) the opportunities and support they need to engage in international trade and strengthen their efforts towards sustainable development. Linking the GSP to certification systems would create a strong economic incentive for manufacturers in our partner countries to embrace sustainable production. At the same time, it would support the efforts of European businesses to make their products more sustainable, particularly given the increasing importance of this issue within Europe in the competition for markets. For producer countries, the benefit would lie in guaranteed and increased demand for sustainably produced goods over the long term, thus providing a greater incentive to invest in sustainability. It is also reasonable to assume that a large share of the preferences would benefit producers and could therefore be used to finance investment.

The chapter shows that through state-level recognition of certification systems, it would be possible to exploit synergies between those systems and the GSP. This would be an effective way for the EU to support the efforts of the private sector in this area while improving the efficiency of its own set of regulatory tools. Recognition would be subject to compliance with established sustainability criteria and minimum requirements on the proof/certification systems. The sustainability criteria would be based on international agreements, like e.g. ILO conventions (ILO 2013) and the corresponding system requirements would be underpinned by internationally recognised ISO standards covering, for example, the details of certification. This would guarantee credibility and legitimacy of the approach and lead to international acceptance.

As shown in this chapter, certification systems already exist in the main sectors covered by the GSP (agriculture, fisheries, flowers and textiles) and could be used to measure compliance with sustainable production criteria. The ongoing costs of developing certification systems are already financed by the systems themselves. This would also contribute to keeping time and costs involved within acceptable limits. In terms of trade and development policy, integrating sustainability criteria into the GSP through state-level recognition of certification systems would be a feasible and attractive means of promoting sustainable business activity in developing countries, and of expanding the market for sustainably produced goods.

The ideas and proposals laid out in this study could also be transferred to bilateral trade agreements or incorporated into other preference systems. The core principle, i.e. state recognition of certification systems, could also be transferred to other areas such as public procurement. This study therefore recommends that the approach be examined in greater detail for possible application in other contexts like trade agreements or public procurement.

However, the proposal is merely offered as an outline and different aspects such as e.g. the potential impact on competition and the credibility of systems should be discussed in greater detail.

References

BMWI – Bundesministerium für Wirtschaft und Technologie (2013) Die Abkommen der WTO: Übereinkommen über technische Handelshemmnisse – TBT (WTO Agreements; Convention on Technical Barriers to Trade – TBT). http://www.bmwi.de/DE/Themen/Aussenwirtschaft/Handelspolitik/wto,did=217768.html. Accessed 02 Sept 2013

CARIS – Centre for the Analysis of Regional Integration at Sussex (2010) Mid-term evaluation of the EU's generalised system of preferences: Final report. http://trade.ec.europa.eu/doclib/docs/2010/may/tradoc_146196.pdf. Accessed 02 Sept 2013

CARIS – Centre for the Analysis of Regional Integration at Sussex (2011) A preliminary investigation into the effects of the changes in the EU's GSP'. http://www.tradesift.com/Reports/GSP_Report_June2011.pdf. Accessed 02 Sept 2013

Charnovitz S, Earley J, Howse R (2008) An examination of social standards in biofuels sustainability criteria. IPC discussion paper – standards series. International Food & Agricultural Trade Policy Council. http://www.agritrade.org/documents/SocialStnds_Biofuels_FINAL.pdf. Accessed 02 Feb 2013

EU – European Union (2010) Communication from the commission on voluntary schemes and default values in the EU biofuels and bioliquids sustainability scheme (2010/C 160/01). In: Official Journal of the European Union C 160, Vol. 53 of 19 June 2010, pp 1–7

FSC – Forest Stewardship Council (2013) Trademark assurance. http://www.fsc.org.vm-fsc-entw.tops.net/trademarkassurance.html. Accessed 02 Sept 2013

GIZ – Deutsche Gesellschaft für Internationale Zusammenarbeit GmbH (2012) Tariff preferences for sustainable products: an examination of the potential role of sustainability standards in generalised preference systems based on the European Union model (GSP). Report published on behalf of the Federal Ministry for Economic Cooperation and Development (BMZ). GIZ, sector project on trade policy, trade and investment promotion, Bonn and Eschborn

ILO – International Labour Organization (2013) NORMLEX – information system on international labour standards. http://www.ilo.org/dyn/normlex/en/f?p=NORMLEXPUB:1:0. Accessed 02 Sept 2013

ISO – International Standardization Organization (2006) ISO 17021:2006; Conformity assessment – requirements for bodies providing audit and certification of management systems

MSC – Marine Stewardship Council (2013) Report ecolabel misuse. www.msc.org/get-certified/use-the-msc-ecolabel/report-ecolabel-misuse. Accessed 02 Sept 2013

UNCTAD – United Nations Conference on Trade and Development (2011) Generalised system of preferences. List of beneficiaries. www.unctad.org/en/docs/itcdtsbmisc62rev5_en.pdf. Accessed 02 Sept 2013

Chapter 28
The Political Challenge of Voluntary Standard Systems

Günther Bachmann

28.1 Introduction

On the agenda for sustainable development strategies, voluntary standard systems, VSS, are an emerging governance feature. In both the public and private sector, their relevance is growing, although it is nowhere near what is needed and should be made possible. VSS respond to the multiple crises of the environment, the economy and what is traditionally understood as development, as well as to the failure of institutions. In Germany, and presumably in other developed and emerging economies also, people are demanding sustainable products and services, demonstrating lasting changes in their attitudes towards enterprises and in their personal consumption behaviour. Those changes might not yet be mainstream, but the move from 'mindless' to 'mindful' consumption lets consumers and parts of the corporate sector take action, with one type of action centred in standards, labels and the idea of stewardship of the global commons.

There is a need to challenge conventional patterns of production and consumption and to foster 'green' or 'sustainability standards' where they exist and to establish those standards where they do not yet exist. With this task, new governance features arise. The governance of VSS is under constant pressure to deliver increasingly reliable, meaningful and inclusive solutions. The single most important step would be to fill the leadership vacuum. This chapter describes the political setting of VSS in Sect. 28.2. The following sections describe VSS as a learning experience and discuss stakeholder views before, in Sect. 28.5, showcasing examples of strategic relevance, including those that comprehensively focus on supply chain patterns. The next two sections discuss the limits of VSS and the conundrum of what is seen as 'voluntary' before Sect. 28.8 closes with conclusions and recommendations.

G. Bachmann (✉)
German National Council for Sustainable Development, Geschäftsstelle c/o GIZ GmbH, Potsdamer Platz 10, 10785 Berlin, Germany
e-mail: guenther.bachmann@nachhaltigkeitsrat.de

C. Schmitz-Hoffmann et al. (eds.), *Voluntary Standard Systems*, Natural Resource Management in Transition 1, DOI 10.1007/978-3-642-35716-9_28,

28.2 Setting the Scene

Today, mankind has become a quasi-geological force. The change of watershed regimes, the warming of the atmosphere, the depletion of the soil's natural carrying capacities, add up to a level of human interference with nature that is unprecedented. Also, a couple of other important issues such as marine littering or the barring off of migrating species that follows infrastructural site development are expected to have long term negative impacts. That raises questions as to the near future of some nine billion people with (hopefully) decent living conditions on a carbon restrained planet that faces severe resource restrictions. Energy-intensive and resource-depleting economies will cease to prevail. Growing the output by simply growing the input into the economy—the notion that still dominates politics and economic thinking—leads to a no-solution ecological no man's land. This is a well-known and widely shared diagnosis. But therapies are missing. Some alternative solutions are in the making, but this process is far too slow, too oriented towards niches and small market shares, and, most importantly, hampered by complexities and misconceived concepts. There are ecological boundaries such as the stability of earth's climate, the availability of soils for the purpose of food production, access to freshwater resources, and the ecological equity of consumers. For 40 years, environmentalists have been discussing the idea of limits to growth, and they find that this debate has had almost no impact on real-world growth strategies. However, to this author's mind, those limits to growth must not be understood as physical limits, in the same sense that ecological science is different from physics. A limited planet obviously establishes final limits, but so far humankind makes those limits a moving target. This, of course, makes transforming ecological frontiers and limits into new patterns of production and consumption even more complicated and demanding. Fortunately, there are concrete experiences with propulsive transformational processes. With the rise and growing appreciation of the terms *sustainability*, *green economy*, *sustainable development strategies*, one may detect interesting new practices (Bachmann 2012a) and tools as well as emerging new roles (Bachmann 2013b)—where voluntary green standards for production and consumption play an increasingly important role. For the majority of business cases it is fair to say that their global dimension increased, and that most of the supply and value chains have to cope with an extraordinary complexity. This is true for hundreds and thousands of commodities, resources and almost all semi-finished products, with, so far, only a couple of those being covered by VSS.

The quest for *sustainable development* is evident and manifest. Sustainability is not a buzzword, nor is it one of those trends that are transitory and will soon allow a return to the old order. People who assume this are deluding themselves and dangerously exposed to becoming obsolete.

Today's leadership is about finding new models for a new age (Bachmann 2012b, 2013a). Consumer behaviour is not just about getting angry and complaining against corporate brands (although this is still often very appropriate and necessary), but will increasingly emphasise their relevance through selective choice making and requiring businesses to meet demands for sustainably managed products. Already today using the possibilities of social media and fostering core values of self-reliance and thrift is very important for consumers. Practicality and collective leadership, transparency and credibility, kindness and respect, fairness and ecological equity clearly offer many very favourable alternatives to the trend of mindless, parsimonious and close-fisted consumption.

Once green standards are established (or as is often the case, with the process of discussing and establishing, and strengthening standards) it becomes clear that user-defined action is voluntary, but not arbitrary in character. This action might not be legally mandatory and "binding", but none the less it effectively earmarks corporate responsibility and civil society's commitment. It is evident that with the growing success of voluntary benchmarking tools, in number and in quality, coherence and compliance (verification, traceability, and transparency) come into focus. Policies that assure and warrant standards are often needed, mostly for the provision of a level-playing-field. National jurisdiction or stakeholder based governance is also often needed to vouch for and authenticate voluntary standards.

VSS, therefore, demonstrate that governance towards sustainable development must be more than just (a) another piece of legislation, and (b) another piece of corporate communication or adding another one to the list of corporate brands. It is rather a regulatory frame that encompasses schemes of voluntary action, verification of ambitious action by social partners, encouragement, preferential choices, and where necessary legal frames for objectives, creating level playing fields, and authorising vouching structures.

28.3 A Learning Experience

The influence of enterprises on global material flows is increasing, just as the material basis of production and consumption are having a much larger impact on enterprises and production. This reciprocal relationship will get stronger with the possible necessary transformation towards a sustainable economy (green economy) and will, in any event, not diminish.

With it, the pressure on companies to legitimise their activities is also increasing: publicly, on the market, with regard to the "global commons" and common property and beyond mere compliance with legal rules.

Often, there are circumstances which make unusual collaborations necessary. This is the case when only the collaboration of a company with other companies along its supply chain and, if need be, with competitors, too, can ensure that fish stocks are not overfished, cotton is not contaminated or farmers can benefit from fair trade.

VSS are instrumental to strengthening cross departmental collaboration. Thus, departments such as product development, design, sourcing, branding and accounting have to work together with procurement, compliance and controlling and, in part, also with a company's strategic unit, of course. They also—and this is critical—have to create and ensure a common basic understanding with their suppliers. The development of and compliance with environmental and social standards affects the entire supply chain. Customers demand of their suppliers' evidence of compliance with environmental and social standards. This may take the form of specific questionnaires, a standardised exchange of data on online platforms or contractual commitments. In the German agricultural sector, forms of contract farming such as in the production of sugar beets and potatoes, which also include quality requirements regarding product features and their measurement and monitoring, have been practised for decades. The additional expense incurred by the supplier is pitted against the advantage of reliable distribution channels.

28.4 Requirements of the Stakeholders

Unsustainable farming practices occur when the cumulative effects of separate interests cannot guarantee the condition of their own existence. That is at the heart of the problems regarding food security, genetic diversity, human rights and a stable climate, and it is ultimately the reason for the fact that a fundamental shift towards sustainable development should be demanded across all social groups.

Apart from civil society, lenders of capital and investors in new markets are increasingly demanding the good professional management offered by a sustainability strategy. This applies to globally operating corporations as well as to small and medium-sized enterprises insofar as they are reliant on the buying and processing of raw materials and primary products from critical sources and fragile natural resources.

Enterprises are faced with the requirement to run their core business in such a way that they also take the environmental and social impacts of their production or purchasing, as well as good corporate governance, into account.

The awareness and recognition of the environmental and social challenges associated with modern production methods and logistics chains are making significant, albeit slow, progress.

Thus, for example, companies are increasingly being assigned responsibility for the observance of human rights by suppliers (e.g. the UN Guiding Principles on Business and Human Rights or the "Ruggie Framework") or the mining conditions of raw materials (see, e.g., traceability of conflict minerals as required by the Dodd-Frank Wall Street Reform and Consumer Protection Act of the USA – H.R. 4173 von 2009).

28.5 Strategic Examples

There are relatively long-standing examples of VSS, such as those with regard to the environmentally responsible and socially fair production of coffee, but also those relating to bananas or wood (Forest Stewardship Council). In addition, climate friendly as well as socially responsible and environmentally friendly production methods for other products and commodities, such as for textiles, cotton, ornamental flowers, cocoa and vanilla, are also the subject of VSS, although their breakthrough into a larger market segment has yet to succeed. Furthermore, there are also more complex areas of need which have voluntary standards and certification systems in place, such as those applying to buildings (sustainability standards for new buildings, conservation), sustainable investments, tourism services, etc. They are presented in a comprehensive and easy-to-communicate way by the "Sustainable Shopping basket", a project run by the Council for Sustainable Development (Nachhaltigkeitsrat 2013).

VSS are only just starting to evolve with the first corporate activities of a small number of front runners globally. In an initial attempt, PUMA SE started accounting its environmental and social profits and losses in full. Internalising and reducing external costs is a cornerstone of a green economy. Of course, this triggers innovative standardisation issues and topics for further research to assess the costs of ecological impacts that the mainstream economy does not account for. Thus, comprehensive profit and loss accounting is possibly one of the next stage challenges for VSS concepts.

In addition to those requirements that specifically address products and services, the German Council for Sustainable Development has established a voluntary standard that can be applied to an enterprise as a whole. The German Sustainability Code aims to provide guidance for entrepreneurial sustainability management as the part of its core business case.

This chapter uses the case of palm oil, one of the most discussed commodities in recent time, to exemplify the business of VSS.

Palm oil is a component of many foods and detergents. It is considered essential. The predominant type of palm oil production is irresponsible and damaging to the environment. Although the extent to which oil palm has been a direct cause of past deforestation is difficult to quantify, its potential as a future agent of deforestation is enormous (Fitzherbert et al. 2008). In addition, it is often at odds with human rights.

The World Wildlife Fund, WWF, has published a so-called scorecard of companies that makes business practices with regard to palm oil transparent and comparable (WWF 2011). It is clear that far from all companies use sustainably produced palm oil. Even very well-known companies operating on a global scale still shirk this responsibility. Many companies do not even pay heed to the most basic safeguards against exploitation and injustice in, for example, land use conflicts. The WWF calls on all enterprises that use palm oil to source 100 % RSPO-certified palm oil (RSPO Roundtable on Sustainable Palm Oil) by 2015. In the future, only fully traceable certified palm oil should be obtained and used. In principle, VSS are an appropriate way to do so. According to its own explanation, the RSPO label ensures that, for the quantity of palm oil purchased and certified, an equivalent volume of sustainable palm oil is produced. This is a "book and claim" principle. Companies who use it may say "we support the production of sustainable palm oil", while the use of certified and fully traceable palm oil enables a tougher statement to be made, namely "we use sustainable palm oil in our products".

Fourteen percent of the global palm oil market is now RSPO-certified. This is still a marginal amount, judged from what must and could be done. The ratio has increased rapidly compared with other products, and it is still on the rise. But only half the palm oil generated according to RSPO criteria is also actually sold as such. Regarding the social and environmental impacts of conventionally grown palm oil, the raising of public awareness and the educating of consumers needs to be made part of corporate responsibility, in cooperation with social stakeholders. The public sector could also contribute by earmarking sustainability criteria in procurement and by assuring fair competition against green washing.

Various sectors continue to criticise RSPO, the main issues being environmental concerns as to the impact of palm oil plantation expansion on peat swamp forest and on tropical forest, in particular those with Orang Utan population. In the RSPO, there were initially serious problems and abuses that are now, partly, the subject of dispute settlement procedures with the involvement of the World Bank.

As a minimum standard, the RSPO has urgent need for improvement. The credibility of all certification labels is markedly influenced by the fact that (a) a wide reach and a substantial level of effectiveness can be counted on, (b) there are strategies to solve conflicts of interest, (c) they are sensibly flanked by the nation-state regulatory policy, and (d) there is monitoring which demonstrates the options for substantial further development. What appears particularly necessary is the tightening of RSPO criteria with regard to the logging practice and the greenhouse gas emissions taking also into account the indirect effect caused by land use. The establishment of plantations on peat soils (or rather on what is now left of them) has to be an exclusion criterion. Respect for human rights must be more firmly anchored. The quality of audits by independent service providers should also be improved.

As the international negotiation process is usually lengthy, we must act more quickly on the national side in Germany. That is an argument that is frequently challenged by what is superficially referred to as globalisation. Globalised products and brands are said to be beyond the range of influence of national jurisdiction. That may or may not be true, and has to be seen in detail. But what is more important is the fact that all consumer markets, at the end of the day, are local and regional markets, and products are sold to people that will be reached by localised communication, or not at all. That is why market based improvements of production and consumption patterns, in the case of palm oil and generally, necessarily has to be made in issue of collective action of corporate companies and social organisations.

The establishment of a (German) Forum for Sustainable Palm Oil in 2013 is intended to accelerate the market mechanism of supply and demand. Experience from other industries, particularly those of chemicals and paper recycling, confirms the significant effect, above all, of alliances of partners from manufacturing, distribution, retail, services as well as from local authorities, educational institutions and NGOs. In the case of palm oil, therefore, sustainability means (a) reducing the demand for palm oil, and (b) substituting its use, wherever possible. The idea of sufficiency must basically be kept in mind when pursuing the paths towards socially and environmentally responsible palm oil production. We also need, therefore, to talk about using less of it. Finally, (c) the residual demand for palm oil has to be extracted in an environmentally sustainable and socially fair way. The need for fundamental changes is, therefore, obvious.

28.6 Limits of Application

Relevant stakeholder processes regarding social and environmental standards show the high normative power of leading companies and the enforceability of sustainability criteria along the value chain. Enforcement is often made more difficult, particularly in the case of large single markets (China) or oligopolistic structures of government-subsidised commodity policy on the part of developing countries.

No VSS exist yet for many value chains and commodities. Herein lies the largest deficit if one just considers the field of medical products and their use of genetic resources from ecosystems or losses in the food supply chains after the harvest and before consumption. Other examples with a huge impact (detrimental to the environment and, ultimately, the net value added) are phosphorus resources as well as the mining of rare earth metals and industrial metals.

Only for the sake of transparency should the fact be added that completely illegal poaching and the receiving of rare or endangered species in order to, for example, obtain ivory, alleged medical products, trophies and, in the case of finning, destructive (supposed) gourmet products cannot be the subject of VSS. In this respect, police protection rules have to be enforced by the state and flanked by companies in order to exert economic pressure on market participants.

In the field of VSS, there are striking examples of how limitations of governance have been able to be shifted, resulting in new solutions. Thus, in part, this has led to changes in supplier and/or the adaptation of contractual conditions away from commodity exchanges to individual contractual relationships with better quality control of the products. Some VSS correctly provide rules for filing complaints and settling disputes.

Meanwhile, however, there are also limits for the use of VSS, which have not yet been satisfactorily resolved. Those products and raw materials whose extraction often leads to conflicts over land use are particularly affected by this. Land use conflicts are the order of the day in countries that have weak governance or simply need to be described as failed states. Even in the case of democratic constitutions, there may well be conflicts if land rights have been unclearly defined or the indigenous people have been robbed of their traditional rights. In other lands though, the country has been destabilised by warlords and subliminal wars. Then it is above all cultivation methods based on decentralised structures and those based on family structures, such as cocoa in West Africa, that suffer.

Another issue, as mentioned above, is the growing incoherence of VSS. Incoherence has resulted from a severe lack of public policies, and the non existence of a concerted action on the international level. The mislabelling of species, in particular regard to seafood, is another problem.

Governments often insufficiently honour their responsibility to safeguard the legal framework used to recognise land use rights and to regulate land-use changes and the compensation claims of local and indigenous populations.

Often, establishing transparency in sourcing procedures is felt to be of high risk and painful, both by the corporate community and by the political class of developing countries. Likewise, the same people often misinterpret VSS and other supply chain requirements and transparency standards as neo-imperialism. The truth is that they are not. Instead, VSS and similar requirements show how it is possible to combine the need to take serious action to safeguard the environment and social habitats with the desire to maximise the chances of a green economy and create decent green jobs. Avoiding risks (material, fiscal or reputational) was the concept of the 'command and control' approach of emerging environmental legislation, and this concept holds true still today. But globalisation increasingly requires additional modalities to cope with the challenges of modern societies, such as the 'commit to manage' approach as exemplified by VSS.

28.7 Voluntary: Profit Before Principle?

Undeniably, there are an exceptionally high number of good examples of corporate sourcing based on voluntary sustainable standards. One might be led to believe that those examples would meet with undivided approval. This, however, is not the case. The voluntary character of standards and their contribution towards adding value are moot points.

Opinions are divided as to regulatory standing. Active enterprises emphasise positive examples where they assume responsibility of their own accord. Their refuting defence of the principle of voluntary will rather than binding regulations threatens to fail to exhaust the opportunities for a new responsibility-based culture as well as for solving the problems in this matter. For those companies already actively involved, a legal rather than a voluntary standard would represent a subsequent regulation which would (mostly) be of no benefit. It would bring with it the risk of competitors keen to duck standards or the competitive advantage of first movers being levelled by focusing on minimum standards.

Critics from civil society are sceptical as to whether the actions are truly and honestly intended or whether they are merely greenwashing. To them, voluntary measures without any legal obligation cannot make any lasting contribution to solving the global issues of sustainability. Ultimately, voluntary standards would all boil down to profit before principle. From the perspective of (certain sections of) civil society, the voluntary commitment of enterprises results from the responsibility of shareholders and executives who continue the maxim of ‘property entails responsibility’.

‘Voluntary will’ is accepted as a high commodity. However, the objections raised against it relate to only a minority of enterprises committing to it. Those taking a sceptical view of the economics of sustainable standards ask how personal and corporate responsibility can best be implemented, made credible and, ultimately, controlled for the benefit of society? Would not only a binding regulation provide for an inclusive green economy that serves to eradicate poverty?

These standpoints illustrate the differing basic understanding of how regulatory policy should govern corporate activity and the nation’s responsibility to protect nature and mankind. The fundamental standpoints outlined need to be called into question in order to make adequate and informed choices to foster a green economy further.

The principle of voluntary will, however, must not be misconstrued as being completely random. A commitment to sustainable supply chains often derives more or less mandatorily from the conditions to act within an enterprise’s political environment. For example, an enterprise in regions with failing state order and structure, pandemics or a destroyed environment needs to do far more than merely comply with environmental and social standards, if only out of pure self-interest. It appears justified not to speak fundamentally of a “voluntary” commitment in such instances but to consider the framework conditions which suggest that it is necessary to assume responsibility (“enforced voluntary will”). If the life expectancy among the workforce decreases dramatically as a result of epidemics and a lack of healthcare, an enterprise is often forced to commit to healthcare which goes beyond what is available in the local communities and often far beyond the enterprise’s direct interests.

The decision as to whether an enterprise becomes involved with sustainable supply chain management most frequently depends on which time perspective defines its own interests.

28.8 Conclusions and Recommendations

The tendency towards short-term thinking is one of the most dangerous trends. The management of natural resources must be transformed into a (more) sustainable modality. No business should be run on the stock market 90-day time horizons. But still, market requirements, public awareness and much of the media are more or less bound to the short term. This contradiction calls out for ambitious and constructive leadership, and new governance modalities, one of them being voluntary standard schemes.

There are four political challenges of VSS: Ignorance (where no VSS solutions are in place so far), confusion (where green standards and brands are numerous but incoherent and intransparent), reluctance (where the implementation of VSS has not yet found sufficient traction), and non-exposure-policies (where companies join VSS but deny leadership). They may be impacting each upon one another, and not for the better.

The entrepreneurial community involved in VSS is dealing with this approach mostly on a step-by-step and case-by-case basis, which is not enough. There is a need and an opportunity to frame VSS as a generically new governance feature. Hesitation provides no guidance and will not improve the business case nor will it help the public sector to implement good practice in transformational governance, but a keen leadership will.

VSS basically is a tool for transforming the way we traditionally produce and ship goods around the world. It should combine the establishment and strengthening of standards for the current use of commodities with the innovation of production and consumption patterns in a way that would end or minimise the use of problematic resources. Introducing sustainable patterns is nothing else than looking ahead to new opportunities for prosperity, new jobs, new career tracks, new markets.

The key responsibility for this lies with the private sector. A keen leadership is also expected to help remedy today's lack of expertise and social competence in dealing with the complexities of a green economy.

References

Bachmann G (2012a) Accelerating implementation by scaling up of good practices, keynote, UN office for sustainable development: International Workshop on Strengthening Planning and Implementation Capacities for Sustainable Development in Post Rio Context, Incheon, 16 Nov 2012. http://www.nachhaltigkeitsrat.de/uploads/media/Bachmann_Keynote_UNOSD_2012-11-16.pdf. Accessed 03 June 2013

Bachmann G (2012b) A sustainable resource strategy, key note low carbon earth summit 2012, October 2012, Guangzhou. http://www.nachhaltigkeitsrat.de/uploads/media/Bachmann_Key_Note_LCES_2012-10-19.pdf. Accessed 03 June 2013

Bachmann G (2013a) Die Idee der Frontiers in der heutigen Nachhaltigkeitsdebatte, Frontiers in Sustainability Science Lecture anlässlich der anstehenden Tätigkeit als Gastprofessor an der Fakultät Nachhaltigkeit der Leuphana Universität Lüneburg am 30 January 2013. http://www.nachhaltigkeitsrat.de/uploads/media/Bachmann_Frontier-Vorlesung_Leuphana_30-01-2013.pdf. Accessed 03 June 2013

Bachmann G (2013b) Emergency response: clustering change. In: Meuleman L (ed) Transgovernance: advancing sustainability governance. Springer, Heidelberg, pp 235–254

Fitzherbert EB, Struebig MJ, Morel A, Danielsen F, Brühl CA, Donald PF, Phalan B (2008) How will oil palm expansion affect biodiversity? Trends Ecol Evol 23(10):529–566

Nachhaltigkeitsrat (ed) (2013) The sustainable shopping basket, Berlin. http://www.nachhaltigkeitsrat.de/uploads/media/Brochure_Sustainable_Shopping_Basket_01.pdf. Accessed 03 June 2013

Round Table Sustainable Palm Oil, RSPO, Internet address: http://www.rspo.org/

WWF – World Wildlife Fund (2011) Scorecard palm oil buyers, 2011. http://wwf.panda.org/what_we_do/footprint/agriculture/palm_oil/solutions/responsible_purchasing/scorecard2011. Accessed 03 June 2013

About the Authors

Eike Albrecht is head of the Department for Civil and Public Law with References to the Law of Europe and the Environment since 2004, between 2004 and 2010 in the position of a junior professor, since 2010 as full professor. His scientific focus is on environmental law, energy law and the law of technology.

Ni An has completed her BSc at the University of Shanghai for Science and Technology. She continued her studies for MSc in Environmental and Resource Management (ERM) at the Brandenburg University of Technology (BTU) in Cottbus in 2006, and afterwards joined the PhD Programme ERM in October 2010. Her PhD research is conducted at the Department of Environmental Meteorology. In her research she analyses the extreme climate events in the federal state Brandenburg, Germany using climate simulations and corrects the BIAS of the applied model with statistical methods.

Miguel Araque Vera is a process engineer at ThyssenKrupp Industrial Solutions, Bad Soden, Germany. He holds a degree in Chemical Engineering from the Rafael Urdaneta University, Maracaibo, Venezuela (2009) and a MSc in Environmental and Resource Management from the Brandenburg University of Technology (BTU) Cottbus, Germany (2012). As a chemical engineer Miguel has experience in the design, evaluation and selection of process technologies for the energy sector. For his Master thesis at Siemens Corporate Technology he evaluated the applicability and influence of environmental databases in life cycle assessment. His areas of interests include sustainable processes and energy technologies and environmental management systems.

Eric Ambe Asoh is a PhD candidate at the Department of Environmental Planning, Brandenburg University of Technology (BTU) Cottbus, Germany. He did his bachelors on Environmental and Resource Management at the BTU Cottbus before moving to the Humboldt University in Berlin to complete his MSc degree in "Integrated Natural Resource Management" (2010). He has worked with the Cameroon Development Corporation (CDC). His work there included establishing

C. Schmitz-Hoffmann et al. (eds.), *Voluntary Standard Systems*, Natural Resource Management in Transition 1, DOI 10.1007/978-3-642-35716-9,

partnership between local population and local government institutions in the field of agriculture and environment. His research interest includes natural resource management and sustainable development.

Günther Bachmann is the Secretary General of the German Council for Sustainable Development. He was born in Berlin, studied landscape planning at the TU Berlin until 1978 and obtained PhD in 1985, and in 2014 the Leuphana University Lüneburg appointed him Honorary Professor. Günther Bachmann took grants from the European University Institute (Jean Monnet scholarship) and the German Marshall Fund of the US. He worked in the Federal Environmental Agency (*Umweltbundesamt*) with the focus on soil protection legislation. Since April 2001 he acted as the Director of the Administration Office of the German Council for Sustainable Development, and in June 2007 Günther Bachmann became its Secretary General.

Julia Baumhauer is a MA Social-Anthropologist. Currently she works as project coordinator for Give Something Back To Berlin, a German NGO which develops strategies and models for sustainable urban integration. With relation to development cooperation, she was working and researching in the field of legal pluralism, social security systems and private sector promotion with a regional focus on sub-Saharan Africa and Southeast Asia. She was lately engaged in the GIZ Programme for Social and Environmental Standards.

John Blewitt is Director of the MSc Social Responsibility and Sustainability at Aston University, Birmingham, UK. He is also Chair of Aston Sustainability Working Group and Chair of the University Academic Assembly. He has long worked in adult community and higher education and has for many years been interested in sustainable community development. He is a Distinguished Schumacher Institute Fellow and a Fellow of the Royal Society of Arts. He is author of Understanding Sustainable Development (Earthscan, 2008), co-author of Searching for Resilience in Sustainable Development (Routledge-Earthscan, 2013) and editor of the four-volume anthology: Sustainable Development: Critical Concepts in the Environment (Routledge, 2013).

Ambe Emmanuel Cheo has joined the PhD Programme Environmental and Resource Management at the Brandenburg University of Technology (BTU) Cottbus, Germany in April 2012. At BTU, he does his PhD thesis at the Department for Environmental Geology, at which he also completed both his BSc and MSc Degrees in Environmental and Resource Management (ERM). Using his additional academic background in Water Management, he is currently working on developing an Integrated Water Resource Management model for Northern Cameroon.

Pia Dewitz has completed her BSc in Environmental and Resource Management (ERM) at the Brandenburg University of Technology (BTU) in Cottbus (2013). Title of her thesis: “Emission Balance and Cost Efficiency of Solid Biofuels: The Example of Co-Firing of Residues from Palm Oil Plantations in Cameroon”. Before, she has gained experience in certification and standards in the agricultural sector during her studies of "Environmental Management" at the University of Technology in Pereira, Colombia. She has further worked for the programme "Social and Environmental Standards" in the Rural Development Division of the Deutsche Gesellschaft für Internationale Zusammenarbeit (GIZ) GmbH. Together with GIZ and BTU Cottbus she is involved in the promotion of voluntary sustainability standards and sustainability issues in higher education through eLearning. The application of standards across various sectors, for example forestry, clothing industry, cocoa and coffee, as well as fishery and aquaculture is a focus of both her work and studies. Currently she is enrolled in the Master's programme "Sustainability Economics and Management M.A." at the Carl-von-Ossietzky University in Oldenburg, Germany.

Mercy Nambu Diangha is a PhD candidate at the chair of General Ecology at BTU Cottbus. She obtained a BSc in Biological sciences 'Zoology' from Ahmadu Belo University (ABU) Zaria, Nigeria and an MSc in Tropical Forestry and Management from Technical University Dresden (TUD), Germany. Mercy worked for the Wildlife Conservation Society (WCS) Cameroon programme, in the Banyang-Mbo Wildlife Sanctuary (BMWS) and the Cameroon-Nigeria transboundary projects, where she served as conservation biologist. She is passionate about conserving biodiversity and is working in her academic research towards contributing to the management of the biodiversity of the Deng Deng National Park in Cameroon.

Stefan Essel is advisor on forest-related sustainability standards in the Programme for Social and Environmental Standards at the Deutsche Gesellschaft für Internationale Zusammenarbeit (GIZ) GmbH, assigned by the German Ministry for Economic Cooperation and Development (BMZ). His main focus is on shaping favourable framework conditions for the implementation of sustainable tropical forest management and supply chains through strengthening synergies between regulatory approaches (e.g. public procurement, FLEGT, EU-Timber Regulation) and voluntary forest standard systems. Before joining GIZ (previously GTZ) in 2009, Stefan has been working in Nepal as advisor to the Federation of Community Forest Users on value chain development and in the Gambia/West Africa as advisor to the Forestry Department in a Community Forestry Project. Stefan holds a degree in forest science from the Technical University of Munich.

Martina Gaebler is project manager at the Programme for Social and Environmental Standards at GIZ Eschborn. Martina joined the Programme in 2009 and has since been working on sustainability standards systems. Her focus areas have been sustainability standards for agricultural commodities and the extractive industries, as well as the recognition of private sustainability certification schemes for liquid biofuels under the EU Renewable Energies Directive. She is currently responsible for a project dedicated to the development of an IT-based comparison tool for sustainability standards systems. Martina holds a Master's degree in International Development from the University of Manchester.

Robert B. Gibson is a professor in the Department of Environment and Resource Studies at the University of Waterloo in Canada. He mostly works on environmental policy issues and broader sustainability imperatives related to land use decision making in urban growth management, corporate greening initiatives, special area governance and environmental assessments at the project and strategic levels. Most recent projects focus on application of the principles set out in his book on *Sustainability Assessment*, published by Earthscan in 2005.

Daniele Giovannucci is President of the Committee on Sustainability Assessment (COSA)—a global consortium of more than three dozen institutions advancing innovative ways to credibly measure the sustainability of our food and agriculture systems. COSA partners with leading firms and agencies for a clear understanding of the impacts of agriculture projects and initiatives around the world. Mr. Giovannucci is a former food company executive and an early advocate of the use of market mechanisms as drivers of sustainability. He serves as advisor to a number of major corporations and institutions such as the World Bank Group and the United Nations Forum on Sustainability Standards. He chaired the global teams formulating the strategic input on Food and Agriculture for the United Nations Division for Sustainable Development ("Sustainable Development in the 21st century") and for the landmark Rio+20 Earth Summit.

Joel Goebelbecker has joined the PhD Programme Environmental and Resource Management at the Brandenburg University of Technology (BTU) Cottbus, Germany in April 2011. At BTU, he does his PhD thesis at the Department for Civil and Public Law with References to the Law of Europe and the Environment, at which he also did his Master's Degree in Business Law (MBL) in October 2010. Using his additional academic background in the fields of chemistry and life sciences he attempts to regulate the risks of nanotechnologies within the German law.

Xiaoying Gu is a PhD candidate in the Environmental and Resource Management PhD Programme at the Brandenburg University of Technology (BTU) Cottbus, Germany. Her research is focused on carcass ecology, biodiversity conservation and food web restoration based on carcasses. Her academic background includes the following subjects: Environmental Engineering (BSc degree from the Liaoning Technical University, China) and Environment and Resource Management (MSc degree from the BTU Cottbus). Her BSc thesis was about the design of a wastewater treatment plant in China, while in the MSc thesis she dealt with the operation models of wastewater treatment plants.

Oliver von Hagen is a Market Analyst at the International Trade Centre (ITC–UN/WTO) in Geneva, Switzerland. He has over 10 years work experience in international trade, export promotion and development. Oliver works on and researches voluntary sustainability standards and their role and impacts in international trade, sustainable sourcing strategies, global supply chains, and export promotion. Prior to arriving at the ITC, Oliver worked for the Growing Sustainable Businesses (GSB) project at the United Nations Development Program (UNDP) in Malawi and for the Malawi Coffee Association developing growth and export strategies for the

Malawian coffee industry. Oliver has extensive experience in working in developing countries delivering trainings to SMEs, NGOs, trade support institutions and policy maker and providing technical advice on the ITC market analysis tools, sustainability standards, development of sustainable value chains, export strategy development and market access barriers. Oliver published numerous ITC papers, peer-reviewed papers in reputable academic journals, book chapters and a wide range of practitioner-oriented articles and business guides. He has been invited to present his research at various international conferences and works as reviewer for international academic conferences, most recently for the Academy of International Business (AIB) conference 2014 in Vancouver.

Berthold Hansmann has 29 years of experience in development cooperation with emphasis on natural resource management, integrated water resource management (IWRM), promotion of UNCCD implementation and projects on adaptation to climate change. He holds a Master's of Science in Geology. Prior to joining GTZ, Mr. Hansmann lectured and managed fieldwork at the 'International Institute for Aerospace Survey and Earth Sciences (ITC)', Enschede, The Netherlands. Following that, he joined the Deutsche Gesellschaft für Internationale Zusammenarbeit (GIZ) GmbH (previously GTZ) in Eschborn as Technical Advisor in the Rural Development Division in the headquarters and as a project manager in the Middle East.

Ulrich Helberg is holding a BSc in International Agricultural Development and an MSc in Environmental Protection. He has worked in the international development projects in Zambia, Tanzania, Mali and Ghana for about 10 years and, as a freelance consultant, in about 30 countries worldwide for over 15 years. From 2010 until early 2013, Ulrich has worked as GIZ project manager for the Certification Capacity Enhancement project (CCE), a Development Partnership with the Private Sector for sustainable cocoa production in West Africa. He still coordinates this partnership.

Callum Hill is a director of JCH Industrial Ecology Limited, an environmental consultancy and acts as a consultant for Renuables and for the Norsk Institutt for Skog og Landskap. From 2007 to 2013 he was the Edinburgh Research Partnership chair of Materials Science at the Forest Products Research Institute of Edinburgh Napier University. Prior to that he was Senior Lecturer in Renewable Materials at Bangor University. Callum Hill has widely published, having contributed to over 150 research papers, 2 books and 3 book chapters. He has been conducting research into sustainable and renewable materials for 20 years. He holds a first class honours degree in chemistry from the University of Bristol and a PhD in organometallic gas sensors from the UK Council for National Academic Awards.

Amy Jackson is the Senior Credibility Manager at the ISEAL Alliance, where she led the global consultation to develop the ISEAL Credibility Principles, a set of core values and actions that ensure the credibility and effectiveness of standards systems. She is leading the work to determine and clarify good practice regarding claims made by and about sustainability standards systems, and works to drive collaboration among members on various technical topics. Prior to her work at

ISEAL, Amy was with the Marine Stewardship Council (MSC) for 11 years working to increase the uptake of the MSC programme. She was responsible for MSC's global chain of custody programme and introduced the 'Traceability and Assurance in the Supply Chain' (TASC) programme where she oversaw the development of innovative approaches to traceability assurance including DNA testing and online assessment.

Sophie Klose is scientific assistant at the Department of Environmental Planning, Brandenburg University of Technology (BTU) Cottbus. She is currently involved in the project 'Establishment of a Long Term Cooperation and Dialog', partnering two Lebanese Universities and the BTU. Thematically the cooperation focuses on the research and communication of "Voluntary Sustainability Standards (VSS)" as an instrument for measuring sustainable development, production and consumption. Sophie holds a Bachelor Degree in Environmental and Resource Management. Prior to joining the BTU as a scientific assistant she deepened her knowledge of VSS during an internship in the Deutsche Gesellschaft für Internationale Zusammenarbeit (GIZ) GmbH in the Programme for Social and Environmental Standards.

Kristin Komives is ISEAL's Senior Monitoring & Evaluation Manager. In this capacity, she manages ISEAL's work on the impacts of sustainability standards. This includes supporting ISEAL members in the development of their own monitoring and evaluation systems and facilitating collaboration between standard systems on the development of indicators, methodologies, and research studies that will help members measure and understand their sustainability impacts. Before joining ISEAL in 2010, Kristin worked as a Senior Lecturer in environment and sustainable development at the International Institute of Social Studies, Erasmus University, Rotterdam. She taught graduate courses on research methodology, survey design, sustainable and social development and environmental policy, and was engaged in applied research related to water and electricity services, pro-poor subsidy design, environmental planning and poverty reduction in Latin America and sub-Saharan Africa. As a faculty member and an independent consultant, Kristin has conducted large-scale household surveys, policy and project evaluations and impact evaluations for a wide variety of international organisations. She obtained her Master's in Regional Planning and PhD at the University of North Carolina, Chapel Hill.

Eberhard Krain has been working for more than 20 years in long- and short-term assignments in technical and research cooperation projects in several countries in Africa and Asia, mostly for GIZ. He is an agriculturalist by profession, well conversant with value chain promotion, rural development and decentralization and a seasoned trainer cum moderator. Since August 2010 he works at GIZ headquarters as advisor for sustainability and standards in global agricultural value chains.

Christine Lemaitre is CEO of the German Sustainable Building Council (Deutsche Gesellschaft für Nachhaltiges Bauen—DGNB). Christine Lemaitre was born in Gießen, Germany and studied structural engineering at the University of Stuttgart. After working in the USA, in 2003 she started working at the Institute of Lightweight Structures Design and Construction at the University of Stuttgart. In 2007 she started working at Bilfinger Berger AG as a project manager for R&D. In January 2009 she took on the role as director system of the German Sustainable Building Council, and since February 2010 Christine Lemaitre was appointed the CEO of the German Sustainable Building Council.

Kerstin Linne is a freelance consultant working under the name of Green Line Consulting. She started working on the topic of climate change in the coffee sector in early 2008 as an employee at the Deutsche Gesellschaft für Internationale Zusammenarbeit (GIZ) GmbH. Kerstin successfully implemented two climate change projects and supported several other initiatives on this topic during her time at GIZ. In late 2011 she became an independent consultant with a focus on climate change challenges in tropical agriculture. She is closely working with voluntary standards systems and has also developed a participatory approach to manage climate risks at the level of the affected population—Managing Climate Risks and Opportunities (MRO).

Oscar David Matallana-Tost is a chemical engineer with a Master's degree in Environmental and Resource Management from the Brandenburg University of Technology (BTU) in Cottbus. He is an alumnus of the scholarship program of the Heinrich-Böll-Foundation. Between 2011 and 2013 he worked as scientific researcher for the "Necros Project: Carcass and Biodiversity". In 2012 he received the prize for the best scientific presentation at the sixth International Conference for Young Water Professionals for his paper "Diquís Dam in Costa Rica: Should Bio and Cultural diversity be sacrificed in favour of a lucrative and clean energy production?", and started his work on energy price assessment and model based projection at the Fraunhofer Institute for Systems and Innovation Research ISI. Nowadays he is working as lecturer and researcher in Chemical Engineering and Environmental Management at the University of Engineering and Technology (UTEC) in Lima, Peru. He has a significant work experience in rubber, chemical, food and vegetable oil refining industries, as well as in educational institutions in Colombia, Germany and Peru. His main research fields are environmental management in mining, methodologies for decision making in the context of sustainability, indicators for measuring sustainability, development cooperation and energy management.

Carolin Möller's academic path is grounded in Water Management at the University of Technology Dresden and Kwame Nkrumah University of Technology Kumasi (Ghana) and in Environmental and Resource Management at the Brandenburg University of Technology (BTU) Cottbus and De La Salle University Manila (Philippines). Her involvement at the Department of Environmental Issues in Social Sciences (BTU Cottbus) enabled the two degree theses: on "Sustainability as

Construct – Sustainable Practices on Beninese Small Scale Farms" (MSc) and on "Empowerment Strategies for Environmental Sustainability – Proposal of Concepts to master Social and Environmental Impacts of the Large-Scale Gold and Copper Mining Project in Didipio, Philippines" (BSc). These academic works embrace her decade of experiences of working for science-based NGOs such as IWMI and LRC-Philippines at the interface of farmers' as well as indigenous people's practical solutions and politics. Working in various countries in sub-Saharan Africa, as well as Southeast Asia, furnished her experience on stakeholder demand according to natural resource use and livelihood maintenance. Her focus is aligned to innovative farming and promoting adaptive farming ideas in a local context while supporting actors to claim their socio-environmental rights to maintain their lives. Currently she is working for two small scale farmers in West Africa by implementing gained expertise in a real life context as a mediator among stakeholder groups based on understanding the emerging positions on socio-economic and socio-environmental concerns and empowering the affected vulnerable groups towards balanced and efficient solution findings.

Vera Morisse works as advisor at the Programme for Social and Environmental Standards in the Rural Development Division of the Deutsche Gesellschaft für Internationale Zusammenarbeit (GIZ) GmbH. There, she works in the secretariat of the German Initiative on Sustainable Cocoa (GISCO). Her work focuses on the management of multi-stakeholder processes, as well as on sustainability and standards in global value chains. Vera holds a Master's degree in political sciences from the Freie Universität Berlin.

Andrew Norton is director of Renuables, a Life Cycle Assessment practitioner and consultant specialising in renewable and sustainable materials. Whilst providing LCA and Carbon Footprinting advice to major international clients, he has also worked on assessing and developing many agricultural and forestry-derived products. His recent clients include: Marks and Spencer, Ernst & Young, Saint Gobain, the Belgian Federal Government, UPM Tillhill, DECC, Imperial College of Science Technology and Medicine and the World Bank. He holds a degree in Forestry and Forest Products, an MSc in Biocomposite Materials and a PhD in life cycle assessment, all from Bangor University.

David Ovando Jeria is a geographer, candidate to the MSc degree in 'Environmental and Resource Management' at the Brandenburg University of Technology (BTU) Cottbus. He has over 15 years of work experience in areas of regional planning, environmental impact assessment and geographic information systems, in the public and private sectors in the region of Magallanes, Chile.

Dmitry Palekhov is a lecturer and research associate at the Department of Environmental Planning, Brandenburg University of Technology, Cottbus, Germany. He obtained PhD in environmental law from the V.M. Koretsky Institute of State and Law of the National Academy of Sciences of Ukraine (2009), PhD in Environmental and Resource Management from the BTU Cottbus (2014), and holds an

LL.M. degree from the National Mining University, Dnepropetrovsk, Ukraine. His research interests include environmental assessment (EIA and SEA), environmental law, environmental planning and sustainable development.

Ludmila Palekhova is associate professor at the Department of Marketing, National Mining University (NMU), Dnepropetrovsk, Ukraine. At the same time, for the last 15 years she has been a director of the Consulting Centre which prepares and supports the implementation of development programmes for Ukrainian industrial enterprises under conditions of the post-Soviet transition economy. Her research scope includes the issues of strategic management and marketing based on the principles of sustainable development, as well as transition to green economy. In particular, in her latest publications she discussed problems related to: marketing and investment mechanisms in the context of the Kyoto Protocol; green innovations in the development of industrialised regions; potential of environmental certification for marketing of industrial products.

Gennady Pivnyak is Rector of the National Mining University, Dnepropetrovsk, Ukraine and member of the National Academy of Sciences of Ukraine. He is a member of various national and international professional and academic organisations, including the Ukrainian Committee on State Prizes in the Field of Science and Technology, State Commission for Academic Degrees and Titles, International Society for Engineering Education, European Society for Engineering Education, expert on sustainable energy at the United Nations Economic Commission for Europe. He holds Dr. h.c. titles from the AGH University of Science and Technology, Krakow and Moscow State Mining University.

Mark Prein is Fisheries Biologist and is presently head of an internationally operating Programme in support of sustainable fisheries and aquaculture at the Deutsche Gesellschaft für Internationale Zusammenarbeit (GIZ) GmbH, assigned by the German Ministry for Economic Cooperation and Development (BMZ). Earlier he was with the WorldFish Center for 15 years leading the research program on inland fisheries and integrated aquaculture-agriculture. Subsequently he had engagements at the Food and Agriculture Organisation (FAO) of the United Nations supporting aquaculture development in Africa and at the Aquaculture Department of the University of Hohenheim coordinating research collaboration between Palestine, Israel and Germany. He continues to teach courses at the University of Hohenheim. In his role in German development cooperation he has conducted numerous missions overseas for project design, technical and operational backstopping and progress review.

Franziska Rückert completed her BSc in Environmental and Resource Management at the Brandenburg University of Technology in Cottbus in 2013 with the thesis "Tropical Timber Trade: Analyzing Certification Schemes and Legal Requirements as Tools for Sustainable Forestry". Currently she is enrolled in the Joint Degree Master of Science in Sustainable Agriculture and Integrated Watershed Management in cooperation between Chiang Mai University, Thailand and Universität Hohenheim, Germany.

Elena Rueda holds Bachelor's degree in International Business Administration and MBA in Sustainability Management. Since June 2012 she is heading the Secretariat to the German Initiative on Sustainable Cocoa (GISCO). She joined GIZ in June 2009 as an advisor within the Programme for Social and Environmental Standards, being responsible for private sector cooperation, policy advice, steering and backstopping GIZ activities in agricultural commodities, including coffee and cocoa. Between 2009 and 2010 she acted as GIZ policy advisor to the 4C Association. Before entering GIZ she worked as project manager and consultant in the fields of business development, corporate sustainability management and fair trade.

Jenny Rust works as a freelance consultant with Como Consult GmbH. Her main fields of expertise relate to sustainability standards in the forestry and agricultural sector, natural resources management and sustainable economic development. Before she started working as an independent consultant in mid-2011, she was employed as an expert at the GIZ Programme for Social and Environmental Standards for 4 years where she also gained extensive experience in working on projects with the private sector. Jenny holds a Bachelor's degree in International Forest Ecosystem Management from the Eberswalde University of Sustainable Development (HNEE) and a Master's degree in Environmental Management from the University of London (SOAS).

Eberhard Schaller is a professor and the director of the chair of Environmental Meteorology at the Brandenburg University of Technology (BTU) Cottbus since 1996. He is now also acting as the vice-dean of Faculty 4 of Environmental Sciences and Process Engineering. His research is focused on improving the understanding and quantitative description of the atmospheric processes and both short-term and long-term impact on air quality. The consequences of a change in air quality on humans, fauna and flora are also included. The research is mainly conducted through a number of different simulation models to study the structure of the near-surface atmosphere. He initiated the Climate Limited-Area Modelling Community, which is an open international network of scientists applying and developing the COSMO-CLM or CCLM.

Michael Schmidt is a professor and Head of the Department of Environmental Planning at the Brandenburg University of Technology (BTU) Cottbus. At the BTU Cottbus, he initiated the international study programmes "Environmental and Resource Management" and "World Heritage Studies". In 2002 he received the "Award for Excellence in International University Cooperation" by the German State Federal Ministry of Education and Research, and in 2005 the Dr. h.c. of the National Mining University, Dnepropetrovsk, Ukraine. His scientific research and lecturing fields include environmental planning with particular focus on climate change adaptation, environmental assessment, setting standards and thresholds for sustainable development.

Evita Schmieg is researcher at the Stiftung Wissenschaft und Politik (SWP, German Institute for International and Security Affairs). She is on leave from the German Federal Ministry for Economic Cooperation and Development, where she was head of the Globalisation, Trade and Investment Division for the last 12 years. She has more than 20 years of experience in the area of trade, globalisation and development in the Ministry as well as in the Commission of the European Communities in Brussels. She has been involved in the Ministry's work on strategic issues related to trade, globalisation and private voluntary standards as well as public–private partnerships. She did her PhD at the University of Leipzig on regional integration and EU trade policy. Her academic career also included a post-graduate course at the German Development Institute. She is an economist by profession.

Carsten Schmitz-Hoffmann is Executive Director for Private Sector Cooperation at GIZ. He is responsible for partnerships and projects with companies, private foundations and business associations. Prior to this function he was Head of the GIZ Competence Center "Agriculture, Trade, Standards" and Director of the GIZ Programme for Social and Environmental Standards. Carsten holds a degree in political science from the University of Marburg. Before joining GIZ in 2001, Carsten gained first experience in working in international cooperation at the German Federal Ministry for Economic Cooperation and Development (BMZ) and the Berlin-based Carl-Duisberg-Gesellschaft. Scope of his work to date focuses on the role of corporations in shaping global supply chains towards more sustainable and responsible business practices.

Uwe Scholz is Fisheries Biologist and Senior Fisheries Advisor in GIZ since 1987 with long term postings in Africa (Zambia, Malawi, Cape Verde) and Asia (Philippines). Numerous missions continue to bring him to South Asia, Asia-Pacific, the Middle East, Africa, the Caribbean and Latin America in support of German development cooperation programs in the fields of marine and inland fisheries governance, coastal zone management, coastal biodiversity conservation through protected areas, sustainable aquaculture and climate change adaptation and mitigation in aquatic environments. He is a member of the Developing World Working Group of the Marine Stewardship Council (MSC) and closely cooperates with the fisheries unit at the NEPAD Planning and Coordinating Agency (NPCA) in support of fisheries development in Africa.

Philipp Schukat is Programme Director of GIZ's Programme on Environmental and Social Standards. He holds a degree in International Economics of the Eberhard Karls University, Tuebingen. GIZ's Programme supports development and improvement of standard systems, advises governments around the globe on policy strategies using standard systems and supports in alliances with private sector conversion of value chains into sustainable value chains. Philipp has been with the Programme since 2006 and previously worked as senior advisor at the German Ministry for Economic Cooperation and Development.

David Smyth qualified with a BSc (Hons) in Microbiology from Queens University Belfast (Northern Ireland) in 2006. During his studies, David availed of the opportunity of paid summer placements within the Environment and Heritage Service (now renamed 'Northern Ireland Environment Agency'), specialising in water microbiology. Prior to this he was active in volunteering for numerous charitable organisations with ties to sustainability and environment conservation, such as: Royal Society for Protection of Birds (RSPB), Wildfowl and Wetlands Trust (WWT), Ulster Wildlife Trust (UWT), World Wildlife Fund (WWF), Concern Worldwide and Traidcraft. After university, David worked within the quality testing and assurance department of Coca-Cola HBC as a microbiologist. Later he migrated to pursuits within criminal justice, fulfilling a personal ambition to contribute as a member of scientific staff within Forensic Science Northern Ireland (FSNI), the sole provider of forensic services in the province. Upon arrival to Germany in late 2010, he continued his passion for sustainability by enrolling for the MSc in Environmental and Resource Management (ERM) at the Brandenburg University of Technology (BTU). Recent work at BTU has included his project entitled "Arthropod Succession Study in Summer 2011 on Deer Carcass in Eastern Brandenburg Open Landscape at Lieberoser Heide (Federal State of Brandenburg, Germany)", with his pending MSc thesis featuring the observation of vertebrates attending large exposed bovine carcasses.

Hans-Jürgen Voigt is professor and Head of the Department of Environmental Geology at Brandenburg University of Technology (BTU) Cottbus, Germany. The Department of Environmental Geology is also part of the Institute of Soil, Water and Air. He was the Chairman of the EU-Cost 629 Group of groundwater monitoring. Courses offered by the department include: Water Resources Management, Applied Geology, Applications of Hydrogeology, Geological Heritage, Environmental Geophysics and Geological Practical. The department has also experience in water management in conflict areas such as Thailand and Vietnam.

Gerhard Wiegleb is visiting professor of ecology at BTU Cottbus. His current research focuses on ecological and socio-economic driving forces of biodiversity change in disturbed landscapes. Recent research was dealing with the legal framework of biodiversity protection under the EU liability directive. His most recent research is focused on statistical analysis and modelling of long-term development of aquatic ecosystems under climate and land use change.

Joseph Wozniak is the Manager of the Trade for Sustainable Development (T4SD) programme at the International Trade Centre (UNCTAD/WTO) in Geneva, Switzerland where he is responsible for the development of the Standards Map website and its related capacity building projects. Prior to arriving at ITC, he worked in government and consulting on the development of analytical tools for the impact assessment of private sector investment projects in Africa, Asia and Latin America.

Index

A

Accreditation, 7, 9, 100, 122, 256–257, 267, 338, 426
Africa, 119, 190, 196, 207, 213, 215, 222, 230, 288–289, 320, 324, 405–408, 413, 438. *See also* Sub-Saharan Africa; West Africa
Agroforestry, 279, 281, 282
American National Standards Institute (ANSI), 85, 93
ANSI. *See* American National Standards Institute (ANSI)
Aquaculture, 315–319, 328–337
Aquaculture Stewardship Council (ASC), 10, 315, 328–333
ASC. *See* Aquaculture Stewardship Council (ASC)
Association of German Engineers (VDI), 66, 69, 71
Assurance, 93, 109, 369, 424
  legality, 187, 190
  quality, 122, 263, 351, 353
  sustainability, 101
  system, 5, 105–107, 109–111
  trends in, 9–10
Atmospheric carbon storage, 156–158

B

Baseline, 45, 155, 219
BAT. *See* Best available technology (BAT)
B2B. *See* Business-to-business (B2B)
Berlin, 200–201
Best available technology (BAT), 63, 68, 92
Biofuels, 10, 101–102, 184, 186
Black tiger shrimp. *See* Shrimp *(Penaeus monodon)*
Brandenburg, 201–202
Brazil, 102, 124, 183, 289, 372
BREEAM. *See* Building Research Establishment's Environmental Assessment Method (BREEAM)
British Standards Institute (BSI), 54, 396
BSI. *See* British Standards Institute (BSI)
Building Research Establishment's Environmental Assessment Method (BREEAM), 346, 393
Built environment, 30, 145–162, 391
Business-to-business (B2B), 5, 85, 151, 309, 330, 338, 368, 369, 411
Business-to-consumer, 338

C

4C. *See* Common Code for the Coffee Community (4C)
Cameroon, 207, 211–225, 230, 234–237, 275–284, 289
Capacity building, 5, 13, 115, 122, 124, 125, 192, 207, 270, 272, 292, 294, 310, 388, 406
Capture fisheries, 315–318, 321, 338
Carbon
  footprint, 44, 162, 216, 314
  sequestration, 43, 157–159, 231, 281, 312, 378
  stores, 145, 148
  valuation, 146, 160–162
Carrying capacity, 23
CCRF. *See* Code of Conduct for Responsible Fisheries (CCRF)
CEN. *See* European Committee for Standardization (CEN)
CENELEC. *See* European Committee for Electrotechnical Standardization

C. Schmitz-Hoffmann et al. (eds.), *Voluntary Standard Systems*, Natural Resource Management in Transition 1, DOI 10.1007/978-3-642-35716-9,

Certification, 4–5, 8, 13, 32, 134, 135, 140, 165–174, 229, 246, 258–259, 261–273, 292, 294, 373, 379, 409–410
bodies, 9, 169, 178, 256–258, 264, 265, 267, 268, 338, 399, 413
capacity, 294
chain of custody, 159–160, 190
costs, 109, 413, 415
credible, 224
fisheries and aquaculture, 315–339
independent, 256, 292, 366, 370
real time information, 10
sustainable buildings, 345–355
systems and schemes, 99–111, 145, 200, 207, 212, 216, 219, 225, 269, 278, 318, 397, 424–430, 435
timber products, 145, 162, 187, 196–198, 255
CFCC. *See* Chinese Forest Certificate Committee (CFCC)
Chain of custody, 5, 10, 102, 111, 146, 150, 426
certification, 159–160, 190, 234–235, 256
Child labour, 52, 117, 137, 282, 290–291, 296, 330
China, 93, 181, 243–251, 253–259, 372, 386, 425, 437
economic development and its consequences, 244–245
opportunities of implementing VSS in, 246–248
paper production in Liaoning Province, 253–259
Chinese Forest Certificate Committee (CFCC), 255–259
CIS. *See* Commonwealth of Independent States (CIS)
Civil law, 66
Claims, 5, 10, 14, 18, 32, 135, 145, 149, 169, 180, 267, 271, 368, 405
false, 10, 14
Climate change, 12, 27, 54, 64, 139, 145–149, 159, 182, 195, 288, 305–314, 362, 387–390, 400, 425
mitigation, 148–149, 157, 216, 219, 306–307, 309–310, 378, 388
Coastal zone management, 326–328
Cocoa
beans, 278, 283, 288–290
certified, 290–293, 363, 379
farms, 277, 279, 290, 296, 382
supply chains, 291–296, 298
Cocoa production
in Cameroon, 275–284
conventional methods of, 278–280, 283
environmental impacts of, 280–282
global challenges and opportunities, 290
main producing countries, 289
sector, 288–290, 438
social impacts of, 282–283
sustainable, 287–302
Code for Sustainable Homes (UK), 393
Code of Conduct for Responsible Fisheries (CCRF), 322–324
Coffee production, 118–119, 305–314, 334
emissions in, 312–313
Co-firing, 212, 218–219, 224
Committee on Sustainability Assessment (COSA), 365, 375–381
Common Code for the Coffee Community (4C), 118
association, 8–9, 119, 305, 307, 309
climate module, 305–311
code of conduct, 306, 308–309
Commonwealth of Independent States (CIS), 166, 172, 265
Competitive strategies. *See* Market strategies
Compulsory certification, 166, 262
Conformity assessment, 9, 180, 262–264, 337
Congo Basin, 179–180, 229–239
Contracts, 36, 66
Co-regulation, 99–111, 422–423
pathways of, 100
Corporate social responsibility (CSR), 53, 126, 198, 359–382
roots of, 360–371
Corporation standards, 68
COSA. *See* Committee on Sustainability Assessment (COSA)
Cost sharing, 411, 413, 427–428
Côte d'Ivoire, 288–289, 294, 298
Credibility, 101, 108, 110, 120, 360–371, 426
CSR. *See* Corporate social responsibility (CSR)

D

Data, 14, 31, 35, 40–41, 81, 153, 374, 378–381
collection, 40–41, 153, 155, 297, 311
environmental, 146, 405
formats, 137, 191
quality, 150
reliable, 377–378
secondary, 156
usefulness of, 137
Development policy, 115–131, 142–143, 316, 429
objectives of, 116

DGNB. *See* German Sustainable Building Council
DGNB certification system, 347–349
  criteria, 349–351
  weighting of topics, 348
DIN. *See* German Institute for Standardization

E
Ecolabelling, 166, 258, 270–271, 318, 321, 324, 330, 404–406. *See also* Labels
Economic impacts, 16, 138, 140–141, 223
Ecuador, 122, 289, 334
EIA. *See* Environmental impact assessment (EIA)
EITI. *See* Extractive Industries Transparency Initiative (EITI)
Embedded carbon, 145–162
Embedded energy, 44
Environmental impact assessment (EIA), 41, 280
Environmental impacts, 88, 138, 140, 149, 153, 155, 200, 221, 280–282, 349, 351, 364, 394, 395, 405
Environmental management systems, 169, 254, 267
EU. *See* European Union (EU)
European Committee for Electrotechnical Standardization (CENELEC), 67
European Committee for Standardization (CEN), 67, 81, 397
European Union (EU), 11, 42, 85, 117, 146, 160, 197, 206, 211, 217, 235, 370, 419–420
EU Timber Regulation, 100, 160, 183, 187–188, 190, 197, 203, 207
Extractive Industries Transparency Initiative (EITI), 248

F
Fairtrade. *See* Fairtrade International
Fairtrade International, 6, 50–52, 140–141, 248, 276, 279, 292, 294, 314, 366, 369
Farmer Business School (FBS), 292
FLEGT. *See* Forest Law Enforcement, Governance and Trade (FLEGT)
Forest Law Enforcement, Governance and Trade (FLEGT), 160, 180, 183, 186–187, 189, 197, 203, 204, 207, 235
Forest management, 52, 159, 179–193, 196, 198, 229–239, 255, 266
  regulatory approaches relevant to, 189–191
Forestry, 45, 52, 100, 138, 146, 148, 159–162, 186, 205, 212
  in the Congo Basin, 229–239
Forest Stewardship Council (FSC), 7, 53, 115, 141, 159, 160, 172, 181, 183, 197–204, 216, 229–239, 246, 249, 256, 257, 263, 435
  adoption in the Congo Basin region, 233–234
  impact in the Congo Basin region, 234–237
  principles and criteria, 232–233, 266
Formal regulations, 77
Frozen fish, 12
FSC. *See* Forest Stewardship Council (FSC)

G
GEN. *See* Global Ecolabelling Network (GEN)
Generalised System of Preferences of the EU (GSP), 419–430
  grades of, 420–422
Generally accepted codes of practice, 65, 67, 68, 70
GEN Internationally Coordinated Ecolabelling System (GENICES), 169, 270
German Civil Code, 70, 72–73
German Constitution, 63
German development policy, 115–131, 142, 316
  contribution towards the promotion of VSS, 121–123
German Industrial Norms, 66, 67, 70
German Initiative on Sustainable Cocoa (GISCO), 287–302
German Institute for Standardization (DIN), 62, 67, 85
German Sustainable Building Council (DGNB), 346–355
Ghana, 187, 190, 207, 289, 290, 294, 298
GISCO. *See* German Initiative on Sustainable Cocoa (GISCO)
Global Ecolabelling Network (GEN), 261, 270
GLOBALG.A.P., 122, 315, 328–333, 338, 359, 369
Green procurement, 170, 195–208, 272. *See also* Public procurement
GSP. *See* Generalised System of Preferences of the EU (GSP)
Guidance for Sustainable Community Development (BS 8904), 395–399

H
HCV. *See* High conservation value (HCV)
Heiligendamm process, 124
High conservation value (HCV), 181, 183, 185, 233

I

IASB. *See* International Accounting Standards Board (IASB)
IEC. *See* International Electrotechnical Commission (IEC)
IFOAM. *See* International Federation of Organic Agriculture Movements (IFOAM)
IFRS. *See* International Financial Reporting Standards (IFRS)
ILO. *See* International Labour Organisation (ILO)
Indicators, 22, 31, 39–46, 138, 188, 276, 308, 309, 377, 378, 392, 393, 395
  analytical aids and tools for presentation of, 44–46
  usefulness of, 42–44
Indonesia, 207, 215, 258, 289, 317, 325, 362, 372
Industrial installations, 63
Industry, 5, 30, 66, 71, 84, 135, 170
  chocolate, 290, 293, 296
  food, 215, 296, 300
  palm oil, 212, 215, 217, 223
  paper-making, 177, 253–259
  self-regulation of, 84
  shrimp, 337
Initiative for Responsible Mining Assurance (IRMA), 171, 270
(In)justice, 391
Integrated Coastal Zone Management, 326–328
Integrated pollution prevention and control, 63, 92
Internal Management System (IMS), 293
International Accounting Standards Board (IASB), 64
International Electrotechnical Commission (IEC), 62, 67, 78, 85
International Federation of Organic Agriculture Movements (IFOAM), 6, 7, 11, 50, 51, 276
International Financial Reporting Standards (IFRS), 64
International Labour Organisation (ILO), 53, 115–117, 124–127, 246, 425
International Organization for Standardization (ISO), 54, 62, 67, 78, 81, 135, 180, 397, 399, 404
  risk management process of, 89, 90
  standards, 50, 53, 54, 79, 109, 149, 151, 167, 169–170, 263, 422, 429
International trade agreements, 205–206
International Trade Centre (ITC), 15, 128, 129, 288, 375, 422
IRMA. *See* Initiative for Responsible Mining Assurance (IRMA)
ISEAL Alliance, 7, 50, 54, 55, 135
  Codes of Good Practice, 15, 109, 292
  Credibility Principles, 15–19
ISO. *See* International Organization for Standardization (ISO)

K

Kenya, 305–315, 320, 426
Kompass Nachhaltigkeit internet platform, 128
Kyoto Protocol, 52, 55, 160

L

Labels, 5, 11, 53, 85, 87, 88, 117, 123, 134, 140, 149, 165, 171, 198, 236, 255, 264, 270–271, 291, 309, 328, 338, 369, 375, 396–398, 436
Lake Victoria, 320–324
LCA. *See* Life cycle assessment (LCA)
Leadership in Energy and Environmental Design (LEED), 11, 346, 393
Legal enforcement, 189
Legality assurance system, 187, 190
Legal standards, 61–73, 82, 403
  function of, 64
  inconsistencies, 203–206
  limitations of, 64–65
  in private law relations, 72–73
Life cycle assessment (LCA), 53, 150–156, 217, 219, 349, 405
Liquid biofuels, 100, 101. *See* also Biofuels

M

Marine Aquarium Council (MAC), 7, 325–326
Marine protected areas (MPAs), 327
Marine Stewardship Council (MSC), 7, 12, 53–55, 140, 171, 248, 263, 315, 320–322, 331, 404, 424
Market strategies, 170–173
  differentiation strategy, 171
  focus strategy, 172–173
MDGs. *See* Millennium Development Goals (MDGs)
Measuring impacts of VSS, 133–143
  methodological approaches for, 136
  motivation for, 135–136
Milkfish *(Chanos chanos)*, 315, 337
Millennium Development Goals (MDGs), 45, 116, 133, 142
MSC. *See* Marine Stewardship Council (MSC)

Multi-stakeholder initiatives, 50, 53, 118, 124, 134, 294, 363. *See also* Stakeholder co-operation

N

Nanotechnologies, 77–95
 ethical, legal and societal aspects of, 79
National accreditation bodies, 9, 100, 122, 267
Natural attenuation, 64
Naturland e.V., 321
Naturland Wildfish, 238, 315, 321–322, 333–339
New Legislative Framework of the EU, 262
New Limits, 362
Newly independent states, 166. *See also* Commonwealth of Independent States (CIS)
NGOs. *See* Non-governmental organisations (NGOs)
Nigeria, 289, 294, 414
Non-governmental organisations (NGOs), 4, 6, 49, 52, 73, 106, 116–118, 125, 130, 135, 198, 327, 364, 407
North Rhine-Westphalia, 202

O

Occupational safety, 85, 329
Oil palm, 184, 435
 energetic potential of residues, 221
 residues, 213–215
 trunks, 214
Oversight. *See* Accreditation
Ozone depletion (stratospheric), 27, 155

P

Palm oil, 10, 180, 196, 369, 435–437
 development of plantations area in Cameroon, 213
 plantation residues, 211–225
Pangasius *(Pangasianodon hypophthalmus)*, 315, 330–333
Paper-making industry
 in Liaoning Province, China, 253–259
 in Russia, 172
PEFC. *See* Programme for the Endorsement of Forest Certification (PEFC)
Philippines, 317, 325–326, 337
Post-Soviet states, 165–178, 261, 270
Power imbalances, 121
Pre-industrial times, 146
Product packaging, 5, 271
Programme for the Endorsement of Forest Certification (PEFC), 3, 53, 159, 181, 183, 191, 197–200, 202–204, 216, 255, 263
Proof of compliance, 9, 32, 188, 424
Public procurement, 11, 15, 100, 128, 180, 183, 188–190, 192, 195–207, 430. *See also* Green procurement
 German requirements for, 198–203

R

Rainforest Alliance (RA), 5–9, 52, 223, 258, 292, 314, 366, 368, 371
RED. *See* Renewable Energy Directive of the EC
Reduced Emissions from Deforestation and Degradation (REDD+), 182, 183, 186, 231, 236
Renewable Energy Directive of the EC (RED), 10, 99–114, 182, 186, 212, 217, 219, 223, 224, 422
 co-regulation approach, 103
Resilience, 34, 56, 137, 387–390
Roundtable on Responsible Soy (RTRS), 6, 8, 54, 102, 182, 184
Roundtable on Sustainable Biomaterials (formerly Biofuels) (RSB), 9, 102, 182, 216, 219, 221, 223–225
Roundtable on Sustainable Palm Oil (RSPO), 7–10, 54, 102, 182, 184, 216, 220–221, 223, 225, 370, 435–436
RSB. *See* Roundtable on Sustainable Biomaterials (RSB)
RSPO. *See* Roundtable on Sustainable Palm Oil (RSPO)
RTRS. *See* Roundtable on Responsible Soy (RTRS)

S

Shared cropping, 288
Shrimp *(Penaeus monodon)*, 315, 335, 337
Shrimp-turtle case, 206
Siemens-Norm, 62, 68
Small and medium enterprises (SMEs), 243, 248, 319, 335, 424
Smallholder producers, 13, 141, 213–214, 223, 280, 288–290, 305, 307, 332, 403–416
Social cohesion, 122, 386, 390, 396
Social contract, 366, 392–395
Social impacts, 45, 122, 138–139, 141–142, 237, 275–284, 318, 434
Solid biofuels, 211–225. *See also* Biofuels
Stakeholder co-operation, 120, 397, 406–407, 434

Standards
  aquaculture, 328–337
  capture fisheries, 319–328
  corporation, 68
  in the development context, 123–125
  environmental and social, 52–54, 119, 130, 410, 434, 439
  in international trade and investment agreements, 127
  labour, 117, 127, 419
  legal, 61–73, 82, 403
  limitations of, 93
  nanotechnologies, 80–81
  organic aquaculture, 333, 337
  performance-based, 8
  practice-based, 8
  private, 4, 50, 61–62, 65, 69, 70, 72, 81–82, 93, 115, 125, 126, 128, 369–371, 423
  sources of, 63–64
Standard setting, 7, 8, 13, 30, 124, 181
  bodies, 8
  initiatives, 36
  organisation, 5, 93, 296, 299
  procedures, 28, 64
Strategic management theory, 171
Sub-Saharan Africa, 403, 407
Supply chains, 9, 10, 18, 39, 55, 99, 111, 115, 134, 142, 171, 189, 217, 290, 360, 370, 389, 424, 439
  cocoa, 291, 293, 296, 299
Supply shortages, 291
Sustainability
  applications, 27–28
  in cocoa production, 291
  of communities, 385–400
  concerns, 4, 16, 34
  in construction and real estate sector, 345–358
  criteria, 11, 34, 35, 102, 127, 186, 216, 225, 350, 353, 366, 369, 398, 422, 424, 437
  of fisheries and aquaculture, 315–339
  indicators for measuring, 39–46
  initiatives, 21–22, 28–30, 49, 51, 297, 300, 388
  liquid biofuels, 100
  measuring, 39–46, 190, 345–358
  movements, 51–52
  objectives, 14–16, 345
  principles, 35, 317, 397, 399
  requirements for progress towards, 28–29
  solid biofuels, 225
  in supply chains, 9, 115, 439
  transition to, 24–32
Sustainable communities, 385–400
Sustainable Development Maturity Matrix, 398
Sustainable Shopping basket, 435

T
Tanzania, 315, 320–322
Targets, 8, 44, 65, 102, 167, 261, 297, 347, 386, 420
Tariff preferences, 419–430
TBT. *See* Technical barriers to trade (TBT)
Technical barriers to trade (TBT), 135
Technical regulations, 65, 94, 166, 263
Thailand, 188, 335, 337, 354
Thresholds, 44, 62, 64, 66, 82
Tilapia *(Oreochromis niloticus)*, 307, 309, 321, 324, 328
Timber products. *See* Wood and wood based products
Traceability, 5, 9–10, 171, 187, 189, 283, 293, 296, 299, 326, 329, 364, 369, 370, 381, 410, 433
Transition economy, 165
Transition Town, 399
Transparency, 9, 14–16, 34, 67, 103–106, 120, 126, 129, 142, 151, 152, 182, 190, 198, 206, 250, 283, 289, 352–354, 360, 437
Type III environmental declarations and standards, 149, 405

U
UK Code for Sustainable Homes, 393
Ukraine, 167, 261–273
Ukrainian State Certification System (UkrSEPRO), 264–270
Undefined legal terms, 70
United Kingdom (UK), 6, 107, 346, 395
Unsustainability, 23–24
Upward drift, 353
Urban advantage, 387
Urbanisation, 244
UTZ Certified, 9, 10, 182, 183, 292–294, 314, 366, 368, 371

V
VDI. *See* Association of German Engineers (VDI)
Vietnam, 188, 207, 330–331, 336–337
Virtual water, 44, 45
Voluntary environmental certification, 165–174
  in Ukraine, 261–273
Voluntary Partnership Agreement (VPA), 160, 186–191
Voluntary Standard Systems. *See* Voluntary Sustainability Standards (VSS)
Voluntary Sustainability Standards (VSS), 3–19, 22, 50, 65–69, 91, 142, 158, 246, 255, 276, 432
  added value of, 73

and carbon sequestration, 158
in China, 243–251, 253–259
in cocoa sector, 292–294
for cocoa supply chains, 292
in coffee sector, 305–314
contribution of German development policy towards promotion of, 121
contribution to sustainable development, 142
driving forces of, 50, 167–170
in fisheries and aquaculture, 315–338
forest related, 181–189
foundations for, 21–37
in German development policy, 115–131
history and evolution of, 6–7, 49–56, 371–372
implications for, 32–36
for interpretation of undefined legal terms, 70
in law, 69, 70
limitations of, 68, 437
market strategies based on, 170–173
measuring the impact of, 133–143
models of, 8
objectives of, 366
political challenge of, 431–440
procedure to develop, 67
their relationship with legal standards, 61–73
as a social contract, 382
sources of, 66–67
structure of, 4–5
for sustainable communities, 385–400
terminology, 4
trends in the uses of, 11–12
VPA. *See* Voluntary Partnership Agreement
VSS. *See* Voluntary Sustainability Standards

W
West Africa, 288, 291, 294–295, 296, 413–414, 438
White leg shrimp *(Litopenaeus vannamei)*, 334
Wood and wood based products, 8, 145, 148, 158, 160–162, 170, 172, 180, 188, 195–208, 234, 404
World Health Organization (WHO), 63, 81
World Trade Organization (WTO), 124–125, 128, 165, 205–206, 412

Z
Zero burning technique, 221
Zero emission, 65
Zero use of cyanide, 325

MIX
Papier aus verantwortungsvollen Quellen
Paper from responsible sources
FSC® C105338
www.fsc.org

If you have any concerns about our products,
you can contact us on
ProductSafety@springernature.com

In case Publisher is established outside the EU,
the EU authorized representative is:
Springer Nature Customer Service Center GmbH
Europaplatz 3, 69115 Heidelberg, Germany

Printed by Libri Plureos GmbH
in Hamburg, Germany